World Sustainability Series

Series Editor

Due to its scope and nature, sustainable development is a matter which is very interdisciplinary, and draws from knowledge and inputs from the social sciences and environmental sciences on the one hand, but also from physical sciences and arts on the other. As such, there is a perceived need to foster integrative approaches, whereby the combination of inputs from various fields may contribute to a better understanding of what sustainability is, and means to people. But despite the need for and the relevance of integrative approaches towards sustainable development, there is a paucity of literature which address matters related to sustainability in an integrated way.

Notes on the quality assurance and peer review of this publication

* * *

Prior to publication, the works published in this book are initially assessed and reviewed by an in-house editor. If suitable for publication, manuscripts are sent for further review, which includes a combined effort by the editorial board and appointed subject experts, who provide independent peer-review. The feedback obtained in this way was communicated to authors, and with manuscripts checked upon return before finally accepted. The peer-reviewed nature of the books in the "World Sustainability Series" means that contributions to them have, over many years, been officially accepted for tenure and promotion purposes.

Shruti Kanga • Suraj Kumar Singh
Khetan Shevkani • Vamdev Pathak
Bhartendu Sajan
Editors

Transforming Agricultural Management for a Sustainable Future

Climate Change and Machine Learning Perspectives

Springer

Editors
Shruti Kanga
Department of Geography, School of Environment and Earth Sciences
Central University of Punjab
Ghudda, Punjab, India

Khetan Shevkani
Department of Applied Agriculture
Central University of Punjab
Ghudda, Punjab, India

Bhartendu Sajan
Centre for Climate Change and Water Research
Suresh Gyan Vihar University
Jaipur, Rajasthan, India

Suraj Kumar Singh
Centre for Sustainable Development
Suresh Gyan Vihar University
Jaipur, Rajasthan, India

Vamdev Pathak
Department of Geology
Central University of Punjab
Ghudda, Punjab, India

ISSN 2199-7373 ISSN 2199-7381 (electronic)
World Sustainability Series
ISBN 978-3-031-63429-1 ISBN 978-3-031-63430-7 (eBook)
https://doi.org/10.1007/978-3-031-63430-7

This Springer imprint is published by the registered company Springer Nature Switzerland AG
The registered company address is: Gewerbestrasse 11, 6330 Cham, Switzerland

Contents

Chapter 1
The Impact of Climate Change on Crop Productivity and Adaptation and Mitigation Strategies in Agriculture

Yedid Guadalupe Zambrano-Medina, Evangelina Avila-Aceves, Lidia Yadira Perez-Aguilar, Sergio Alberto Monjardin-Armenta, Wenseslao Plata-Rocha, Cuauhtémoc Franco-Ochoa, and Osiris Chávez-Martínez

Abstract Climate change is one of the most pressing challenges facing humanity today. Its effects extend to numerous sectors of society, and agriculture is one of the most vulnerable to its consequences. Changes in weather patterns threaten global crop productivity, giving rise to serious sustainability concerns. Climate change significantly negatively impacts global crop productivity. Research has consistently shown that increased temperatures, climate variability, and extreme events are associated with decreased crop yields, as well as conditions conducive to the spread and proliferation of agricultural pathogens. These effects are of particular concern in regions that are already experiencing adverse climatic conditions, mainly droughts, and floods. This chapter reviews the literature on the influence of climate change on crop productivity worldwide, analyzes the observed effects on agriculture, and explores adaptation and mitigation strategies used to address these challenges.

Keywords Climate Change · Crop Productivity · Drought · Adaptation · Mitigation

1.1 Introduction

1.1.1 Climate Change and Agriculture

Large-scale changes in weather patterns are intensifying with global warming. The Intergovernmental Panel on Climate Change (IPCC) sixth assessment report, titled

Y. G. Zambrano-Medina (✉) · E. Avila-Aceves · L. Y. Perez-Aguilar
S. A. Monjardin-Armenta · W. Plata-Rocha · C. Franco-Ochoa · O. Chávez-Martínez
Facultad de Ciencias de la Tierra y el Espacio, Universidad Autonoma de Sinaloa, Culiacan, Sinaloa, Mexico
e-mail: yedidzambrano@uas.edu.mx; evangelinaavila.facite@uas.edu.mx; lidiaperez@uas.edu.mx; sa.monjardin12@info.uas.edu.mx; wenses@uas.edu.mx; cfrancoo@uas.edu.mx; osirischavez.facite@uas.edu.mx

S. Kanga et al. (eds.), *Transforming Agricultural Management for a Sustainable Future*, World Sustainability Series, https://doi.org/10.1007/978-3-031-63430-7_1

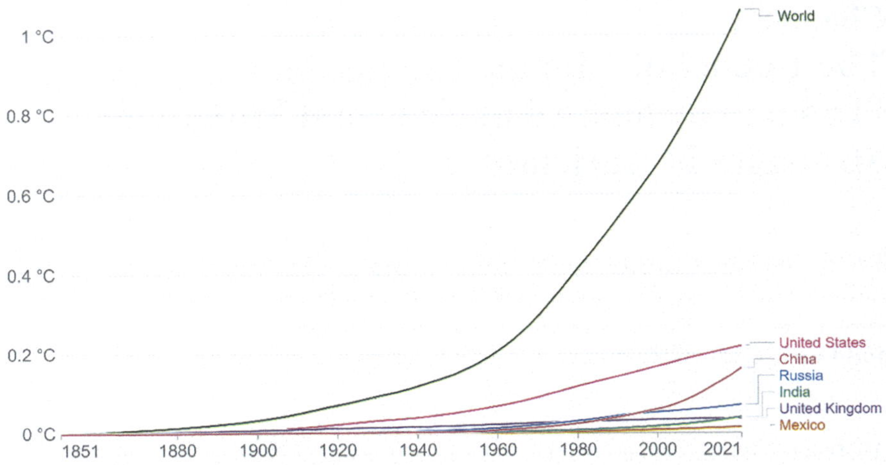

Fig. 1.1 Contribution to global mean surface temperature rise from fossil sources, 1851 to 2021. This refers to the alteration in the global average surface temperature caused by the cumulative emissions of carbon dioxide, methane, and nitrous oxide originating from a particular country or region. These emissions pertain solely to fossil fuels and industrial activities, excluding any contributions from land use or agriculture. Source: (Jones et al. 2023)

"Climate Change 2021: the Basis of Natural Science", published in 2021 and highlights that the climate has experienced unprecedented warming by human activities (Wang et al. 2023). The burning of fossil fuels, along with agriculture, deforestation, and manufacturing, is causing an increase in Greenhouse Gas (GHG) (Fig. 1.1) among other drivers of climate change (Balasundram et al. 2023).

Global warming directly affects extreme changes in both temperature and the global hydrological cycle (Wang et al. 2023). This results in an increase in the frequency, intensity, and severity of weather phenomena, such as temperature, intense rainfall, humidity, heat waves, droughts, and floods (Islam et al. 2023; Pathak 2023; Zhang 2021). Another impact is to consider sea level rise, which mainly affects low-lying coastal areas, as well as hurricanes and storms that erode shorelines (Zambrano-medina et al. 2023).

Climate change characterizes the regional redistribution and intensity of precipitation, as well as its lack of seasonality (Rani et al. 2023). This increases droughts and floods, mainly in the form of droughts and floods (Wang et al. 2023). Changes in historical weather patterns include more intense precipitation events interrupted by more prolonged drought conditions (Joyal et al. 2023). Droughts and floods are major natural disasters globally with devastating impacts on agricultural systems (Qian et al. 2023), bringing large monetary losses in agricultural productivity and forestry (Islam et al. 2023).

Climate change has been intensifying rapidly (IPCC 2021), and as it does, the interaction between climate regimes is becoming less and less stationary (Joyal et al. 2023) affecting crop yields, soil processes, water availability, and pest dynamics (Pathak 2023). Crop production is therefore very sensitive to climate change and in

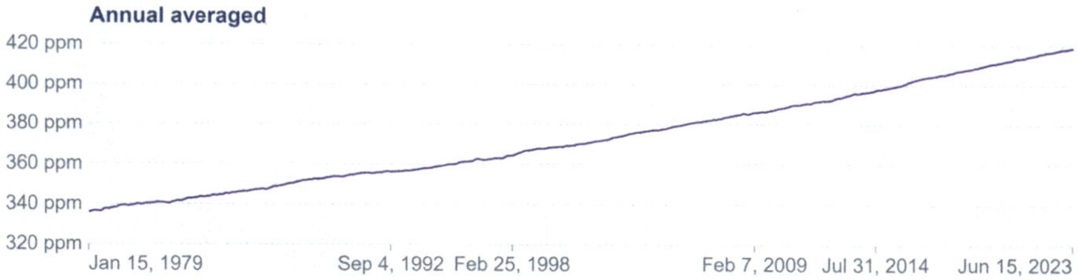

Fig. 1.2 Global atmospheric CO_2 concentration. Atmospheric carbon dioxide (CO_2) concentration is measured in parts per million (ppm). Source: Administración Nacional Oceánica y Atmosférica (NOAA)

recent decades, agricultural regions around the world have experienced rapid climatic changes and an increase in carbon dioxide (CO_2) and ozone (O_3) (Lobell and Gourdji 2012). This causes losses in the productivity of food crops and in the agricultural sector (Muneer et al. 2023) (Fig. 1.2).

Climate change causes changes in the timing of planting and harvesting, as well as altering the growing conditions of crops (Lavane et al. 2023). These changes can lead to decreased food security and economic losses for farmers (Balasundram et al. 2023). Therefore, climate change could pose a severe challenge to agricultural systems, as they will need to adapt to ensure food security (Straffelini and Tarolli 2023).

1.2 Climate Trends

The interaction between climate trends and crop productivity is a growing scientific concern. Climate variability and trends impact global crop yields and depend heavily on factors such as location, crop type, and irrigation (Kukal and Irmak 2018). Different aspects of climate variability (temperature, precipitation, and their interaction) can disproportionately affect growth and resulting productivity (D. K. Ray et al. 2015). Droughts are the most affected, for a prolonged period and have a direct impact on crops.

1.2.1 Rising Temperatures

Rising temperatures because of climate change have significantly increased the numerical values of global temperature. During the twentieth century, the average temperature of the Earth's surface increased by ~1 °C (Fig. 1.3). In addition, in recent decades, we have to observe a faster increase in temperatures, with a warming rate that has exceeded the historical average (Islam et al. 2023; Singh et al. 2023a, b).

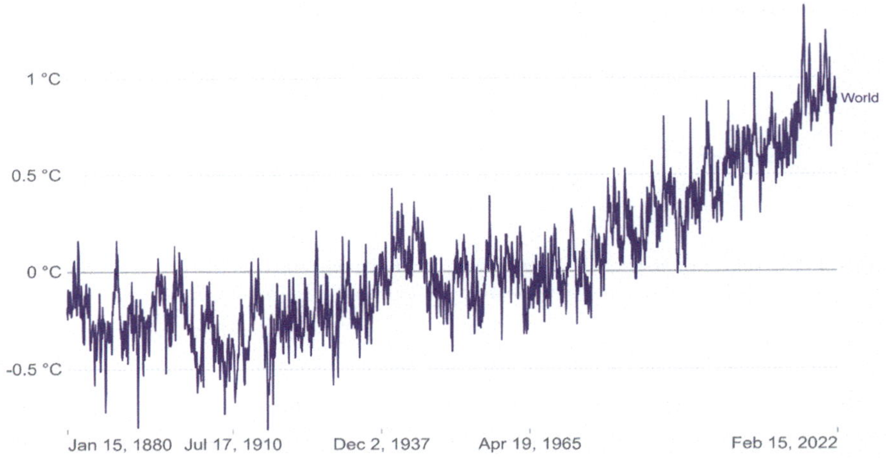

Fig. 1.3 Global warming: monthly temperature anomaly. The combined land-surface air and sea-surface water temperature anomaly is given as the deviation from the 1951–1980 mean. Source: Source: National Aeronautics and Space Administration (NASA), Goddard Institute for Space Studies (GISS)

According to the IPCC, rising temperatures might hasten the rate of plant development and may alter growth cycles, which may affect the yield of crops (IPCC 2021).

Weather and climate are the most important drivers that are influenced by changing trends in climatic variables, thus affecting crop yields even with advances in today's technologies. Increasing atmospheric temperatures can have a direct impact on reducing crop duration, photosynthesis, and ultimately crop yield (Pathak 2023). Air temperatures have been warming in most major cereal-growing regions around the world (Lobell and Gourdji 2012). In addition, the frequency and intensity of extreme weather events, such as droughts and floods, can also have a devastating impact on crops. According to the Food and Agriculture Organization of the United Nations report, protracted droughts and flash floods can cause significant losses in agricultural production (FAO 2020).

Research shows that high temperatures have the potential to expedite plant development, resulting in a decrease in the time available for photosynthesis and biomass accumulation (Li et al. 2023; Lobell and Gourdji 2012). This process can compromise the quality and quantity of crop yields (Tripathi et al. 2023). In addition, heat strain can negatively affect grain formation and grain filling in cereals, such as rice, maize, and wheat, which reduces final yields (Prasad et al. 2017; Zhao et al. 2017). Wheat and corn crop yields may increase because climate change in colder regions and may decrease in countries near the equator (Farooq et al. 2023).

The growth stage of the crop can also influence how the crop responds to heat. Studies estimate that high temperatures during flowering may cause flower loss and a reduction in fruit and seed formation (Matiu et al. 2017; Zhao et al. 2017). Also, high temperatures during the grain-filling stage can decrease grain size and affect food quality (Lu et al. 2020).

1.2.2 Droughts

At present, drought is recognized as among the most severe, recurrent, and hazardous natural disasters, capable of enduring for several months and lasting up to several years, with a progressive cumulative impact (Jabalameli et al. 2023; Yan et al. 2016; Yuan and Quiring 2014). It occurs with below-average rainfall, periods of above-average temperatures and high evapotranspiration that can induce hydrological stress over a prolonged period (Carroll et al. 2021; Kuwayama et al. 2018; Yuan and Quiring 2014).

Drought events are mainly characterized by duration, severity, and intensity. Therefore, the negative effects of this condition can be of great magnitude (Afshar et al. 2020). Droughts can cause a decline in crop sales income, an increase in production costs, reduced net farm income, and negative impacts on agriculture-dependent communities (Kuwayama et al. 2018). For a better analysis of droughts, they can be classified into three main types: meteorological, agricultural, and hydrological (S. Cao et al. 2022; Liu et al. 2017; Maity et al. 2016; Waseem et al. 2022).

Meteorological drought is the driver of agricultural drought where precipitation over a period is less than the average precipitation (Cao et al. 2022), can lead to a significant decrease in agricultural yields, mainly in arid regions (Waseem et al. 2022). Agricultural drought occurs at the critical time of crop growth, resulting in decreased soil moisture and crop failure (Lu et al. 2020). Soil moisture and precipitation during the growing season are inadequate to support crop growth to maturity (Cao et al. 2022). Therefore, agricultural drought is the consequence of inadequate soil moisture during the crop growing season caused mainly by low rainfall (Li et al. 2017). Hydrological drought occurs when there is a significant depletion of surface water and groundwater resources (Cao et al. 2022; Malik et al. 2021; Ojha et al. 2021).

Globally, droughts cause billions of dollars in damages (Dai 2011; Yuan and Quiring 2014). However, it is difficult to define and monitor drought, so drought indices have been developed to evaluate its condition and impacts for better decision-making (Lu et al. 2020). Drought indices are the integration of climatic or hydrological variables (precipitation, temperature, soil moisture, groundwater levels, etc.). They are used to analyzing the occurrence and severity of drought in various investigations (Table 1.1) to study its spatial and temporal effects (Sun et al. 2011).

Some of the most used drought indices include the standardized Precipitation Index (SPI) is an indicator that determines the probability of occurrence of precipitation in a region over a certain period. Standardized Precipitation Index (SPI) is an indicator that determines the probability of occurrence of precipitation in a region within a certain period. The calculations are simple and can be applied to extensive regions on different time scales (C. Liu et al. 2021; Malik et al. 2021; Pramudya and Onishi 2019). The Standardized Precipitation Evapotranspiration Index (SPEI) is calculated with the difference between monthly precipitation and evapotranspiration.

Table 1.1 Research on the impact of droughts on crops

Study Area	Year	Crop	Index	Reference
Canada	2012	Wheat	A multi-index drought model	Sun et al. (2011)
USA	2016	Corn and Soy	SPEI	Zipper et al. (2018)
USA	2017	Corn, Coy, Wheat, and Rice	SPEI	Matiu et al. (2017)
USA	2018	Wheat, Corn, and Cotton	PDSI, Palmer's Z-index, precipitation percent normal, precipitation percentiles, SPI and SPEI	Tian et al. (2018)
USA	2019	Barley, Winter wheat, Soy, Corn, and Cotton	SPEI	Peña-Gallardo et al. (2019)
USA	2020	Corn, Soy, Hay, Wheat, Cotton, Sorghum, Barley, and Rice	SPEI SPI	Lu et al. (2020)
Pakistan	2022	Corn	SPEI	Waseem et al. (2022)
Pakistan	2023	Corn, Rice, Sugar cane, Tobacco, and Wheat	SPEI	Hussain et al. (2023)

It also considers the temperature, which makes it more sensitive to the reaction caused by drought (Gao et al. 2017; Liu et al. 2021).

Palmer's Drought Severity Index (PDSI) is another indicator that provides robust estimates of drought severity. It studies current and past changes in aridity by considering the deficit between actual precipitation and the precipitation needed to maintain climatic moisture conditions (Dai 2011; Palmer 1965; Serrano et al. 2011). The moisture anomaly index (Z-index) is mainly used to calculate short-term moisture deficits and better estimate agricultural moisture shortage, percentage of normal, and forest fire danger (Pandžić et al. 2022).

Given the significant effects of drought on agriculture, several studies have been conducted using different drought indices to assess the impacts on agriculture (Singh et al. 2023a, b). One of the most widely used indexes is SPEI, which has been used extensively, e.g. Zipper et al. (2018) used SPEI to estimate spatiotemporal patterns of drought effects on corn and soybean yields in the USA. Matiu et al. (2017) also used SPEI to determine the interaction and intensification effects of climate variability affecting yields of corn, soybean, wheat, and rice crops in the USA. Peña-Gallardo et al. (2019) used SPEI at different drought timescales to analyze annual yields of barley, wheat, soybean, corn, and cotton in areas in the USA. Hussain et al. (2023) assessed the impact of agricultural drought using the SPEI Index for maize, rice, sugarcane, tobacco, and wheat crops in Pakistan at the national and provincial levels.

Identified the spatial pattern of agricultural sensitivity to drought for corn, soybean, hay, wheat, cotton, corn silage, sorghum, barley, and rice crops in the USA using a correlation analysis between crop yield anomalies and the SPEI and SPI indices. Waseem et al. (2022) quantified the impacts of meteorological drought on maize crop production using yield and rainfall data and the SPI index in Pakistan. Tian et al. (2018) evaluated the performance of six drought indices to analyze agricultural drought in the south-central USA and determine the yield of wheat, corn, and cotton crops.

Sun et al. (2011) developed a drought model combining the strengths of several drought indices for the assessment of wheat production risk affected by agricultural drought in Canada. Daryanto et al. (2016)conducted meta-analysis studies on maize, wheat, legumes, root and tuber crops, wheat, rice, barley, sorghum, millet, and oats globally, considering environmental variables that intensity with climate change (precipitation, temperature, aridity, etc.). Ray et al. (2018) using satellite imagery and spatial analysis tools in Geographic Information Systems (GIS), analyzed wheat, corn, cotton, and sorghum crops before, during, and after drought.

Drought is undoubtedly a devastating natural hazard that severely threatens agriculture. Therefore, identifying the spatial pattern of agricultural sensitivity to drought can provide scientific information for decision-makers to prepare for, allocate resources to, and mitigate the impacts of drought (Lu et al. 2020). Drought risk assessment is of great importance in formulating disaster prevention and increasing farmers' income, while agricultural production is often more susceptible to these conditions as it uses natural resources and depends on the natural environment (Liu et al. 2019). Therefore, droughts have a significant impact on agricultural production security (Yan et al. 2016). Likewise, It is crucial to analyze and create a precise technique for evaluating the temporal and spatial patterns of droughts, which can aid agricultural growth and lead to successful policies for preventing and managing agricultural disasters. (Liu et al. 2019).

1.2.3 Floods

Floods are common and damaging natural events worldwide, caused mainly by heavy rainfall. There are two types of floods: flash floods and slow floods, which give the population more time to prepare for their effects (Avila-Aceves et al. 2023). This situation may have the potential to cause damage to the population, agriculture, livestock, and infrastructure. The frequency and severity of its effects increase with climate change (Ellis et al. 2021; Li et al. 2021; Wijayarathne and Coulibaly 2020). As a result, farmers in Bangladesh believe that "Climate change means frequent flooding" (Islam et al. 2023).

Because of climate change, the inhomogeneous temporal distribution of precipitation has led to abrupt alternating cycles of droughts and floods (Qian et al. 2023). In addition, changes in vegetation and soil conditions associated with increased agricultural and pasture cover become a threat to the water resilience of watersheds (Joyal et al. 2023).

Atmospheric anomalies modify the characteristics, such as frequency, intensity, and impact of floods, within the context of climate change (Gu et al. 2023). The rainfall regime in tropical areas is characterized by high-magnitude and high-intensity events that occur frequently. Unlike higher latitudes, these regions experience relatively low annual hydrological variability. However, this annual climatic stability makes them vulnerable to climate change, as any alteration in these areas can have a significant impact on their climate (Joyal et al. 2023). In addition, climate change will cause more frequent flooding in floodplains in the near term, making sustainability and agricultural production less assured (Islam et al. 2023). Water resources are a crucial issue in the fight against climate change and adaptation measures to be taken (Asmael et al. 2023).

Floods can have a devastating impact on agriculture and, ultimately, on food security. The spatio-temporal distribution of heterogeneous rainfall characteristics provides baseline information for agricultural development (Wang et al. 2023). Irrigated agricultural production allows the production of crops where water is a limiting factor (Asmael et al. 2023). However, reliance on a single seasonal crop makes smallholder farmers vulnerable to seasonality, characterized by the vagaries of nature (Islam et al. 2023), It increases even more if they depend on rainfall as a source of water for irrigation (Yang et al. 2023). Floods, which are a consequence of climate change, have the potential to affect various crops globally, and the magnitude of their impact will depend on the severity of floods and local characteristics. In this context, research related to the impact of floods on different crops will be presented in Table 1.2 below.

The main challenge in flood management is to understand past and future disaster trends and patterns. This is achieved by preparing flood maps and event projections. Geospatial flood modeling is presented as the key flood prevention strategy that provides the ability to identify flood-prone areas. Geospatial flood modeling gives

Table 1.2 Research on the impact of floods on crops

Study Area	Year	Crop	Study	Reference
México	2015	Sugar cane	Flood hydraulic model	Wurl et al. (2015)
China	2021	Rice	Influence of flooding on production efficiency	Zhang (2021)
Myanmar	2021	Rice	Flood damage function, using surveys	Shrestha et al. (2021)
Ecuador	2022	Rice and Soy	Geospatial modeling of agricultural impacts	Urrutia et al. (2022)
Vietnam	2023	Rice	Drought and flood cycle	Cao et al. (2023)
Bangladesh	2023	Rice	Survey of farmers on the effects of climate change	Islam et al. (2023)
China	2023	Cotton	Drought and flood cycle	Qian et al. (2023)

important results in flood prevention and control decision-making (Ben Khalfallah and Saidi 2018). They are tools used to predict flood extent, useful in disaster prevention and scientific research (Dong et al. 2021).

These models mainly address the directives of hydrological and hydraulic models. The former allows the rainfall-runoff system to be established, analyzing the various mechanisms involved in the process (Tamiru and Dinka 2021). While in the hydraulic models, the hydrodynamic behavior of the water during the flood event is involved (Petroselli et al. 2019).

1.3 Impact of Climate Change on Agriculture and Crop Productivity

Agriculture has been a crucial part of human civilization since its emergence and continues to play an important role in our lives. Approximately 80% of the population in developing countries live in rural areas, where agriculture is the mainstay of livelihood (Wijerathna-Yapa and Pathirana 2022), and also the main sector for food production. However, agriculture and other land use changes account for up to 24% of GHG emissions from approximately 22.2 million km^2 of agricultural land area (Pathak 2023). While climate variability accounts for about one-third (~ 32–39%) of observed yield variability (Ray et al. 2015). So the agricultural sector is not spared from the negative effects of climate change (Akter et al. 2023).

Climate change has affected agricultural productivity through weather pattern changes such as higher temperatures, altered rainfall distribution, and extreme weather events. These challenges have been documented in numerous studies (Afshar et al. 2020; Kuwayama et al. 2018; Parida and Roy Chowdhury 2021; Peña-Gallardo et al. 2019; Qiu et al. 2022; Ray et al. 2015; Waseem et al. 2022; Zhao et al. 2017). The increase in average temperatures has led to changes in crop growth cycles, affecting their development and yields. Agriculture has been affected by climate change, in combination with the impact of GHG, nitrogen (N), phosphorus (P), and pesticides that the same activity generates (FAO 2017, 2018).

The observed variability in cereal crop yields accounts for one-third of global climate variability (Fig. 1.4). Yields of crops such as maize, rice, wheat, and soybean are the main explanations for the variability in temperature, precipitation, and their interaction (Ray et al. 2015). In addition, when extreme temperatures are observed during critical periods, such as flowering and grain filling, lower grain yield and a decrease in final product quality result (Xin and Tao 2019). High temperatures increase the rate of plant maturation, which reduces the time available for photosynthesis and biomass accumulation (Lobell and Gourdji 2012; Zhao et al. 2017). Thus, increasing mean temperature negatively affects corn and wheat production in the United States of America (USA) (Schlenker and Roberts 2009).

The agricultural sector in tropical and subtropical countries in the Americas is vulnerable to the impact of climate change and its variability (Terán-Chaves et al.

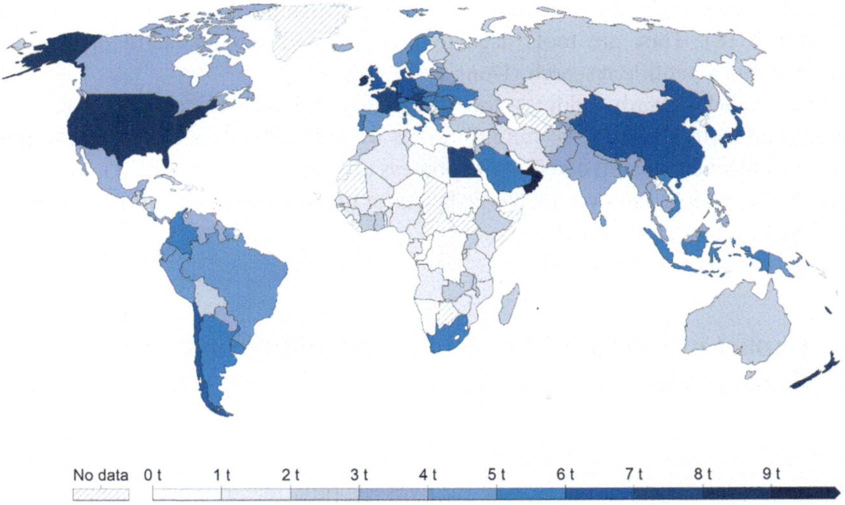

Fig. 1.4 Cereal: Yiels, 2021. Cereals include wheat, rice, maize, barley, oats, rye, millet, sorghum, buckwheat, and mixed grains. Source: Food and Agriculture Organization of the United Nations. Yields are measured in tonnes per hectare

2023), is are some regions in Asia are among the most vulnerable in the world to the impacts of climate change (Bandara and Cai 2014). Rice crop yields may decline in almost all major rice-producing countries due to water shortages, and may be amplified by climate change effects (Farooq et al. 2023). In most South Asian countries, production is expected to decline by up to 4%, and 11% for rice and wheat, because of climate change-induced alteration in land productivity compared to baseline food production (Bandara and Cai 2014).

Drought-flood cycles occur during the growing season of crops. However, the effect on yield differs if the stress produced in the plant is because drought or flooding (Qian et al. 2023). Because both water deficit and saturation point (caused by flooding) can prevent optimal root development of crops, preventing the proper absorption of water and fertilizers, mobilization of fixed phosphorus (P) from the soil by reducing oxides of Fe and Mn (Qian et al. 2023; S. Zhang et al. 2022). Flooding in the later stages of crop maturation and harvest can contaminate yields with mineral oils, polycyclic aromatic hydrocarbons, and wastewater and even promote the emergence of microorganisms and stimulate the development of pests (Glavan et al. 2020).

Extreme events such as floods can damage crops and soil, affecting long-term productivity. Developing countries are vulnerable to agricultural losses resulting from floods, which leads to economic and food security losses (Shrestha et al. 2021). An example of this is Bangladesh and India, which top the list of countries most affected by floods (Parida and Roy Chowdhury 2021)(Zhang 2021).

Rising temperatures and variations in precipitation patterns have led to risks related to water shortages, such as droughts, floods, and droughts (Perez-Aguilar et al. 2021). Since the 1980s, longer durations and higher intensities, primarily by rising temperatures, have characterized droughts (Gu et al. 2023), which is one effect of global warming. Drought can bring losses in agriculture and cause widespread crop mortality and an increase in the frequency of pests and diseases, further reducing crop yields (Qiu et al. 2022).

Droughts can have a significant environmental, economic, and social impact, but agriculture is the main sector affected by this condition. Droughts reduce the availability of surface and groundwater, which contributes to a reduction in crop yields and thus crop losses (Daryanto et al. 2016; Kuwayama et al. 2018; Lu et al. 2020; Maity et al. 2016; Waseem et al. 2022; Yuan and Quiring 2014).

Direct and indirect consequences of aridity are reflected in plant germination, as shown by studies in Spain and Italy (Estrelles et al. 2015), with consequences on crops. Northwestern Italy is moving towards drier conditions, posing a challenge for agriculture. The Adriatic coast could become arid, a finding in line with historical observations. Rice paddies will be most at risk, with 76% of their area becoming arid in the future, as well as irrigated land (Straffelini and Tarolli 2023).

The scientific community has also documented an increase in the incidence and severity of diseases in crops because of climate change. The intensification of extreme weather events, coupled with variations in land use and global trade, is likely to further exacerbate pathogen transmission and disease incidence (Singh et al. 2023a, b). Rising temperatures and alterations in precipitation patterns have influenced the geographic distribution of pests and diseases, affecting both crop and pathogen populations and increasing pressure on agricultural production (Ayoub Shaikh et al. 2022; Pathan et al. 2020).

Intensified movement of crop pathogens via globalization and international trade due to climate change is projected to raise plant diseases in crops, increasing the transmission risk from disease-prevalent regions to disease-free regions (Singh et al. 2023a, b). Global changes in pathogen distribution are of concern, as growing evidence suggests that pathogens cause more damage in newly invaded regions and new hosts than in their native regions and hosts (Cao et al. 2023; Singh et al. 2023a, b).

In addition, climate change also affects the plant-pathogen interaction, affecting its life cycle, development of host resistance, disease severity, introduction of new types, pathogenicity, etc. (Balasundram et al. 2023). Changes in global temperatures can profoundly affect the emergence of pathogens in agricultural and natural ecosystems, increasing the risk of exposure to new pests and pathogens (Hajjar et al. 2023). Both natural disasters and pest infestations reduce the technical efficiency of crops (technical efficiency is the ability of a farmer to achieve the highest possible production, using existing and accessible technology) (T. M. Cao et al. 2023).

Changes in precipitation patterns also play an important role in agricultural pathology. Irregular rainfall can influence the spread and survival of pathogens in the soil and plants (Angon et al. 2023). Soil moisture is an important factor to consider as it affects the hydrologic cycle and influences plant growth and crop yields. It is a hydrological parameter that influences hydrological, carbon, or climate cycles, as well as crop yields and natural disasters such as droughts and floods (Lee et al. 2023). Excessive humidity resulting from heavy and prolonged rainfall can favor the spread of fungal and bacterial diseases, as well as the proliferation of disease vectors (Hajjar et al. 2023).

1.4 Adaptation and Mitigation Strategies in Agriculture

Climate change has the potential to alter agricultural systems, leading to a decrease in crop yields and, thus, food security. Therefore, adaptation strategies and mitigation efforts are needed to help reduce the negative impacts of climate change on agriculture and crop production (Straffelini and Tarolli 2023; Wijerathna-Yapa and Pathirana 2022). Implementing climate-smart agricultural practices helps mitigate the negative impacts of climate change (Cao et al. 2023; Mutengwa et al. 2023). As well as mitigation of GHG emissions from agriculture can be achieved by changing land use management practices and improving input use efficiency (Pathak 2023).

The adoption of conservation agriculture techniques, such as no-tillage and crop rotation, helps to improve soil quality, reduce erosion, and maintain the health of the agricultural ecosystem (Lal 2015). As well as proper agricultural and livestock waste management can reduce greenhouse gas emissions and improve soil quality (Zhao et al. 2017). Including trees and shrubs in agricultural systems can improve biodiversity, soil resilience, and climate regulation, a technique called agroforestry (Bado et al. 2021; Mbow et al. 2014; Mukoobwa et al. 2023).

The development and cultivation of plant varieties resistant to diseases and climatic stresses can reduce the vulnerability of crops to adverse conditions (Hafeez et al. 2023; Shahzad et al. 2021). Similarly, encouraging crop diversification can reduce dependence on a single crop and increase resilience to extreme weather events (Almagro et al. 2023). Implementing efficient irrigation systems, such as drip irrigation, can help conserve water and improve crop yields (Bhavsar et al. 2023; Vaddula and Singh 2023). Education and training programs are also needed for farmers to promote sustainable practices and the adoption of mitigation techniques (Liu et al. 2023).

One of the most arduous tasks in agriculture is the early detection of diseases in time so as not to affect productivity. This problem can be addressed with Machine Learning and Deep Learning techniques using an automated method to detect plant diseases in large farms. This results in a benefit by reducing supervision time (Ahmed and Yadav 2023). Machine Learning is expected to be an efficient resource for the sustainability of the inputs used, where it can provide a solution to the concerns of the agricultural industry such as lack of proper irrigation systems, weeds, crop yields, plant control problems due to crop height and extreme meteorological conditions (Ayoub Shaikh et al. 2022).

Machine Learning helps to identify plant diseases, improve irrigation, pesticide control, crop phenotyping, and improve crop yields (Ayoub Shaikh et al. 2022; Pathan et al. 2020). As well as the use of integrated pest management practices, which include biological, chemical, and cultural methods, can help control the spread of diseases and pests (Hajjar et al. 2023). Used of natural enemies of pests and diseases, known as biocontrol, provides effective and sustainable methods of disease and pest control (Angon et al. 2023).

A very important technique is precision agriculture, which uses technology to optimize the use of resources such as water and fertilizers, reducing environmental impact and improving crop efficiency (Nath 2023). It aims to improve productivity and resource use, either by increasing yields or by reducing inputs and adverse environmental effects (Balafoutis et al. 2017). Supported by technologies such as remote sensing and satellite navigation of constellations such as Global Navigation Satellite Systems, Geographic Information Systems, Internet of Things, Big Data analysis, and artificial intelligence. With this, precision agriculture can be considered as a tool in issues of food security, natural resource management, and reduction in environmental impact (Pantazi et al. 2020). According to (Radočaj et al. 2023) it leads the list in global research on precision agriculture, followed by China, India, and Brazil.

1.5 Conclusion

Climate trends are playing a crucial role in crop productivity. Rising temperatures, altered rainfall patterns and increased frequency of extreme weather events are negatively affecting agriculture. Addressing these challenges through sustainable agricultural practices and climate change adaptation policies is critical to ensure long-term food security.

Rising temperatures caused by climate change have a detrimental impact on crop productivity, affecting various physiological processes and developmental stages. These findings underscore the urgent need for adaptive and mitigation strategies to ensure food security in an ongoing climate change scenario.

Droughts and floods exert a significant and negative influence on agriculture and crop productivity globally. Droughts, by reducing water availability and increasing water strain, can reduce crop production and affect food quality. Floods, alternatively, directly damage crops, erode soil, and pollute, causing considerable losses in agricultural production. Both extreme weather events are becoming more frequent and intense because of climate change, highlighting the importance of implementing adaptation and risk management strategies to protect food security and the sustainability of agriculture in the future. In addition, greenhouse gas emissions must be addressed to mitigate climate change and reduce the frequency and severity of these extreme weather events.

Climate change is also altering conditions affecting agricultural pathology by increasing the incidence and severity of crop diseases. Changing climatic conditions, including increasing temperatures and erratic rainfall patterns, are favoring the proliferation, reproduction, and spread of pathogens in agricultural fields. These impacts emphasize the importance of developing disease management strategies adapted to a developing climate. The scientific literature consistently shows that climate change has harmed agriculture and food productivity through changes in temperatures, precipitation patterns, and extreme weather events.

References

Afshar MH, Ünal ŞA, Mehr AD (2020) Climate change impact assessment on mild and extreme drought events using copulas over Ankara, Turkey. Theor Appl Climatol 141(1-2): 1045–1055. https://doi.org/10.1007/s00704-020-03257-6

Ahmed I, Yadav PK (2023) A systematic analysis of machine learning and deep learning based approaches for identifying and diagnosing plant diseases. Sustainable Oper Comput 4 (February):96–104. https://doi.org/10.1016/j.susoc.2023.03.001

Akter A, Mwalupaso GE, Wang S, Jahan MS, Geng X (2023) Towards climate action at farm-level: Distinguishing complements and substitutes among climate-smart agricultural practices (CSAPs) in flood prone areas. Clim Risk Manag 40:100491. https://doi.org/10.1016/j.crm.2023.100491

Almagro M, Díaz-Pereira E, Boix-Fayos C, Zornoza R, Sánchez-Navarro V, Re P, Fernández C, Martínez-Mena M (2023) The combination of crop diversification and no tillage enhances key soil quality parameters related to soil functioning without compromising crop yields in a low-input rainfed almond orchard under semiarid Mediterranean conditions. Agric Ecosyst Environ 345:108320. https://doi.org/10.1016/J.AGEE.2022.108320

Angon PB, Mondal S, Jahan I, Datto M, Antu UB, Ayshi FJ, Shafiul Islam M (2023) Integrated Pest Management (IPM) in agriculture and its role in maintaining ecological balance and biodiversity. Adv Agric 2023. https://doi.org/10.1155/2023/5546373

Asmael N, Dupuy A, McLachlan P, Franceschi M (2023) Hydro-geochemical characteristics of the shallow alluvial aquifer and its potential artificial recharge to sustain the low flow of the Garonne river. Water 15(16):2972. https://doi.org/10.3390/w15162972

Avila-Aceves E, Plata-Rocha W, Monjandin-Armenta SA, Rangel-Peraza JG (2023) Geospatial modelling of floods: a literature review. Stoch Env Res Risk A 2023:1–20. https://doi.org/10.1007/S00477-023-02505-1

Ayoub Shaikh T, Rasool T, Rasheed Lone F (2022) Towards leveraging the role of machine learning and artificial intelligence in precision agriculture and smart farming. Comput Electron Agric 198(June):107119. https://doi.org/10.1016/j.compag.2022.107119

Bado BV, Whitbread A, Sanoussi Manzo ML (2021) Improving agricultural productivity using agroforestry systems: performance of millet, cowpea, and ziziphus-based cropping systems in West Africa Sahel. Agric Ecosyst Environ 305:107175. https://doi.org/10.1016/J.AGEE.2020.107175

Balafoutis AT, Beck B, Fountas S, Tsiropoulos Z, Vangeyte J, van der Wal T, Soto-Embodas I, Gómez-Barbero M, Pedersen SM (2017) Smart farming technologies—description, taxonomy and economic impact. https://doi.org/10.1007/978-3-319-68715-5_2

Balasundram SK, Shamshiri RR, Sridhara S, Rizan N (2023) The role of digital agriculture in mitigating climate change and ensuring food security: an overview. Sustainability 15(6):5325. https://doi.org/10.3390/su15065325

Bandara JS, Cai Y (2014) The impact of climate change on food crop productivity, food prices and food security in South Asia. Econ Anal Policy 44(4):451–465. https://doi.org/10.1016/j.eap.2014.09.005

Ben Khalfallah C, Saidi S (2018) Spatiotemporal floodplain mapping and prediction using HEC-RAS – GIS tools: case of the Mejerda river, Tunisia. J Afr Earth Sci 142:44–51. https://doi.org/10.1016/j.jafrearsci.2018.03.004

Bhavsar D, Limbasia B, Mori Y, Imtiyazali Aglodiya M, Shah M (2023) A comprehensive and systematic study in smart drip and sprinkler irrigation systems. Smart Agric Technol 5:100303. https://doi.org/10.1016/J.ATECH.2023.100303

Cao S, Zhang L, He Y, Zhang Y, Chen Y, Yao S, Yang W, Sun Q (2022) Effects and contributions of meteorological drought on agricultural drought under different climatic zones and vegetation types in Northwest China. Sci Total Environ J 821. https://doi.org/10.1016/j.scitotenv.2022.153270

Cao TM, Lee SH, Lee JY (2023) The impact of natural disasters and pest infestations on technical efficiency in rice production: a study in Vietnam. Sustainability 15(15):11633. https://doi.org/10.3390/su151511633

Carroll CJW, Slette IJ, Griffin RJ, Lauren N, Ava EB, Denton EM, Gray JE, Post AK, Johnston MK, Yu Q, Collins SL, Luo Y, Smith MD, Knapp AK (2021) Is a drought a drought in grasslands? Productivity responses to different types of drought. Oecologia 0123456789. https://doi.org/10.1007/s00442-020-04793-8

Dai A (2011) Characteristics and trends in various forms of the Palmer Drought Severity Index during 1900–2008. J Geophys Res-Atmos 116(December 2010):1–26. https://doi.org/10.1029/2010JD015541

Daryanto S, Wang L, Jacinthe P (2016) Global synthesis of drought effects on cereal, legume, tuber and root crops production: a review. Agric Water Manag. https://doi.org/10.1016/j.agwat.2016.04.022

Dong B, Xia J, Zhou M, Deng S, Ahmadian R, Falconer RA (2021) Experimental and numerical model studies on flash flood inundation processes over a typical urban street. Adv Water Resour 147(August 2020):103824. https://doi.org/10.1016/j.advwatres.2020.103824

Ellis C, Visser-Quinn A, Aitken G, Beevers L (2021) Quantifying uncertainty in the modelling process; future extreme flood event projections across the UK. Geosciences (Switzerland) 11(1):1–20. https://doi.org/10.3390/geosciences11010033

Estrelles E, Biondi E, Galiè M, Mainardi F, Hurtado A, Soriano P (2015) Aridity level, rainfall pattern and soil features as key factors in germination strategies in salt-affected plant communities. J Arid Environ 117:1–9. https://doi.org/10.1016/j.jaridenv.2015.02.005

FAO (2017) The future of food and agriculture – Trends and challenges. https://www.fao.org/3/i6583e/i6583e.pdf

FAO (2018) More people, more food, worse water? a global review of water pollution from agriculture. https://www.fao.org/3/CA0146EN/ca0146en.pdf

FAO (2020) Food and Agriculture Organization of the United Nations. http://www.fao.org/3/y5028s/y5028s0a.htm

Farooq A, Farooq N, Akbar H, Hassan ZU, Gheewala SH (2023) A critical review of climate change impact at a global scale on cereal crop production. Agronomy 13(1). https://doi.org/10.3390/agronomy13010162

Gao X, Zhao Q, Zhao X, Wu P, Pan W, Gao X, Sun M (2017) Temporal and spatial evolution of the standardized precipitation evapotranspiration index (SPEI) in the Loess Plateau under climate change from 2001 to 2050. Sci Total Environ 595:191–200. https://doi.org/10.1016/j.scitotenv.2017.03.226

Glavan M, Cvejić R, Zupanc V, Knapič M, Pintar M (2020) Agricultural production and flood control dry detention reservoirs: example from Lower Savinja Valley, Slovenia. Environ Sci Pol 114:394–402. https://doi.org/10.1016/j.envsci.2020.09.012

Gu X, Zhang P, Zhang W, Liu Y, Jiang P, Wang S, Lai X, Long A (2023) A study of drought and flood cycles in Xinyang, China, using the wavelet transform and M-K Test. Atmosphere 14(8):1196. https://doi.org/10.3390/atmos14081196

Hafeez U, Ali M, Hassan SM, Akram MA, Zafar A (2023) Advances in breeding and engineering climate-resilient crops: a comprehensive review. Int J Res Adv Agric Sci 15

Hajjar MJ, Ahmed N, Alhudaib KA, Ullah H (2023) Integrated insect pest management techniques for rice. Sustainability 15(5):4499. https://doi.org/10.3390/SU15054499

Hussain A, Jadoon KZ, Rahman KU, Shang S, Shahid M, Ejaz N, Khan H (2023) Analyzing the impact of drought on agriculture: evidence from Pakistan using standardized precipitation evapotranspiration index. Nat Hazards 115(1):389–408. https://doi.org/10.1007/S11069-022-05559-6/METRICS

IPCC (2021) Summary for policymakers. In: Climate change 2021: the physical science basis. Contribution of working group I to the sixth assessment report of the intergovernmental panel on climate change. In Climate change 2021: The physical science basis.

Islam MM, Sarker UK, Monira S, Kheya SA, Kaysar MS, Ali MI, Salam MU, Hwang H, Hashem A, Abd Allah EF, Uddin MR (2023) Farmers' understanding about impact of climate change on cropping systems and nutrition: a study on Dingaputa Haor of Netrakona District in Bangladesh. Sustainability 15(16):12378. https://doi.org/10.3390/su151612378

Jabalameli A, Saghafian B, Zolfagharpour F (2023) Predicting agricultural water shortage in Karkheh Basin, Iran. Iran J Sci Technol Trans Civ Eng 47(5):3175–3184. https://doi.org/10.1007/S40996-023-01099-6/METRICS

Jones MW, Peters GP, Gasser T, Andrew RM, Schwingshackl C, Gütschow J, Houghton RA, Friedlingstein P, Pongratz J, Le Quéré C (2023) National contributions to climate change due to historical emissions of carbon dioxide, methane and nitrous oxide since 1850. [Data set]. Scientific Data 10(1):155

Joyal T, Fremier AK, Boll J (2023) Modeled forest conversion influences humid tropical watershed hydrology more than projected climate change. Hydrology 10(8):160. https://doi.org/10.3390/hydrology10080160

Kukal MS, Irmak S (2018) Climate-driven crop yield and yield variability and climate change impacts on the U.S. Great plains agricultural production. Sci Rep 8(1):1–18. https://doi.org/10.1038/s41598-018-21848-2

Kuwayama Y, Thompson A, Bernknopf R, Zaitchik B, Vail P (2018) Estimating the impact of drought on agriculture using the U.S. Drought monitor. Am J Agric Econ 101(July):193–2010. https://doi.org/10.1093/ajae/aay037

Lal R (2015) Sustainable intensification for adaptation and mitigation of climate change and advancement of food Security in Africa. Sustainable Intensification Adv Food Secur Enhance Clim Resilience Afr:3–17. https://doi.org/10.1007/978-3-319-09360-4_1

Lavane K, Kumar P, Meraj G, Han TG, Ngan LHB, Lien BTB, Ty TV, Thanh NT, Downes NK, Nam NGD, Minh HVT, Singh SK, Kanga S (2023) Assessing the effects of drought on rice yields in the Mekong Delta. Climate 11(1):13. https://doi.org/10.3390/cli11010013

Lee S-J, Choi C, Kim J, Choi M, Cho J, Lee Y (2023) Estimation of high-resolution soil moisture in Canadian Croplands using deep neural network with Sentinel-1 and Sentinel-2 images. Remote Sens 15(16):4063. https://doi.org/10.3390/rs15164063

Li R, Tsunekawa A, Tsubo M (2017) Assessment of agricultural drought in rainfed cereal production areas of northern China. Theor Appl Climatol 127:597–609. https://doi.org/10.1007/s00704-015-1657-x

Li Z, Zhang Y, Wang J, Ge W, Li W, Song H, Guo X, Wang T, Jiao Y (2021) Impact evaluation of geomorphic changes caused by extreme floods on inundation area considering geomorphic variations and land use types. Sci Total Environ 754(100):142424. https://doi.org/10.1016/j.scitotenv.2020.142424

Li M, Wu X, Zhu Y, Ullah N, Song Y (2023) Assessment of heat risk of winter wheat cropping based on long-term meteorological data. Agronomy 13(8). https://doi.org/10.3390/agronomy13082149

Liu Y, Zhu Y, Ren L, Singh VP, Yang X, Yuan F, 1State (2017) A multiscalar Palmer drought severity index Yi. Geophys Res Lett 44:6850–6858. https://doi.org/10.1002/2017GL073871

Liu Y, You M, Zhu J, Wang F, Ran R (2019) International journal of disaster risk reduction integrated risk assessment for agricultural drought and flood disasters based on entropy information diffusion theory in the middle and lower reaches of the Yangtze River, China. Int J Disaster Risk Reduct 38(May):101194. https://doi.org/10.1016/j.ijdrr.2019.101194

Liu C, Yang C, Yang Q, Wang J (2021) Spatiotemporal drought analysis by the standardized precipitation index (SPI) and standardized precipitation evapotranspiration index (SPEI) in Sichuan Province. Sci Rep 11(1):1–14. https://doi.org/10.1038/s41598-020-80527-3

Liu D, Huang Y, Luo X (2023) Farmers' technology preference and influencing factors for pesticide reduction: evidence from Hubei Province, China. Environ Sci Pollut Res 30(3):6424–6434. https://doi.org/10.1007/S11356-022-22654-0/TABLES/5

Lobell DB, Gourdji SM (2012) The influence of climate change on global crop productivity. Plant Physiol 160(4):1686–1697. https://doi.org/10.1104/pp.112.208298

Lu J, Carbone GJ, Huang X, Lackstrom K, Gao P (2020) Agricultural and forest meteorology mapping the sensitivity of agriculture to drought and estimating the effect of irrigation in the United States, 1950–2016. Agric For Meteorol 292–293(March):108,124. https://doi.org/10.1016/j.agrformet.2020.108124

Maity R, Suman M, Verma NK (2016) Drought prediction using a wavelet based approach to model the temporal consequences of different types of droughts. J Hydrol 539:417–428. https://doi.org/10.1016/j.jhydrol.2016.05.042

Malik A, Kumar A, Rai P, Kuriqi A (2021) Prediction of multi-scalar standardized precipitation index by using artificial intelligence and regression models. Climate 9:1–25. https://doi.org/10.3390/cli9020028

Matiu M, Ankerst DP, Menzel A (2017) Interactions between temperature and drought in global and regional crop yield variability during 1961–2014. PLoS One:1–23. https://doi.org/10.1371/journal.pone.0178339

Mbow C, Smith P, Skole D, Duguma L, Bustamante M (2014) Achieving mitigation and adaptation to climate change through sustainable agroforestry practices in Africa. Curr Opin Environ Sustain 6(1):8–14. https://doi.org/10.1016/J.COSUST.2013.09.002

Mukoobwa D, Zaninka B, Patrick Mugunga C (2023) The assessment of agroforestry practices in Mukura Sector, Huye District, Southern Rwanda. Int J For Res 2023. https://doi.org/10.1155/2023/2288114

Muneer S, Bakhsh K, Ali R, Yasin MA, Kamran MA (2023) Farm households' perception and adaptation to climate change in relation of food crop productivity in Pakistan. Environ Dev Sustain:1–18. https://doi.org/10.1007/S10668-023-03333-7/METRICS

Mutengwa CS, Mnkeni P, Kondwakwenda A (2023) Climate-smart agriculture and food security in southern Africa: a review of the vulnerability of smallholder agriculture and food security to climate change. Sustainability 15(4):2882. https://doi.org/10.3390/SU15042882

Nath S (2023) A vision of precision agriculture: balance between agricultural sustainability and environmental stewardship. Agron J. https://doi.org/10.1002/AGJ2.21405

Ojha SS, Singh V, Roshni T (2021) Comparison of meteorological drought using SPI and SPEI. Civil Eng J 7(12)

Palmer WC (1965) Meteorological drought. United States: U.S. Department of Commerce, Weather Bureau. https://books.google.com.mx/books?hl=en&lr=&id=kyYZgnEk-L8C&oi=fnd&pg=PA6&ots=U4hxhiXGso&sig=uZk03DSONn4Mj_rfGVdhXCRYR-M&redir_esc=y#v=onepage&q&f=false

Pandžić K, Likso T, Pejić I, Šarčević H, Pecina M (2022) Application of the self – calibrated palmer drought severity index and standardized precipitation index for estimation of drought impact on maize grain yield in Pannonian part. Nat Hazards 113(2):1237–1262. https://doi.org/10.1007/s11069-022-05345-4

Pantazi XE, Moshou D, Bochtis D (2020) Sensors in agriculture. Intell Data Mining Fusion Syst Agric:1–15. https://doi.org/10.1016/b978-0-12-814391-9.00001-7

Parida Y, Roy Chowdhury J (2021) An empirical analysis of the effect of floods on rural agricultural wages across Indian states. World Dev Perspect 23:100272. https://doi.org/10.1016/j.wdp.2020.100272

Pathak H (2023) Impact, adaptation, and mitigation of climate change in Indian agriculture. Environ Monit Assess 195(1). https://doi.org/10.1007/s10661-022-10537-3

Pathan M, Patel N, Yagnik H, Shah M (2020) Artificial cognition for applications in smart agriculture: a comprehensive review. Artifi Intell Agric 4:81–95. https://doi.org/10.1016/j.aiia.2020.06.001

Peña-Gallardo M, Vicente-Serrano SM, Quiring S, Svoboda M, Hannaford J, Tomas-burguera M, Martín-hernández N, Domínguez-castro F, El A (2019) Agricultural and forest meteorology response of crop yield to different time-scales of drought in the United States: Spatio-temporal patterns and climatic and environmental drivers. Agric For Meteorol 264(July 2017):40–55. https://doi.org/10.1016/j.agrformet.2018.09.019

Perez-Aguilar LY, Plata-Rocha W, Monjardin-Armenta SA, Franco-Ochoa C, Zambrano-Medina YG (2021) The identification and classification of arid zones through multicriteria evaluation and geographic information systems—Case study: arid regions of northwest Mexico. ISPRS Int J Geo Inf 10(11). https://doi.org/10.3390/ijgi10110720

Petroselli A, Vojtek M, Vojteková J (2019) Flood mapping in small ungauged basins: a comparison of different approaches for two case studies in Slovakia. Hydrol Res 50(1):379–392. https://doi.org/10.2166/nh.2018.040

Pramudya Y, Onishi T (2019) Assessment of the Standardized Precipitation Index (SPI) in Tegal City, Central Java, Indonesia. IOP Conference Series: Earth and Environmental Science PAPER

Prasad PVV, Bheemanahalli R, Jagadish SVK (2017) Field crops and the fear of heat stress—Opportunities, challenges and future directions. Field Crop Res 200:114–121. https://doi.org/10.1016/J.FCR.2016.09.024

Qian L, Meng H, Chen X, Tang R (2023) Evaluating agricultural drought and flood abrupt alternation: a case study of cotton in the middle-and-lower Yangtze River, China. Agric Water Manag 283:108313. https://doi.org/10.1016/j.agwat.2023.108313

Qiu R, Li X, Han G, Xiao J, Ma X, Gong W (2022) Monitoring drought impacts on crop productivity of the U.S. Midwest with solar-induced fluorescence: GOSIF outperforms GOME-2 SIF and MODIS NDVI, EVI, and NIRv. Agric For Meteorol 323(November 2021):109038. https://doi.org/10.1016/j.agrformet.2022.109038

Radočaj D, Šiljeg A, Marinović R, Jurišić M (2023) State of major vegetation indices in precision agriculture studies indexed in web of science: a review. Agriculture (Switzerland) 13(3). https://doi.org/10.3390/agriculture13030707

Rani A, Gupta SK, Singh SK, Meraj G, Kumar P, Kanga S, Đurin B, Dogančić D (2023) Predicting future land use utilizing economic and land surface parameters with ANN and Markov chain models. Earth 4(3):728–751

Ray DK, Gerber JS, Macdonald GK, West PC (2015) Climate variation explains a third of global crop yield variability. Nat Commun 6:1–9. https://doi.org/10.1038/ncomms6989

Ray RL, Fares A, Risch E (2018) Effects of drought on crop production and cropping areas in Texas. Agric Environ Lett:1–5. https://doi.org/10.2134/ael2017.11.0037

Schlenker W, Roberts MJ (2009) Nonlinear temperature effects indicate severe damages to U.S. crop yields under climate change. Proc Natl Acad Sci USA 106(37):15594–15598. https://doi.org/10.1073/pnas.0906865106

Serrano SMV, Beguería S, Moreno JIL (2011) Comment on " Characteristics and trends in various forms of the Palmer Drought Severity Index (PDSI) during 1900 – 2008 " by Aiguo Dai. J Geophys Res Atmos 116:1–9. https://doi.org/10.1029/2011JD016410

Shahzad A, Ullah S, Dar AA, Sardar MF, Mehmood T, Tufail MA, Shakoor A, Haris M (2021) Nexus on climate change: agriculture and possible solution to cope future climate change stresses. Environ Sci Pollut Res 28(12):14211–14232. https://doi.org/10.1007/S11356-021-12649-8/METRICS

Shrestha BB, Kawasaki A, Zin WW (2021) Development of flood damage functions for agricultural crops and their applicability in regions of Asia. J Hydrol Reg Stud 36:100872. https://doi.org/10.1016/j.ejrh.2021.100872

Singh SK, Kotecha MJ, Kanga S, Pankhaniya SK, Agrawal S, Meraj G (2023a) Urban Heat Island (UHI) Resilience plan in varying climatic conditions using geospatial approach: a case study of Rajkot City. Clim Change Agric Soc 1:307–338

Singh BK, Delgado-Baquerizo M, Egidi E, Guirado E, Leach JE, Liu H, Trivedi P (2023b) Climate change impacts on plant pathogens, food security and paths forward. Nat Rev Microbiol 21(October):640–656. https://doi.org/10.1038/s41579-023-00900-7

Straffelini E, Tarolli P (2023) Climate change-induced aridity is affecting agriculture in Northeast Italy. Agric Syst 208(December 2022):103647. https://doi.org/10.1016/j.agsy.2023.103647

Sun L, Mitchell W, Davidson A (2011) Multiple drought indices for agricultural drought risk assessment on the Canadian prairies. Int J Climatol. https://doi.org/10.1002/joc.2385

Tamiru H, Dinka MO (2021) Application of ANN and HEC-RAS model for flood inundation mapping in lower Baro Akobo River Basin, Ethiopia. J Hydrol Reg Stud 36:100855. https://doi.org/10.1016/j.ejrh.2021.100855

Terán-Chaves CA, Duarte-Carvajalino JM, Ipaz-Cuastumal C, Vega-Amante A, Polo-Murcia SM (2023) Assessing the vulnerability of maize crop productivity to precipitation anomalies: a case study in the semiarid region of Cesar, Colombia. Water (Switzerland) 15(11). https://doi.org/10.3390/w15112108

Tian L, Yuan S, Quiring SM (2018) Agricultural and forest meteorology evaluation of six indices for monitoring agricultural drought in the south- central United States. Agric For Meteorol 249(December 2017):107–119. https://doi.org/10.1016/j.agrformet.2017.11.024

Tripathi G, Shakya A, Upadhyay RK, Singh SK, Kanga S, Pandey SK (2023) Landslide Susceptibility Mapping of Tehri Reservoir Region Using Geospatial Approach. In: Climate Change Adaptation, Risk Management and Sustainable Practices in the Himalaya, pp 135–156. https://link.springer.com/chapter/10.1007/978-3-031-24659-3_7

Urrutia JM, Scheffczyk K, Riembauer G, Mendoza J, Yanez D, Jímenez S, Ramírez A, Acosta M, Argüello J, Huerta B, Neteler M, Walz Y (2022) A validated geospatial model approach for monitoring progress of the Sendai Framework: the example of people affected in agriculture due to flooding in Ecuador. Prog Disaster Sci 15:100233. https://doi.org/10.1016/j.pdisas.2022.100233

Vaddula Y, Singh K (2023) Progression of drip irrigation and fertigation in cotton across the globe and its future perspectives for sustainable agriculture: an overview. Appl Water Sci 13(9):1–20. https://doi.org/10.1007/S13201-023-01986-3

Wang J, Peng T, Xiang Y, Yin Z, Qi H (2023) Analysis of the characteristics of Uneven Spatio-temporal distribution in Wujiang River Basin over the last 60 years. Atmosphere 14(9):1356. https://doi.org/10.3390/atmos14091356

Waseem M, Khurshid T, Abbas A, Ahmad I, Javed Z (2022) Impact of meteorological drought on agriculture production at different scales in Punjab, Pakistan. Water Clim Change 13(1):113–124. https://doi.org/10.2166/wcc.2021.244

Wijayarathne DB, Coulibaly P (2020) Identification of hydrological models for operational flood forecasting in St. John's, Newfoundland, Canada. J Hydrol Reg Stud 27. https://doi.org/10.1016/j.ejrh.2019.100646

Wijerathna-Yapa A, Pathirana R (2022) Sustainable agro-food systems for addressing climate change and food security. Agriculture (Switzerland) 12(10):1–26. https://doi.org/10.3390/agriculture12101554

Wurl J, García CNM, Lamadrid MÁI (2015) Caracterización del peligro por inundaciones en el oasis La Purísima, Baja California Sur, México. Investigaciones Geograficas 87(87):76–87. https://doi.org/10.14350/rig.41858

Xin Y, Tao F (2019) Optimizing genotype-environment-management interactions to enhance productivity and eco-efficiency for wheat-maize rotation in the North China Plain. Sci Total Environ 654:480–492. https://doi.org/10.1016/J.SCITOTENV.2018.11.126

Yan H, Wang S, Wang J, Lu H, Guo A, Zhu Z, Myneni RB, Shugart HH (2016) Assessing spatiotemporal variation of drought in China and its impact on agriculture during 1982–2011 by using PDSI indices and agriculture drought survey data. J Geophys Res-Atmos 121(5):2283–2298. https://doi.org/10.1002/2015JD024285.Received

Yang W, Zhang L, Gao Y (2023) Drought and flood risk assessment for rainfed agriculture based on Copula-Bayesian conditional probabilities. Ecol Indic 146:109812. https://doi.org/10.1016/j.ecolind.2022.109812

Yuan S, Quiring SM (2014) Drought in the U.S. Great Plains (1980–2012): a sensitivity study using different methods for estimating potential evapotranspiration in the Palmer Drought Severity Index. J Geophys Res-Atmos. https://doi.org/10.1002/2014JD021970.Received

Zambrano-medina YG, Plata-rocha W, Monjardin-armenta SA, Franco-Ochoa C (2023) Assessment and forecast of shoreline change using geo-spatial techniques in the Gulf of California. Land 12(4):782. https://doi.org/10.3390/land12040782

Zhang Y (2021) Influence of frequent flood disaster on agricultural productivity of rice planting and structural optimization strategy. Microprocess Microsyst 82:103863. https://doi.org/10.1016/j.micpro.2021.103863

Zhang S, Chen S, Jin J, Wu G, Bolan NS, White JR, Shaheen SM, Rinklebe J, Chen Q (2022) Incorporation of calcium cyanamide and straw reduces phosphorus leaching in a flooded agricultural soil. Geoderma 428:116,150. https://doi.org/10.1016/j.geoderma.2022.116150

Zhao C, Liu B, Piao S, Wang X, Lobell DB, Huang Y, Huang M, Yao Y, Bassu S, Ciais P, Durand JL, Elliott J, Ewert F, Janssens IA, Li T, Lin E, Liu Q, Martre P, Müller C et al (2017) Temperature increase reduces global yields of major crops in four independent estimates. Proc Natl Acad Sci USA 114(35):9326–9331. https://doi.org/10.1073/pnas.1701762114

Zipper SC, Qiu J, Kucharik CJ (2018) Drought effects on US maize and soybean production: spatiotemporal patterns and historical changes Drought effects on US maize and soybean production: spatiotemporal patterns and historical changes. Environmental Research. https://iopscience.iop.org/article/10.1088/1748-9326/11/9/094021

Chapter 2
Climate Change Impacts on Water Resources and Implications for Agricultural Management

Hafsa Muzammal, Muhammad Zaman, Muhammad Safdar, Muhammad Adnan Shahid, Muhammad Khalil Sabir, Aka Khil, Aamir Raza, Muhammad Faheem, Jahanzaib Ahmed, Jaffar Sattar, Muhammad Sajid, and Aurang Zaib

Abstract Water supplies are severely impacted by climate change, providing serious problems for agricultural management and food security. Water availability is changing as a result of changed patterns of precipitation, more frequent droughts, and higher evaporation rates, which has an effect on crop growth and yield potential. Water shortage becomes greater by glacier retreat and sea level rise, particularly in coastal and glacier-fed areas. The sustainability of agricultural practices is at risk by

---

H. Muzammal · M. Safdar (✉) · M. Adnan Shahid
Department of Irrigation and Drainage, University of Agriculture, Faisalabad, Pakistan

Agricultural Remote Sensing Lab (ARSL), University of Agriculture, Faisalabad, Pakistan

M. Zaman · M. K. Sabir · A. Raza
Department of Irrigation and Drainage, University of Agriculture, Faisalabad, Pakistan

A. Khil
Department of Agronomy, University of Agriculture, Faisalabad, Pakistan

M. Faheem
Department of Farm Machinery and Power, University of Agriculture, Faisalabad, Pakistan

J. Ahmed
Directorate of Agriculture Research (Pulses), Agriculture Research Institute, Sariab Quetta, Pakistan

J. Sattar
Department of Irrigation and Drainage, University of Agriculture, Faisalabad, Pakistan

Department of Agricultural Engineering, Khwaja Fareed University of Engineering and Information Technology, Rahim Yar Khan, Pakistan

M. Sajid
Institute of Micro-Nano Optoelectronics, Optical Engineering, Shenzhen University, Shenzhen, China

A. Zaib
Directorate of Agriculture Research Water Management & High Efficiency Irrigation System, Agriculture Research Institute, Sariab Quetta, Pakistan

S. Kanga et al. (eds.), *Transforming Agricultural Management for a Sustainable Future*, World Sustainability Series, https://doi.org/10.1007/978-3-031-63430-7_2

these changes, which affect existing water supply channels. Adaptive agriculture management techniques need to be implemented for managing these issues. The concentration on water-efficient irrigation techniques, like drip and sprinkler systems, together with rainwater gathering improving the water use efficiency and improves crop water availability. A variety of cropping methods, enhanced soil health management, and the selection of climate- and drought-resistant crop types all contribute to increased climate resilience. Furthermore, stakeholders can effectively predict and respond to threats correlated to climate change according to integrated water resource management and advance alerting systems. To maintain sustainable water management in agriculture, this chapter highlights the need of implementing adaptive techniques into use. Stakeholders may reduce the effects of climate change, increase food production, and increase the resilience of agricultural systems in facing the challenges of a changing climate by adopting these practices.

Keywords Water supplies · Climate change impacts · Agricultural management · Water availability · Drought and evaporation · Adaptive techniques · Irrigation efficiency · Climate resilience

2.1 Introduction

2.1.1 Background of Climate Change and Global Trends

The long-term changing of Earth's temperature patterns that results from human activities including the burning of fossil fuels, deforestation, and industrial operations is known as climate change, and it is an urgent worldwide issue, as explain in Fig. 2.1. As a result of these activities, a significant amount of greenhouse gases (GHGs), such as carbon dioxide (CO2), are released, improving the earth's natural

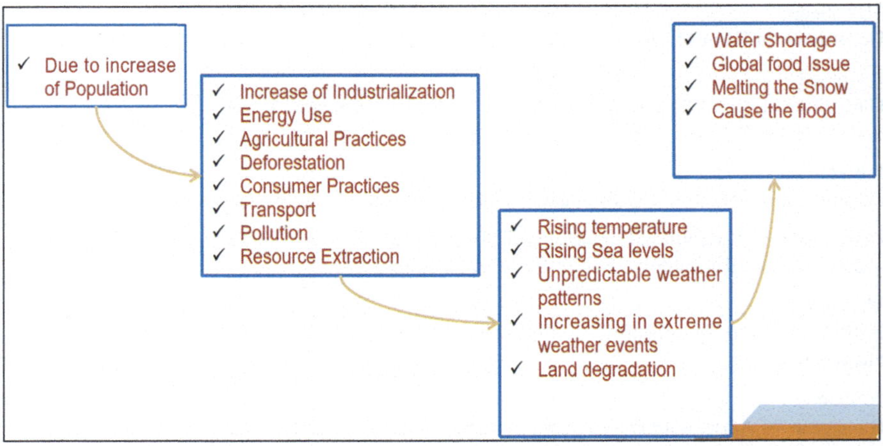

Fig. 2.1 Background of climate change and global trends

greenhouse effect, and reducing global warming (Singh et al. 2023a, b). This phenomenon caused to the rise of several interconnected worldwide trends. Additionally, the rise in average global temperatures over the past century has produced historic the warmth with 2016 referred to as the warmest year yet. As a result of this warming trend, extreme weather events like storms, droughts, and heatwaves have become more frequent and more severe, with disastrous effects on infrastructure and communities. Additionally, the seas' increased acidity because of absorbing too much CO2 has put marine life in threat, particularly sensitive region ecosystems (Singh et al. 2023a, b). The Paris Agreement of 2015, which brought nations together to combat global warming and keep temperature increases to well below 2 degrees Celsius over per-industrial levels, is one example of how these patterns have caused international initiatives. The shift to renewable energy sources has risen at the same time, providing an exciting chance to reduce greenhouse gas emissions. The establishment of effective climate policies and regulations has become crucial to guiding societies towards sustainable and resilient futures as governments and communities adjust to the difficulties caused by climate change (IPCC 2021).

2.1.2 Importance of Water Resources in Agriculture

Water resources are essential to agriculture and act as a condition for the production and security of food worldwide. For irrigation, which is crucial for supporting crop development and yields, irrigation needs to be supported by sufficient and reliable water supply. In addition to ensuring the supply of food, efficient water management techniques in agriculture also support rural communities, economic stability, and overall sociolect-environmental satisfaction (Sqib et al. 2023).

Farmers may grow a larger variety of crops, increase growing seasons, and improve production by using irrigation, which is made possible by water resources. Irrigated agriculture makes up around 20% of the world's cultivated area currently contributes about 40% of food production globally, according to the Food and Agriculture Organization (FAO). Without irrigation, it is predicted that global food production will fall by nearly 50% (FAO 2020), highlighting the important part that water resources play in suffering agricultural productivity.

The management and availability of water also affect the standard of agricultural outputs and crops. Water which is either insufficient or of low quality can affect the health of crops and reduce output. When plants are going through essential developmental stages including germination, flowering, and fruit development, an adequate water supply is very important (Singh et al. 2023a, b). Additionally, water resources are essential for livestock production since they support the growth of forage and fodder as well as provide drinking water.

A solution must be found between the significance of water supplies for agriculture and growing competition for water among different industries. These issues become severe by climate change, pollution, and inefficient water usage practices, highlighting the importance of sustainable water management techniques. The long-

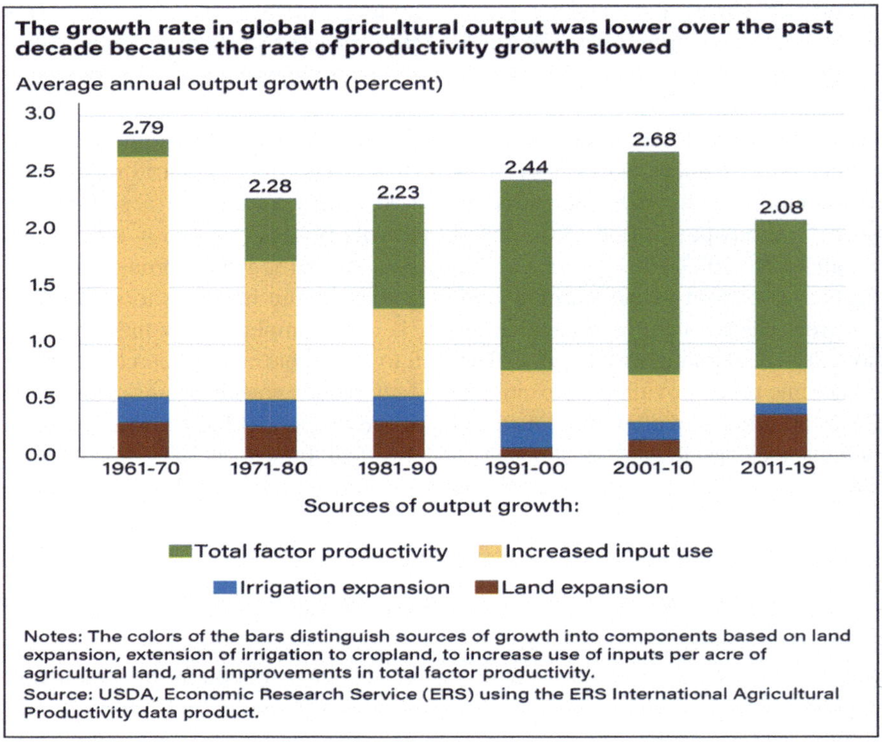

Fig. 2.2 Distinguish sources of growth global agricultural

term viability of water resources in agriculture depends on effective irrigation technologies, water conservation techniques, and integrated water resource management (Sud et al. 2023).

In conclusion, water resources are essential to agriculture to maintain food production, rural economies, and assure the safety of the world's food supply. For societies to address the problems caused by a growing population and a changing climate, effective water management practices are necessary to assure the continuous availability and quality of water for agricultural activities. Figure 2.2 demonstrates an increase in the use of inputs, but a decrease in the growth rate of agricultural products due to insufficient surface water supply for irrigation.

2.2 Change in Precipitation Patterns and Water Availability

Climate change is causing significant changes in precipitation patterns around the world. Some regions are becoming wetter, while others are becoming drier. These changes are having a major impact on water availability, which is essential for agriculture, industry, and human consumption. These changes in precipitation patterns are having a major impact on water availability. In some regions, water scarcity is becoming a serious problem. This is particularly true in arid and semi-arid regions, where water resources are already limited.

2.2.1 *Precipitation Patterns Variations*

Changes in precipitation patterns have significant effects on ecosystems, water supplies, and human cultures. Changes in rainfall timing, intensity, and distribution could have an effect on agriculture, water supply, and environmental stability as a whole (IPCC 2014). Frequently, these variations are associated with more general climate phenomena, such as alterations in air circulation patterns and surface warming.

Several regions have described variations in precipitation patterns over the past few decades, which have affected the hydrological cycle and the availability of water, as depicted in Fig. 2.3. Due to more frequent and intense rainfall events, the risk of flooding and soil erosion is elevated in some areas (Dai 2013). In contrast, other regions experience protracted droughts that impede the availability of water for residential and agricultural purposes (Trenberth et al. 2014).

The impact on agricultural output is one of the most alarming aspects of shifting precipitation patterns. Due to the fact that crops have varying water requirements at various stages of development, agriculture is highly susceptible to variations in precipitation. According to Lobell et al. (2014), an irregular distribution of precipitation can lead to crop failures and decreased yields, which influence food security and livelihoods.

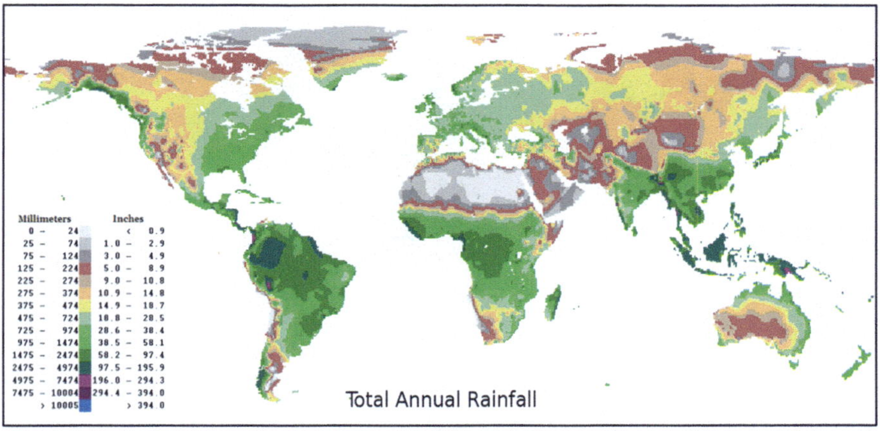

Fig. 2.3 Total annual rainfall of the whole world

Changes in precipitation patterns may also affect the freshwater supply in aquifers, lakes, and rivers. This may affect aquatic habitats and species, with cascading impacts on ecosystems and biodiversity (Vörösmarty et al. 2010). Changes in river flow patterns may influence hydropower production and other water-dependent industries.

Changes in precipitation could necessitate revaluation of urban infrastructure constructed for previous weather patterns. Field et al. (2012) state that urban inundation and property damage can result from increased rainfall intensity phases overworking drainage systems. Such incidents not only disrupt daily life, but also place a strain on emergency response and recovery resources.

Focusing on the problems caused by altering precipitation patterns requires an alternate strategy. Sustainable water management options, such as effective irrigation techniques and water conservation measures, are essential for adapting to changing conditions (Pereira et al. 2012). In addition, actions taken to prevent climate change through the reduction of emissions make it possible to reduce future precipitation unpredictability.

2.2.2 *Reduction in Water Availability*

The diminishing availability of water is posing significant issues for ecosystems, agriculture, and human populations. The decline in water availability, as seen in Fig. 2.4, is particularly evident in regions experiencing water stress. This can be attributed to alterations in precipitation patterns, heightened evaporation rates due to escalating temperatures, and the presence of conflicting demands for water resources (Vörösmarty et al. 2010). To assure continuous availability of this crucial resource,

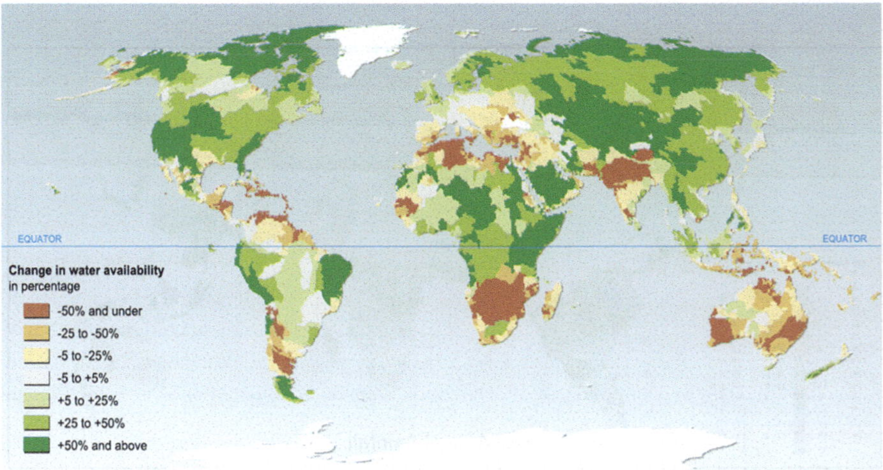

Fig. 2.4 Change in water availability

accurate management strategies are necessary for addressing the important implications it holds for sustainable development.

One consequence of reduced water availability is the increased demand placed on agricultural systems. In order to facilitate the growth and cultivation of crops, it is imperative that an adequate water supply is provided. The impact of water scarcity on agricultural production has been observed to result in decreased harvest yields and an increased likelihood of food shortages (Lobell et al. 2014). Specifically, for small-scale farmers who rely predominantly on rain-fed agriculture, this phenomenon can have adverse implications for both food prices and livelihoods.

Ecosystems and biodiversity are highly vulnerable to alterations in water availability. The phenomenon of decreased water flow in rivers and streams has the potential to disrupt aquatic habitats and jeopardize the well-being of animal species that rely on these water sources for their survival (Vörösmarty et al. 2010). Wetlands and marshes, integral to the process of water purification and the sustenance of diverse animals, exhibit heightened vulnerability to desiccation under conditions of water scarcity.

Adopting a holistic approach is important in order to effectively tackle the issue of declining water supplies. The use of water-efficient farming techniques, investment in water storage and management infrastructure, promotion of water conservation and reuse, and adoption of integrated water resource management strategies are some instances of this (UNEP 2016). In order to safeguard water supplies for current and future generations, it is imperative to address the underlying factors contributing to reduced water availability, including climate change and unsustainable water utilization.

2.2.3 Seasonal Variability in Water Supply

Water supply seasonality is a natural occurrence with substantial effects on ecosystems, agriculture, and human cultures. Throughout the year, variations in precipitation, temperature, and hydrological patterns can affect water availability, affecting many sectors and requiring adaptive measures for long-term water management. When analyzing water supply, the seasonal distribution of precipitation is extremely important (Upadhyay et al. 2023). There are usually distinct wet and dry seasons in many places, which generally correspond to monsoon or rainy seasons. There is a shortage of water resources because the wet season brings more rainfall, replenishing rivers, lakes, and groundwater. This great supply maintains ecosystems, supports farming, and refills soil moisture. The dry season, on the other hand, can result in decreased water availability, affecting crop growth, hydropower production, and other water-dependent organizations (Bates et al. 2008). The dry season is marked by reduced rainfall and higher evaporation rates.

Water supply seasonality in agriculture affects the timing of planting and harvesting. Farmers frequently depend on seasonal rainfall to water their crops, and changes in precipitation can affect crop production. Droughts during the dry season can lead to crop failures and food insecurity in places where rainfall is the primary supply of water for agriculture (Kundzewicz et al. 2007). Seasonal variations in the water supply have a significant impact on ecosystems as well. Water levels and flow rates fluctuate in aquatic ecosystems, affecting the behavior and migration patterns of aquatic animals. Wetlands and marshes are especially vulnerable to changes in water supply because they are essential for preserving biodiversity while providing ecosystem services. During the dry season, decreased water levels can result in habitat loss and the degradation of these important ecosystems (Davidson 2014).

Integrated water resource management is crucial to overcoming the problems caused by seasonal fluctuations in water availability (Tomar et al. 2023). This strategy takes into account the complete water cycle and includes strategies for water distribution, storage, and conservation. Improvements in water facilities, such as irrigation systems and reservoirs, can aid agriculture by reducing the effects of dry seasons and ensuring a more reliable water supply (UN-Water 2018).

2.3 Extreme Weather Events and Agricultural Risks

Extreme weather events, such as droughts, floods, heat waves, and cold snaps, are becoming more frequent and severe due to climate change. These events pose significant risks to agriculture, as they can damage crops and livestock, reduce yields, and disrupt supply chains. Extreme weather events pose a number of risks to agricultural producers. Extreme weather events can also have a negative impact on food security. When crop yields are reduced, food prices can increase. This can make it difficult for people to afford the food they need. we have to reduce the vulnerability of agriculture to extreme weather events and ensure that everyone has access to the food they need.

2.3.1 Frequent Extreme Precipitation Flows and Drought Events

Droughts and frequent extreme precipitation events are important indicators of how the climate is changing and have significant effects on ecosystems, water supplies, and human traditions. These situations highlight the complex relationships between water supply and climatic uncertainty, providing problems for water management and resilience. The impact of drought on food crop farming is seen in Fig. 2.5. A significant proportion of drought occurrences, namely 41%, resulted in suboptimal agricultural yields, while an additional 35% of drought events led to the withering and mortality of crops. Furthermore, it was found that 12% of the participants

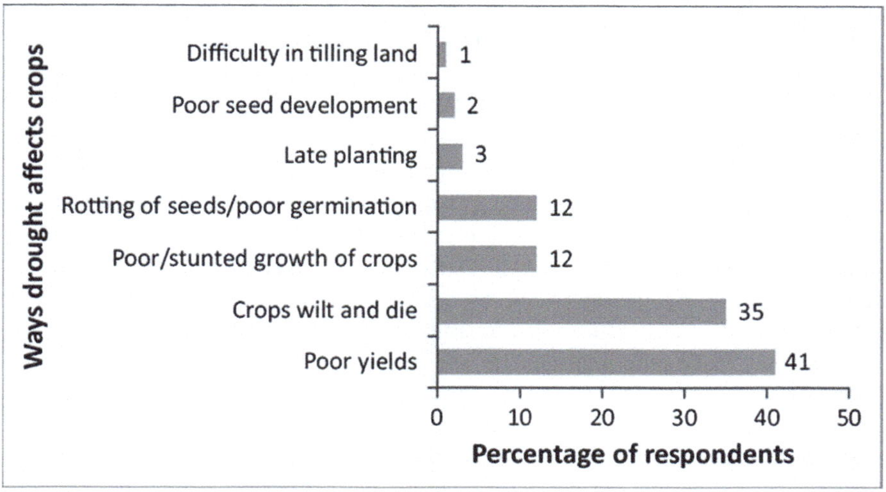

Fig. 2.5 Exploring the consequences of drought on food crop farming. (Source: Derbile et al. 2016)

ascribed the cause of poor germination to drought, while an additional 12% attributed poor growth to the same factor.

Flash floods, landslides, and infrastructure damage can result from extreme precipitation events, which occur when there is excessively high and severe rainfall over a brief period of time. These occurrences have the potential to clog drainage systems and cause severe harm, particularly in large cities that lack the resources to deal with an unexpected flood of water (Hirabayashi et al. 2013). Extreme precipitation can also increase runoff, which can result in soil erosion, sedimentation of water bodies, and problems with water quality (Duan et al. 2016). Such occurrences have wide-ranging implications, alarming communities, uprooting populations, and taxing emergency response mechanisms.

On the other hand, challenging periods of especially low precipitation are referred to as drought events, and they cause water scarcity and decreased soil moisture. Droughts can have a significant negative impact on agriculture, which can lead to crop failure, decreased yields, and financial losses (Wilhite 2000). Droughts may also restrict the availability of water for households, businesses, and energy production, increasing challenges and competition over limited resources (Browne 2019).

2.3.2 *Impacts of Floods on Crops*

Floods have severe negative effects on crops, preventing agricultural systems and resulting in major losses in production and yield. According to Zhang et al. (2015), excessive flooding during floods can result in wet soil conditions that suffocate plant roots and prevent nutrient uptake. In addition, prolonged standing water can encourage the development of infections, causing root and stem disorders that further risk the health of plants (Mohanty et al. 2014).

Additionally, floods frequently cause sediment deposition and soil erosion, which change the soil's structure and nutrient distribution. According to Brady and Weil (2016), this may result in a reduction in soil fertility and crop development. Floodwaters' physical force has the potential to uproot or harm crops, destroy standing vegetation, and lower yields (Jain et al. 2019). Floods can also disrupt critical growth stages and lower crop quality overall by delaying planting and harvesting schedules (IPCC 2014). The destruction of infrastructure like irrigation systems and storage facilities may make crop management and post-harvest handling more challenging (FAO 2016).

As a result of water logging, nutritional depletion, disease transmission, and physical damage to crops, floods pose a serious danger to agricultural productivity. Particularly in areas that are at risk for flooding, these effects contribute to shortages of food and economic losses. Figure 2.6 illustrates the impact of flood disturbance on the production of crops. The red region delineates the impact of flood-induced disturbance on agricultural growth, while the green region signifies the eventual recuperation of the crop.

2.3.3 Impacts of Droughts and Water Stress on crops

Crops are significantly impacted by droughts and water stress, which prevent photosynthesis, slow growth, make it difficult to absorb nutrients, and cause decay and early maturity. According to Rizhsky et al. (2004) and Tardieu et al. (2018), these stresses risk agricultural yields, quality, and reproductive development, increasing the danger of food shortages and financial losses. The negative impacts are increased by nutrient deficits, greater susceptibility to pests and diseases, and soil

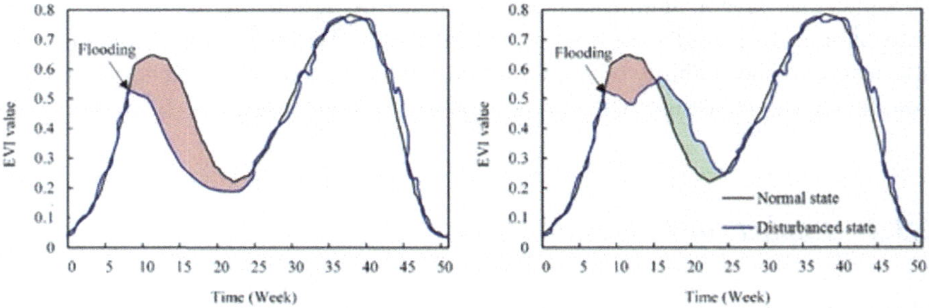

Fig. 2.6 Visualizing the influence of flood disturbance on crop production. (Source: Wang et al. 2022)

degradation (Liu et al. 2019; Hochman et al. 2020). To increase resilience in the face of shifting water supplies, mitigation options include adopting drought-tolerant crop varieties, optimizing irrigation management, and advancing sustainable agricultural practices (Kumar and Sharma 2018).

2.4 Melting Glaciers and Snow-pack

Glaciers and snowpack are important reservoirs of freshwater. They provide water for drinking, irrigation, hydropower, and other human uses. They also play an important role in the water cycle, regulating river flows and replenishing groundwater. Climate change is causing glaciers and snowpack to melt at an accelerated rate. This is due to rising global temperatures. As the Earth warms, glaciers and snowpack melt faster, releasing more water into the environment. It is important to take action to mitigate the negative impacts of melting glaciers and snowpack. This is because these reservoirs of freshwater are essential for human life and ecosystems.

2.4.1 Role of Glaciers and Snow-Pack in Freshwater Supply

Glaciers and snowfall play a crucial role in ensuring the availability of freshwater for various ecosystems and human populations. These cryosphere formations serve as reservoirs that accumulate water in the form of ice and snow from precipitation, retaining it until the arrival of warmer seasons, when it is progressively discharged as meltwater. The rivers, streams, and aquifers that sustain ecosystems, agriculture, and human consumption serve as inherent hydrological systems, acting as natural reservoirs that enable the movement of water. In arid and semi-arid environments, the gradual melting of glaciers and the annual melting of snowpack play a crucial role in providing a reliable water supply. The ice and snow formations serve to refill rivers and groundwater, thus functioning as a protective measure against water scarcity during periods of low precipitation according to Immerzeel et al. (2010).

2.4.2 Impacts of Glacier Retreat on Downstream Water Resources

There have been significant effects on downstream water resources as a result of glacial retreat. In fact, the decline in glacier volume has caused up to a 50% decrease in river flow during the dry season in many glacier-fed river basins, such as the Indus in South Asia (Immerzeel et al. 2010). The decrease in flow has a direct impact on the communities and ecosystems downstream that rely heavily on the steady supply of meltwater. Studies have shown a 20% decline in annual discharge from glacier-fed rivers due to retreat in areas like the Andes, where glacier melt considerably contributes to water supply (Vuille et al. 2008). By competing with irrigation plans and hydropower production, this decline in river flow can have an effect on both agricultural productivity and energy supplies.

Additionally, glacial retreat has an impact on the temporal distribution of water supplies. Peak flows in the Western United States have been observed to be higher than usual due to earlier spring melting caused by declines in snowpack resulting from increasing temperatures (Mote et al. 2005). Strategies for managing water resources that are based on previous hydrological patterns may be evaluated by this change in time. These situations highlight the significance of taking glacial retreat into account when estimating and planning for downstream water availability, especially in regions that largely rely on glacier meltwater.

2.5 Rising Temperatures and Evaporation

Rising global temperatures are leading to increased evaporation of water from the Earth's surface. This is because warmer air can hold more moisture. Increased evaporation can lead to several negative consequences, reduced water availability, increased risk of drought, increased risk of wildfires, sea level rise. We must reduce the negative impacts of rising temperatures and evaporation on water resources and ecosystems.

2.5.1 Higher Temperatures Leading to Increased Evaporation Rates

Research has shown evidence that the escalation of temperatures resulting from climate change leads to heightened rates of evaporation, consequently impacting the supply of water in various regions. For example, research findings depicted in Fig. 2.7 indicate that a rise in temperature by 1 °C can lead to an elevation in evapotranspiration rates ranging from 5% to 15% (Allen et al. 2017). This fast evaporation is responsible for a greater depletion of water from both soil and surface water bodies.

Elevated temperatures have the potential to induce soil desiccation in agricultural regions, hence impacting the crucial moisture levels necessary for optimal plant development. The phenomenon of temperature elevation leads to an augmented water consumption by crops via evapotranspiration, potentially exacerbating strain on the available water resources. The aforementioned phenomenon is particularly concerning in regions where irrigation is already necessary to compensate for insufficient natural precipitation. The decrease in soil moisture due to the escalating

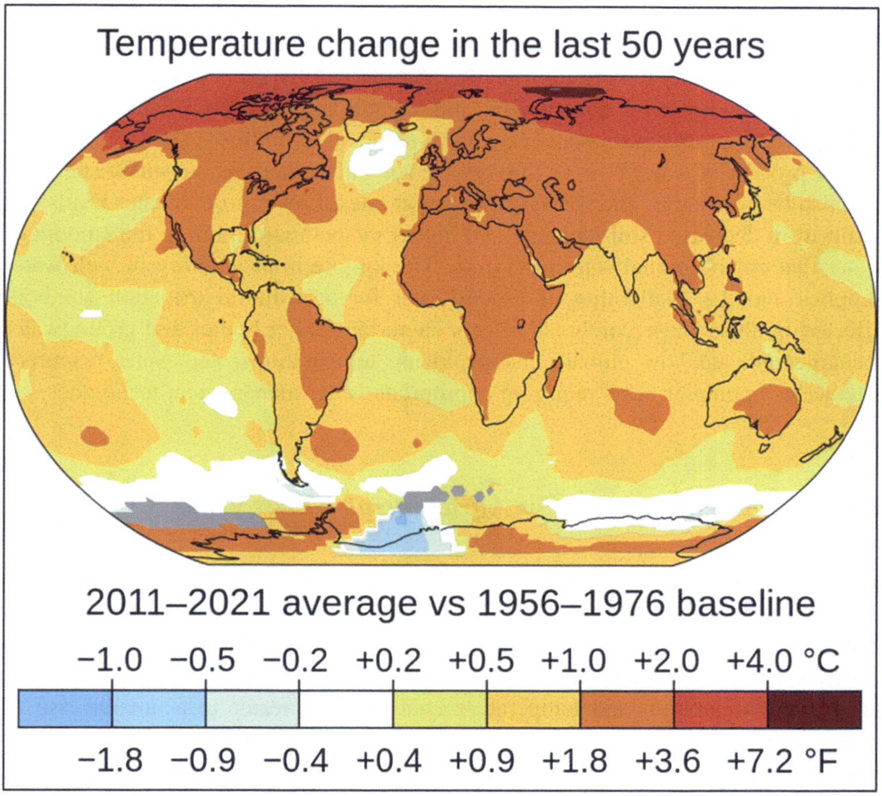

Fig. 2.7 Changes in temperature

rates of evaporation has implications for the uptake of water by crops. This particular circumstance has the potential to lead to diminished crop quality, reduced crop yields, and heightened irrigation requirements. In regions characterized by water scarcity, it is imperative to adapt agricultural practices in order to enhance soil water retention and optimize water utilization efficiency.

2.5.2 Effects on Surface Water Bodies and Groundwater Recharge

The effects of warmer temperatures and higher evaporation rates brought about by climate change have a significant impact on both surface water bodies and groundwater recharge. According to studies, increasing temperatures can cause surface reservoir water levels to reduce by up to 16% and streamflow to decline by around 5–10% for every degree Celsius of warming (IPCC 2018). Surface water bodies, including lakes, rivers, and ponds, continue to experience increased water loss as evaporation rates increase. Reduced water volumes, shallower water depths, and reduced water quality may result from this. Additionally, populations that depend on downstream ecosystems, aquatic life, and streams for a variety of functions may experience downstream consequences from reduced streamflow.

The process of water flowing through the soil and recharging subsurface aquifers can also be affected by increasing evaporation rates. Lower groundwater levels can result from decreased soil moisture and higher evaporation, limiting the amount of water that enters groundwater reservoirs. The long-term availability of well water supplies and the availability of groundwater for agricultural irrigation are both affected by this. These combined effects on surface water bodies and groundwater recharge indicate how climate change, rising temperatures, and water resources interact in a complex way, requiring appropriate water management techniques.

2.5.3 Challenges in Suffer Agricultural Water Needs

In order to ensure the long-term viability of agricultural water requirements, it is imperative to address substantial challenges arising from escalating evaporation rates and a dynamic environmental landscape. As exemplified by a research conducted in the Mediterranean region, it has been determined that the combined impact of evaporation and temperature changes may result in a notable rise of approximately 20–30% in the demand for water in agricultural activities by the year 2050 (Fader et al. 2016). The escalating demands have the potential to exacerbate the scarcity of water resources.

The escalation of evaporation intensifies the struggle for valuable freshwater resources in regions where irrigation plays a vital role in sustaining food production, such as certain areas within sub-Saharan Africa and the Middle East. The likelihood of water scarcity escalates with the augmentation of crop water demands, potentially leading to crop failure, diminished yields, and food insufficiency. The task of forecasting and effectively handling water resources in regions experiencing fluctuations in precipitation patterns, such as heightened occurrences of droughts or unpredictable rainfall, becomes increasingly challenging. The (Fig. 2.8) is depicting the water requirements for food production from 1960 to projected needs in 2050, measured in cubic kilometers per year. The graph shows a steady increase in water demand from 1960 to 2002, with substantial additional increases required by 2015, 2030, and 2050 to meet hunger and poverty eradication targets. The 2015 target indicates a significant increase over 2002 levels, with further substantial increases needed by 2030 and 2050 to achieve poverty eradication. The data underscores the growing water demands tied to food production and the challenges of addressing global hunger and poverty. The aforementioned challenges underscore the pressing need for agricultural practices that possess the ability to adapt and mitigate the impacts of increasing evaporation rates, hence ensuring the sustainability of food production.

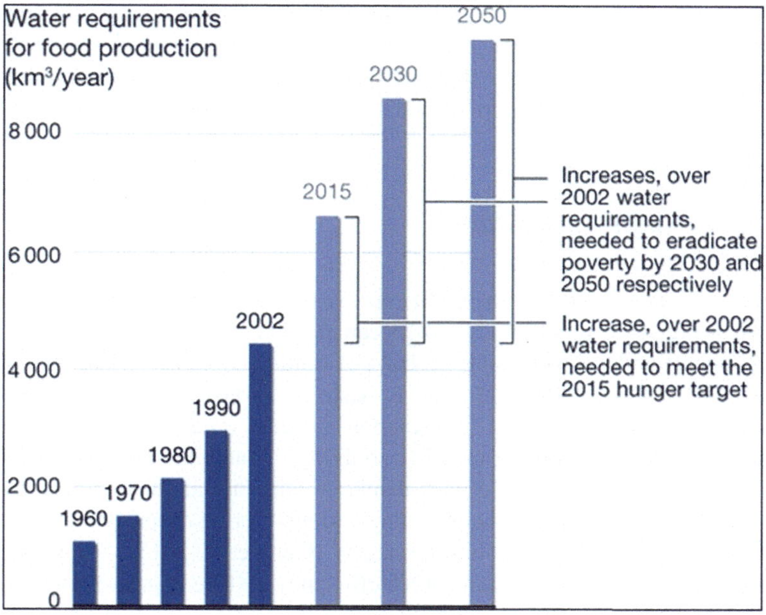

Fig. 2.8 Water requirement for food production 1960–2050. (Hugo Ahlenius et al. 2017)

2.6 Water Use Efficiency and Sustainable Agricultural Practices

Water use efficiency (WUE) is a measure of how much crop yield is produced per unit of water used. Sustainable agricultural practices are those that meet the needs of the present without compromising the ability of future generations to meet their own needs. Sustainable agricultural practices can help to improve WUE in a number of ways. For example, these practices can help to Improve soil health, Reduce water losses, Increase crop yields. Sustainable agricultural practices are essential for creating a more sustainable food system. By adopting these practices, farmers can help to protect the environment and produce healthy food for everyone.

2.6.1 Advanced Irrigation Techniques

Modern irrigation methods are essential for improving agriculture's water-use effectiveness and sustainability. In this case, drip irrigation provides water directly to plant roots, reducing loss of water and evaporation (Kumar et al. 2018). In order to optimize water application based on current plant needs, precision irrigation uses sensors and technology (Pannkuk et al. 2019). This encourages water usage that is efficient. Water is irrigated below the soil's surface to reduce evaporation and prevent weed growth (Yazar et al. 2008). In addition, graywater reuse and rainwater collection offer additional water sources, reducing the demand on freshwater resources (Di Francesco et al. 2021). Utilizing these advanced techniques may increase crop growth and support sustainable water management.

2.6.2 Precision Agriculture

Utilizing information and technology to improve different aspects of crop production is known as precision agriculture. Farmers can monitor fields and obtain information on crop health and growth using global positioning systems (GPS) and remote sensing (Keesstra et al. 2016). In order to make decisions about planting, fertilization, and irrigation, geographic information systems (GIS) analyze spatial data (Liakos et al. 2018). According to field variability, inputs like fertilizers and insecticides can be applied at specific rates thanks to variable-rate technology (Sui and Wang 2018). For the purpose of controlling and monitoring crops, unmanned aerial vehicles (UAVs) offer high-resolution imagery (Gago et al. 2015). Using precision agriculture methods can increase resource efficiency, reduce negative effects on the environment, and increase farm productivity.

2.6.3 Rainwater Harvesting and Water Storage

Water storage and harvesting of rainwater are sustainable practices that play a role in protecting water supplies for a variety of uses, such as irrigation, home requirements, and agriculture. Using tanks or cisterns, rainwater from roofs and other surfaces is collected and stored as part of the rainwater harvesting process (Alaerts et al. 2019). By using the water that has been conserved, less freshwater will be needed when it is dry (Ali and Abas 2015).

Reservoirs and ponds are examples of water storage systems that offer a way to store extra water during wet seasons for use in the future (Davies and Simonovic 2015). According to Rahman et al. (2015), these systems contribute to groundwater recharge, flood control, and agricultural irrigation. In areas at risk of water scarcity and seasonal rainfall variations, implementing rainwater collection and water storage practices can increase water availability, improve water use efficiency, and contribute to sustainable water management.

2.7 Integrated Water Resource Management

Integrated Water Resource Management (IWRM) is a holistic approach to water management that considers all aspects of the water cycle, including surface water, groundwater, and rainwater. It also considers the social, economic, and environmental impacts of water use. IWRM is important because water is a finite resource that is essential for life. It is also a resource that is under increasing pressure from climate change, population growth, and economic development. IWRM aims to ensure that water resources are used sustainably and equitably. It does this by promoting coordination and cooperation between different sectors, such as agriculture, industry, and energy. It also promotes the participation of stakeholders, such as communities, businesses, and NGOs.

IWRM is being implemented in a number of countries around the world. It is a complex process, but it is essential for ensuring sustainable water management in the face of growing challenges. IWRM is an essential tool for sustainable water management in the face of growing challenges. By adopting an integrated approach to water management, we can ensure that water resources are used sustainably and equitably for all.

2.7.1 Balancing Water Needs Across Sectors

Sustainable water allocation requires integrated management solutions to address the complicated task of balancing water needs across sectors. Effective solutions are essential to prevent conflicts and maintain equitable distribution as the demand for water rises in sectors like agriculture, industry, and urban areas (United Nations 2018). In order to optimize water distribution, integrated water resources management (IWRM) principles emphasize an integrated approach that takes into account social, economic, and environmental issues (Kalenikova 2009). In addition, demand management, efficient technology, and water pricing systems all contribute to maximizing water consumption (Brouwer et al. 2019).

Additionally, precision agriculture and remote sensing provide methods for maximizing resource utilization and raising agricultural productivity (Lobell et al. 2018). These developments in technology and data-driven techniques To create and execute effective climate-smart agricultural policies and strategies, interaction between governments, researchers, and communities will be necessary (FAO 2021).

2.7.2 Collaboration and Stakeholder Involvement

In order to ensure the efficacy of water resource management and agricultural adaptation strategies in the face of climate change, it is imperative to incorporate stakeholder participation and foster collaborative efforts. The collaboration of diverse stakeholders is crucial in the formulation of novel solutions and the promotion of sustainable utilization of water resources for agricultural purposes in a global context characterized by escalating challenges arising from water scarcity, erratic weather patterns, and environmental pressures.

In order to formulate inclusive policies and strategies that effectively address the unique challenges faced by different regions, it is imperative to foster collaboration among governmental bodies, local communities, agricultural practitioners, researchers, non-governmental organizations, and private sector entities. According to Pahl-Wostl (2006), stakeholders have the ability to provide solutions that are tailored to the specific situation, with the aim of promoting water conservation, enhancing agricultural yields, and mitigating climate change concerns. This can be achieved through the collaborative pooling of their knowledge, resources, and expertise.

The involvement of stakeholders extends beyond the mere involvement in decision-making processes, but also include their active participation in the planning, implementation, and monitoring of water management programs. Farmers, being the primary beneficiaries of water resources, possess distinct viewpoints regarding the local opportunities and limitations that prevail. The active participation of individuals ensures that the approaches employed are aligned with typical circumstances and yield discernible benefits (Prakash and Singh 2019).

2.8 Climate-Resilient Crops and Innovation in Agriculture

Climate change is posing a major challenge to agriculture. Extreme weather events, such as droughts, floods, and heat waves, are becoming more frequent and severe. This is making it difficult for farmers to grow crops and produce food. One way to address this challenge is to develop and grow climate-resilient crops. Climate-resilient crops are crops that can tolerate and thrive in a variety of climate conditions. They are essential for ensuring food security in the face of climate change.

Innovation in agriculture is also essential for adapting to climate change. This includes developing new technologies and practices that can help farmers to produce more food with less water and other resources. By developing and growing climate-resilient crops and adopting innovative agricultural practices, farmers can adapt to climate change and produce food for a growing population. By taking these steps, we can help to create a more climate-resilient food system that can meet the needs of a growing population in the face of climate change.

2.8.1 Research and Development of Climate-Resistant Varieties

The most important strategy for addressing the problems caused by climate change is research and development (R&D) of various kinds that are resistant to the effects of the environment. In order to achieve this, crop types that are resilient to elements like heat, drought, pests, and diseases must be produced and selected (Dwivedi et al.

2017). Scientists are working to improve features including heat tolerance, water use efficiency, and insect resistance through conventional breeding and advanced biotechnological methods (Varshney et al. 2019). In a time of climate uncertainty, these initiatives help to maintain agricultural productivity and provide food security.

2.8.2 Adoption of Innovative Agricultural Practices

The deployment of novel agricultural practices is crucial for bolstering output, sustainability, and resilience in response to climatic change and environmental challenges. The aforementioned techniques encompass a range of methodologies, namely precision farming, integrated pest management, conservation agriculture, and agroforestry. According to Midingoyi et al. (2019), farmers can benefit from advantages such as enhanced crop yields, improved soil health, and reduced input costs. Furthermore, the implementation of agroecological techniques such as the utilization of cover crops and the diversification of crops has been shown to enhance ecosystem services and contribute to the conservation of biodiversity (Gliessman 2015). The widespread use of these approaches may yield significant benefits in terms of achieving global food security and mitigating the impacts of climate change.

2.9 Policy and Governance for Climate Adaptation in Agriculture

Policy and governance play a critical role in supporting climate adaptation in agriculture. Governments can develop and implement policies that promote climate-smart agriculture practices, such as water-efficient irrigation, crop rotation, and cover cropping. Governments can also invest in research and development of climate-resilient crops and agricultural technologies. In addition, governments can create an enabling environment for climate adaptation in agriculture by providing financial assistance to farmers, educating farmers about climate change, and promoting sustainable agricultural practices.

It is important to note that climate adaptation in agriculture is a complex issue, and there is no one-size-fits-all approach. The best policy and governance measures will vary depending on the specific context of each country or region. However, the measures described above can provide a starting point for developing and implementing effective climate adaptation policies and governance for agriculture. By taking these steps, governments can develop and implement effective policy and governance measures to support climate adaptation in agriculture. This will help to ensure that the agricultural sector is resilient to climate change and can continue to produce food for a growing population.

2.9.1 Government Initiatives and Support

The promotion of sustainable agricultural practices and the effective management of environmental challenges necessitate robust government backing and proactive measures. The inclusion of measures such as organic farming, agroecology, and efficient water management in various efforts is a common occurrence, often facilitated through the enactment of legislation, provision of subsidies, and implementation of programs (Huang et al. 2020). The promotion of innovative crop varieties and current technology is facilitated by the provision of financial and technical support (Morton et al. 2011). Furthermore, it has been suggested that farmers have the potential to improve their agricultural practices and make informed decisions through the utilization of research and extension programs supported by the government (Erenstein and Laxmi 2008). Governments have a crucial role in enhancing food security, promoting sustainable resource management, and bolstering agricultural resilience through the establishment of a conducive and supportive environment.

2.9.2 Climate-Smart Agricultural Policies

Climate-smart agriculture policies enjoy robust backing due to their endorsement of sustainable practices aimed at mitigating the impacts of climate change and enhancing agricultural resilience. The strategies under consideration encompass a range of techniques, including agroforestry, sustainable land management, promotion of climate-resilient crop varieties, and facilitation of efficient water utilization (FAO 2013). According to Lipper et al. (2014), the implementation of policies that facilitate the accessibility of climate information and technology, along with the provision of financial and technical assistance to farmers, is imperative. Agricultural development plans at the national level that incorporate climate considerations align with global climate objectives, consequently enhancing environmental sustainability and ensuring food security.

2.10 Conclusion

Changes in precipitation patterns, an escalation in the frequency of extreme weather events, and fluctuations in temperature regimes are exerting an influence on crop yields and water availability. The agricultural sector encounters significant challenges, particularly in regions characterized by constrained water resources resulting from shifts in precipitation patterns and escalating rates of evaporation. Droughts and floods exert detrimental effects on crop production, resulting in diminished yields, compromised crop quality, and heightened vulnerability of crops to pest infestations and diseases. The implementation of climate-resilient crop varieties, along with the adoption of sophisticated agricultural practices such as precision farming and conservation agriculture, has the potential to enhance agricultural resilience. Governments can facilitate the adoption of climate-smart agricultural practices and enhance adaptation efforts by enacting supportive legislation, offering incentives, and providing agricultural extension services. The implementation of successful climate adaptation strategies necessitates active engagement from local communities, the incorporation of indigenous knowledge systems, and the facilitation of inclusive participatory approaches.

In order to foster the cultivation of crop varieties capable of adapting to the dynamic conditions of a shifting climate, additional research and development efforts are necessary. Enhancing capacity would facilitate the adoption of adaptive practices among farmers and other agricultural professionals through the establishment of platforms for information exchange and training. The implementation of efficient water management strategies, such as the utilization of rainwater harvesting and advanced irrigation techniques, has the potential to mitigate the risks associated with water scarcity. In order to further the adoption of sustainable practices, it is imperative for governments to formulate and execute climate-smart agricultural policies that encompass both financial incentives and technical assistance. Engaging in decision-making processes and acknowledging the traditional knowledge of local populations can potentially lead to the development of adaptation methods that are better suited to the specific environmental context and more feasible in practice. International collaboration plays a pivotal role in facilitating the exchange of best practices and knowledge across countries, firms, and academics, hence fostering sustainable adaptation to climate change in the field of agriculture. To ensure the establishment of a robust and enduring agricultural sector, it is imperative to adopt a comprehensive and cooperative strategy that integrates scientific expertise, indigenous wisdom, and enabling governance measures to effectively tackle the challenges posed by climate change.

The effective administration of agricultural activities and water resources, taking into account the impacts of a dynamic climate, presents a range of potential advantages and obstacles. The Intergovernmental Panel on Climate Change (IPCC 2021) predicts that if climate change continues, there will be additional alterations in precipitation patterns and heightened fluctuations in temperature. According to Dai (2013), the increased occurrence and intensity of droughts may have implications for the accessibility of water for agricultural irrigation and crop development. Conversely, it is important to note that the cultivation of crops can be influenced by the variability of planting zones, as alterations in environmental conditions have the potential to generate novel prospects for agricultural productivity (Luedeling et al. 2009). Effectively managing these changes will be crucial. The assurance of water supply for agricultural purposes is achieved by the implementation of integrated water resource management strategies, encompassing the utilization of efficient irrigation methods and the collection of rainfall (Seyoum et al. 2020). Agroecology and conservation agriculture represent two instances of sustainable land management practices that have the potential to enhance soil health and water retention, hence mitigating the impacts of fluctuating precipitation patterns (Pittelkow et al. 2015).

Furthermore, the implementation of precision agriculture and remote sensing techniques offers strategies to optimize resource allocation and enhance agricultural productivity (Lobell et al. 2018). The advancements in technology and the utilization of data-driven methodologies. In order to develop and implement efficient climate-smart agriculture policies and strategies, it will be imperative to foster collaboration among governmental entities, researchers, and local people (FAO 2021).

References

Hugo Ahlenius et al (2017)

Alaerts GJ, van der Sande C, van der Steen P (2019) Design principles for rooftop rainwater harvesting systems: a review. Water 11(2):290

Ali M, Abas N (2015) Rainwater harvesting and its importance for water management and climate change adaptation in rural areas of Pakistan. Geosciences 5(3):332–346

Allen ST, Reba ML, Edwards BL, Keim RF (2017) Evaporation and the subcanopy energy environment in a flooded forest. Hydrol Process 31(16):2860–2871

Bates BC, Kundzewicz ZW, Wu S, Palutikof JP (Eds.). (2008) Climate change and water. Technical paper of the intergovernmental panel on climate change. IPCC Secretariat

Brady NC, Weil RR (2016) The nature and properties of soils. Pearson

Brouwer S, Pieron M, Sjerps R, Etty T (2019) Perspectives beyond the meter: A Q-study for modern segmentation of drinking water customers. Water Policy 21(6):1224–1238

Browne D (2019) The United Nations World Water Development Report. Leaving no one behind. Highways 88:3. https://doi.org/10.5334/bcw.d

Dai A (2013) Increasing drought under global warming in observations and models. Nat Clim Chang 3(1):52–58

Davidson NC (2014) How much wetland has the world lost? Long-term and recent trends in global wetland area. Mar Freshw Res 65(10):934–941

Davies EGR, Simonovic SP (2015) Water storage for managing climate change impacts on water resources: a case study of the Grand River watershed in Ontario. Canada Hydrol Res 46(6):826–843

Derbile EK, Dongzagla A, File DJ (2016) The double tragedy of agriculture vulnerability to climate variability in Africa: how vulnerable is smallholder agriculture to rainfall variability in Ghana? Jàmbá: J Disaster Risk Stud 8(3):1–9

Di Francesco S, Lanza LG, Volpe M (2021) Rainwater harvesting systems for sustainable water management in agriculture: a comprehensive review. Sci Total Environ 750:141534

Duan W, Quan X, Zhao H (2016) Effects of precipitation intensity on soil erosion and hydrological processes on steep slopes. Catena 137:286–297

Dwivedi SL, Sahrawat KL, Upadhyaya HD, Mengoni A, Galardini M, Bazzicalupo M et al (2017) Advances in host plant and rhizobium genomics to enhance symbiotic nitrogen fixation in grain legumes. Adv Agron 144:1–92

Erenstein O, Laxmi V (2008) Assessing the adoption potential of conservation agriculture practices in the Indo-Gangetic Plains of India. Food Policy 33(3):236–246

Fader M, Shi S, von Bloh W, Bondeau A, Cramer W (2016) Mediterranean irrigation under climate change: more efficient irrigation needed to compensate for increases in irrigation water requirements. Hydrol Earth Syst Sci 20(2):953–973

FAO (2013) Climate-smart agriculture sourcebook. Food and Agriculture Organization of the United Nations

FAO (2016) Coping with floods: the role of agricultural practices. Food and Agriculture Organization of the United Nations. Retrieved from http://www.fao.org/3/a-i5289e.pdf

FAO (2020) The global overview of water scarcity: North Africa & Middle East Region. Food and Agriculture Organization of the United Nations. Retrieved from http://www.fao.org/3/cb0593en/cb0593en.pdf

FAO (2021) Climate-smart agriculture sourcebook, 2nd edn. Food and Agriculture Organization of the United Nations

Field CB, Barros VR, Dokken DJ, Mach KJ, Mastrandrea MD, Bilir TE et al (2012) Managing the risks of extreme events and disasters to advance climate change adaptation: special report of the intergovernmental panel on climate change. Cambridge University Press

Gago J, Douthe C, Coopman RE, Gallego PP, Ribas-Carbó M, Flexas J (2015) UAVs challenge to assess water stress for sustainable agriculture. Agric Water Manag 153:9–19

Gliessman SR (2015) Agroecology: the ecology of sustainable food systems, 3rd edn. CRC Press

GRID-Arendal resources library by: Hugo Ahlenius www.grida.no/resources/6806

Hirabayashi Y, Mahendran R, Koirala S, Konoshima L, Yamazaki D, Watanabe S et al (2013) Global flood risk under climate change. Nat Clim Chang 3(9):816–821

Hochman Z, Lilley J, Pratley J, Kirkegaard J (2020) Impact of simulation and decision support systems on sustainable agriculture. In: Kirkegaard J, JP (eds) Australian Agriculture in, pp 345–366

Huang Q, Rozelle S, Vos R (2020) The subsidies and other incentives for pesticide use in China. Glob Food Sec 26:100389

Immerzeel WW, van Beek LPH, Bierkens MFP (2010) Climate change will affect the Asian Water Towers. Science 328(5984):1382–1385. https://doi.org/10.1126/science.1183188

IPCC (2014) Climate change 2014: impacts, adaptation, and vulnerability. Part A: global and sectoral aspects. contribution of working group II to the fifth assessment report of the intergovernmental panel on climate change

IPCC (2018) Global Warming of 1.5°C. An IPCC Special Report on the impacts of global warming of 1.5°C above pre-industrial levels and related global greenhouse gas emission pathways, in the context of strengthening the global response to the threat of climate change, sustainable development, and efforts to eradicate poverty [Masson-Delmotte V, Zhai P, Pörtner H-O, Roberts D, Skea J, Shukla PR, Pirani A, Moufouma-Okia W, Péan C, Pidcock R, Connors S, Matthews JBR, Chen Y, Zhou X, Gomis MI, Lonnoy E, Maycock T, Tignor M, & Waterfield T (Eds.)]. In Press

IPCC (2021) Climate change 2021: the physical science basis. Contribution of working group I to the sixth assessment report of the intergovernmental panel on climate change [Masson-Delmotte V, Zhai P, Pirani A, Connors S, Péan C, Berger S, ... & Zhou X (Eds.)]. Cambridge University Press

Jain V, Beyene M, Varay LS, Wasson RJ, Jain S (2019) Riverine flood hazard: Part A types, processes and causative factors. In Proc Indian Natl Sci Acad 85(1):43–64

Kalenikova I (2009) The integrated water basin approach for the sustainable water management in international and regional legislation (Doctoral dissertation)

Keesstra S, Mol G, De Leeuw J, Okx J, Molenaar C (2016) Soil mapping using a mobile NIR-Sensor. Catena 137:160–169

Kumar V, Sharma SK (2018) Drought management strategies for agriculture: recent trends and future challenges. In: Drought stress tolerance in plants. Springer, pp 105–135

Kumar RR, Singh S, Singh RK (2018) Drip irrigation: a modern technique for maximizing agricultural production and water saving. Int J Mech Eng Technol 9(10):1630–1637

Kundzewicz ZW, Mata LJ, Arnell NW, Doll P, Kabat P, Jiménez B, Stakhiv EZ (2007) Freshwater resources and their management. Climate change 2007: impacts, adaptation and vulnerability. Contribution of working group II to the fourth assessment report of the intergovernmental panel on climate change

Liakos B, Busato P, Moshou D, Pearson S, Bochtis D (2018) Machine learning in agriculture: a review. Sensors 18(8):2674

Lipper L, Thornton P, Campbell BM, Baedeker T, Braimoh A, Bwalya M et al (2014) Climate-smart agriculture for food security. Nat Clim Chang 4(12):1068–1072

Liu F, Jensen CR, Andersen MN (2019) Drought stress effect on carbohydrate concentration in soybean leaves and pods during early reproductive development: Its implication in altering pod set. Field Crop Res 235:52–61

Lobell DB, Burke MB, Tebaldi C, Mastrandrea MD, Falcon WP, Naylor RL (2014) Prioritizing climate change adaptation needs for food security in 2030. Science 319(5863):607–610

Lobell DB, Burke MB, Tebaldi C, Mastrandrea MD, Falcon WP, Naylor RL (2018) Prioritizing climate change adaptation needs for food security in 2030. Science 319(5863):607–610

Luedeling E, Zhang M, Girvetz EH (2009) Climatic changes lead to declining winter chill for fruit and nut trees in California during 1950–2099. PLoS One 4(7):e6166

Midingoyi SKG, Kassie M, Muriithi B, Diiro G, Ekesi S (2019) Do farmers and the environment benefit from adopting integrated pest management practices? Evidence from Kenya. J Agric Econ 70(2):452–470

Mohanty M, Lee SS, Bae H (2014) Impacts of flooding on agriculture and food security. Agric Nat Resour 48(1):109–116

Morton LW, Bitto EA, Oakland MJ, Sand M (2011) Solving the problems of agricultural runoff: does the solution depend on the framers? Ecol Econ 70(11):2278–2287

Mote PW, Hamlet AF, Clark MP, Lettenmaier DP (2005) Declining mountain snowpack in Western North America. Bull Am Meteorol Soc 86(1):39–49

Pahl-Wostl C (2006) The importance of social learning in restoring the multifunctionality of rivers and floodplains. Ecol Soc 11(1):10

Pannkuk CD, Colaizzi PD, Moorhead JE, Gowda PH, Marek GW (2019) Managing and monitoring center pivot irrigation using crop canopy temperature and remote sensing. Trans ASABE 62(4):921–935

Pereira LS, Cordery I, Iacovides I (2012) Improved indicators of water use performance and productivity for sustainable water conservation and saving. Agric Water Manag 108:39–51

Pittelkow CM, Liang X, Linquist BA, van Groenigen KJ, Lee J, Lundy ME et al (2015) Productivity limits and potentials of the principles of conservation agriculture. Nature 517(7534):365–368

Prakash A, Singh R (2019) Farmer participation in water management: a review of evidence and implications for governance and policy. Water Altern 12(2):298–319

Rahman AA, Salehin M, Inamura T (2015) Flood control and water storage in the haor basin of Bangladesh: a combined approach. Int J River Basin Manag 13(2):175–187

Rizhsky L, Liang H, Mittler R (2004) The combined effect of drought stress and heat shock on gene expression in tobacco. Plant Physiol 134(4):1683–1696

Seyoum A, Adamowski J, Ozga-Zielinski B, Gebremeskel Y (2020) Climate-smart agriculture: a review of an approach to improve agricultural resilience to climate variability and climate change impacts. Agriculture 10(9):390

Singh SK, Kanga S, Gulati B, Rai˘c M, Sajan B, Đurin B, Singh S (2023a) Spatial and temporal analysis of hydrological modelling in the Beas basin using SWAT+ Model. Water 15:3338. https://doi.org/10.3390/w1519333

Singh SK, Kotecha MJ, Kanga S, Pankhaniya SK, Agrawal S, Meraj G (2023b) Urban Heat Island (UHI) resilience plan in varying climatic conditions using geospatial approach: a case study of Rajkot City. In: Climate change, agriculture and society, vol 1, pp 307–338

Singh S, Meraj G, Kumar P, Singh SK, Kanga S, Johnson BA (2023c) Decoding Chambal river shoreline transformations: a comprehensive analysis using remote sensing, GIS, and DSAS. Water 15(9):1793

Sqib N, Rai PK, Kanga S, Kumar D, Đurin B, Singh SK (2023) Assessment of ground water quality of Lucknow city under gis framework using Water Quality Index (WQI). Water 15(17):3048

Sud A, Kanga R, Singh SK, Meraj G, Kanga S, Kumar P, Ramanathan AL, Bhardwaj V (2023) Simulating groundwater potential zones in mountainous Indian Himalayas—a case study of Himachal Pradesh. Hydrology 10(3):65

Sui D, Wang J (2018) Big data analytics in geographic information science: a systematic review. ISPRS J Photogramm Remote Sens 140:58–76

Tardieu F, Simonneau T, Muller B (2018) The physiological basis of drought tolerance in crop plants: a scenario-dependent probabilistic approach. Annu Rev Plant Biol 69:733–759

Tomar P, Singh SK, Kanga S, Pattanaik A (2023) Water bodies mapping and monitoring using high-resolution satellite images. Sustain Agric Food Environ Res 11

Trenberth KE, Dai A, Rasmussen RM, Parsons DB (2014) The changing character of precipitation. Bull Am Meteorol Soc 84(9):1205–1218

UNEP (2016) Global environment outlook: water resources. United Nations Environment Programme. Retrieved from https://www.unep.org/resources/report/global-environment-outlook-water-resources

UN-Water (2018) Integrated monitoring guide for SDG 6 – Target 6.4. United Nations. Retrieved from http://www.sdg6monitoring.org/integrated-monitoring-guide-for-sdg-6-target-6-4/

UN-Water (2021) Water scarcity. United Nations. Retrieved from https://www.unwater.org/water-facts/scarcity/

Upadhyay RK, Tripathi G, Đurin B, Šamanović S, Cetl V, Kishore N, Sharma M, Singh SK, Kanga S, Wasim M, Rai PK, Bhardwaj V (2023) Groundwater potential zone mapping in the Ghaggar River Basin, North-West India, using integrated remote sensing and GIS techniques. Water 15(5):961

Varshney RK, Thudi M, Roorkiwal M, He W, Upadhyaya HD, Yang W et al (2019) Resequencing of 429 chickpea accessions from 45 countries provides insights into genome diversity, domestication and agronomic traits. Nat Genet 51(5):857–864

Vörösmarty CJ, McIntyre PB, Gessner MO, Dudgeon D, Prusevich A, Green P et al (2010) Global threats to human water security and river biodiversity. Nature 467(7315):555–561

Vuille M, Francou B, Wagnon P, Juen I, Kaser G, Mark BG, Bradley RS (2008) Climate change and tropical Andean glaciers: past, present and future. Earth Sci Rev 89(3–4):79–96

Wang X, Liu Z, Chen H (2022) Investigating flood impact on crop production under a comprehensive and spatially explicit risk evaluation framework. Agriculture 12(4):484

Wilhite DA (2000) Drought as a natural hazard: concepts and definitions. A Global Assessment, Drought, pp 3–18

Yazar A, Şenay Y, Ekinci K (2008) Subsurface drip irrigation management principles. Int J Agric Biol 10(5):547–553

Zhang L, Xu W, Liu W, Jiang G, Li Y (2015) Effects of waterlogging on plant nutrient concentrations and stoichiometry in herbaceous communities along the Yellow River, China. Ecol Eng 75:70–75

Chapter 3
Managing Water Resources for Sustainable Agricultural Production

Rehan Mehmood Sabir, Abid Sarwar, Muhammad Shoaib, Azka Saleem, Mohamad Hasan Alhousain, Syed Aftab Wajid, Fahd Rasul, Muhammad Adnan Shahid, Lubna Anjum, Muhammad Safdar, Nalain E. Muhammad, Rana Muhammad Waqas, Usman Zafar, and Aamir Raza

Abstract Sustainable agricultural production and maintaining long-term productivity depend on effective water management. This chapter covers multiple aspects of water resource management, such as determining the amount of water needed, effective irrigation methods, controlling soil moisture, choosing sustainable crops, and integrating technology. Various precision irrigation methods such as sprinkler irrigation techniques, drip irrigation systems, and the delicate balance between water efficiency and coverage, emphasize the interaction between water and plant growth.

R. M. Sabir (✉) · M. Adnan Shahid · M. Safdar
Department of Irrigation and Drainage, University of Agriculture, Faisalabad, Pakistan

Agricultural Remote Sensing Lab (ARSL), University of Agriculture, Faisalabad, Pakistan

A. Sarwar · L. Anjum · U. Zafar · A. Raza
Department of Irrigation and Drainage, University of Agriculture, Faisalabad, Pakistan

M. Shoaib
Institute of Agro- Industry and Environment, The Islamia University, Bahawalpur, Punjab, Pakistan

A. Saleem
Department of Botany, University of Agriculture, Faisalabad, Pakistan

M. H. Alhousain
Department of Structures & Environmental Engineering, University of Agriculture, Faisalabad, Pakistan

S. A. Wajid · F. Rasul
Department of Agronomy, University of Agriculture, Faisalabad, Pakistan

N. E. Muhammad
Department of Irrigation and Drainage, University of Agriculture, Faisalabad, Pakistan

National University of Science and Technology, Islamabad (NUST), Islamabad, Pakistan

R. M. Waqas
Institute of Agricultural Extension, Education and Rural Development, University of Agriculture, Faisalabad, Pakistan

S. Kanga et al. (eds.), *Transforming Agricultural Management for a Sustainable Future*, World Sustainability Series, https://doi.org/10.1007/978-3-031-63430-7_3

The significance of assessing soil moisture and including it in irrigation decision-making processes. Discusses drought-tolerant and resistant crop varieties, conservation tillage techniques, and cover crops to demonstrate the symbiotic relationship between sustainable crop selection and water management. The combination of technological innovations and data-driven strategies, such as satellite imaging, GIS, and remote sensing help preserve the resources. This chapter emphasizes the necessity of comprehensive water resource management, including sustainable runoff and drainage, and incorporating natural systems such as artificial wetlands. The chapter ends with a summary of the main ideas and a focus on the relationship between integrated water management and sustainable agriculture. The encouragement to action stimulates the use of knowledge and practices to guarantee a future with safe access to food and water.

Keywords Sustainable agricultural production · Water management · Irrigation methods · Soil moisture · Sustainable crops · Precision irrigation · Technological innovations

3.1 Introduction

The effective management of water resources to ensure sustainable agricultural output is a critical and urgent global concern, given the concurrent factors of population growth, climate change, and escalating food demand. Water plays a crucial role in agricultural systems, providing an essential component for the growth and sustenance of crops and cattle. Nevertheless, the unpredictable availability and distribution of freshwater resources are progressively being influenced by fluctuating weather patterns and the degradation of aquifers (Mishra 2023). Therefore, it is imperative to employ proactive and sustainable strategies to effectively harness water supplies for agricultural purposes, while simultaneously maintaining the fragile equilibrium of ecosystems (Singh et al. 2023a, b, c).

Sustainable water management in the agricultural sector involves the implementation of various techniques and practices to enhance water utilization efficiency, mitigate wastage, and minimize adverse effects on the environment (Irfeey et al. 2023). This involves the utilization of contemporary irrigation methods, like drip or precision irrigation, to deliver an optimal quantity of water with accuracy, targeting specific locations and timing. The implementation of conservation and reclamation strategies, such as the utilization of rainwater harvesting and recycling techniques, has the potential to significantly augment the accessibility of water resources (Singh et al. 2023a, b, c). Furthermore, the implementation of drought-tolerant crop cultivars and the adoption of crop rotation techniques have the potential to alleviate water-related hazards. To maintain agricultural output that is resilient, ecologically responsible, and capable of fulfilling the increasing global food demands, it is imperative to integrate these strategies with novel technologies and comprehensive water governance systems.

3.1.1 Importance of Efficient Water Management in Agriculture

The efficient use of water in agriculture holds significant importance, not only within the agricultural sector but also in the wider context of global sustainability. Agriculture is recognized as one of the primary contributors to freshwater resource consumption, constituting a significant proportion of worldwide water utilization (Sqib et al. 2023). With the ongoing growth of the world population and the increasing impact of climate change on water supplies, the prudent use of water in agriculture becomes a critical necessity (Kim et al. 2016). The potential consequences of inefficient water utilization include the exacerbation of water shortages, which in turn can have detrimental effects on agricultural yields, food security, and the livelihoods of a significant portion of the global population who rely on agriculture for their survival. Hence, the implementation of effective water management methods is of utmost importance to secure sufficient water supply for agricultural irrigation, while simultaneously preserving ecosystem integrity and safeguarding the limited global freshwater supplies (Singh et al. 2023a, b, c).

The agricultural sector's inefficient water management practices can lead to various issues, including waterlogging, soil salinization, ecological imbalances, and water scarcity (Rasul 2016). Over-irrigation can cause waterlogging, soil salinization, and hinder crop growth. Over-extraction of water from rivers and aquifers can cause ecological imbalances, disturbance to aquatic ecosystems, and exacerbate water scarcity issues. Agriculture also contributes to water pollution through agrochemical discharge, deteriorating water quality and posing risks to aquatic ecosystems and public well-being. Implementing effective water management strategies like precision irrigation, rainwater harvesting, and integrated water resource management can enhance productivity while demonstrating environmental conscientiousness and resilience. This aligns with sustainable agriculture and the preservation of Earth's valuable water resources (Sud et al. 2023).

3.1.2 Significance of Sustainable Practices for Long-Term Agricultural Productivity

Sustainable agriculture is crucial for long-term agricultural productivity, addressing global food needs and ensuring future generations can meet them (Wang et al. 2022). It involves practices like soil nurturing, water optimization, and reduced reliance on agrochemicals, which maintain farmlands in a fertile, resilient state. Sustainable agriculture also addresses environmental issues like soil erosion, water pollution, and biodiversity depletion, preserving ecosystem balance and mitigating climate change impacts. It also boosts farmers' economic well-being by reducing input costs, improving crop resilience, and extending market options. As the global population grows and environmental challenges increase, implementing sustainable practices is not just a choice but a necessity to ensure a successful future in harmony with the natural system.

3.2 Assessing Water Needs

3.2.1 Understanding Crop Water Requirements

It is necessary to have a complete understanding of the water needs of the crops in order to put into practice irrigation management practices that are both successful and sustainable. The term "crop water requirements," or CWR for short, refers to the amount of water that is required for a crop to achieve its maximum potential for growth. CWR is affected by a wide range of external circumstances, which include the following:

The water requirements of a crop are significantly influenced by the climatic conditions in which it is cultivated. Crops cultivated in regions characterized by hot, arid, windy, and sun-drenched climates are anticipated to exhibit a greater Crop Water Requirement (CWR) in comparison to crops grown in cooler, more humid, and less windy climatic conditions. Various crops have varying water requirements. As an illustration, it can be observed that alfalfa has a greater crop water need (CWR) in comparison to wheat. Various crops exhibit varying water requirements during different stages of their growth. For instance, it is well observed that crops in the vegetative stage generally exhibit a lower Crop Water Requirement (CWR) compared to crops in the reproductive stage.

The cultivation of a crop can be influenced by the characteristics of the soil in which it is grown, hence impacting its crop water need (CWR). Sandy soils exhibit a reduced capacity to retain water in comparison to clay soils, resulting in crops cultivated in sandy soils displaying an elevated crop water requirement. Farmers may optimize their irrigation schedules by acquiring knowledge about crop water requirements, thereby ensuring efficient water utilization without any wastage. The implementation of this approach has the potential to enhance agricultural productivity, mitigate expenses, and safeguard ecological integrity.

3.2.1.1 Role of Water in Plant Growth and Development

The presence of water is crucial for the optimal growth and development of plants. Cellulose constitutes approximately 90% of the plant's fresh weight and plays a crucial role in various vital plant functions (Manzoor et al. 2022), encompassing:

Water serves as a crucial reactant in the process of photosynthesis, wherein plants harness sunlight, water, and carbon dioxide to synthesize sugar and release oxygen. The process of dissolution in water facilitates the transportation of essential nutrients from the soil to various plant organs, including the leaves. Turgor pressure, facilitated by water, plays a crucial role in maintaining the structural integrity of plant cells. Water serves as a solvent for numerous metabolic activities that take place within the cellular structures of plants. The process of transpiration, wherein water evaporates from the leaves of plants, aids in the cooling of plants.

The significance of water extends to the process of seed germination as well. Upon the process of seed germination, the seed absorbs water from the surrounding soil. The presence of water induces the seed to undergo swelling and subsequent rupture. Subsequently, the embryonic organism enclosed within the seed initiates a process of growth and maturation, ultimately giving rise to a nascent botanical entity. The water requirements of plants vary based on factors such as their species, prevailing climatic conditions, and stage of growth. For instance, agricultural produce cultivated in arid and warm regions necessitates a greater amount of water compared to crops cultivated in colder and more humid regions. Seedlings and juvenile plants have a higher water requirement compared to fully developed plants. Insufficient water supply to plants can lead to wilting and eventual mortality. Water stress has the potential to diminish crop productivity and render plants more vulnerable to the detrimental effects of diseases and pests.

3.2.1.2 Factors Affecting Crop Water Needs: Climate, Soil, Crop Type, and Growth Stage

Farmers may optimize their irrigation schedules by comprehending the various aspects that impact crop water requirements, so ensuring efficient water usage while adequately meeting the crop's needs (Tuninetti et al. 2015). The implementation of this approach has the potential to enhance agricultural productivity, minimize expenses, and safeguard ecological integrity.

- **Climate**

The climate has a crucial role in determining the water requirements of crops. Cultivating crops in hot, arid, windy, and sun-drenched environments necessitates greater water use compared to crops cultivated in cooler, more humid, and less windy climes. The promotion of evaporation from the soil and plant surfaces can be attributed to the prevailing hot, dry, windy, and sunny conditions (Upadhyay et al. 2023).

- **Soil**

The water requirements of crops are influenced by the kind of soil. Sandy soils exhibit a diminished capacity to retain water compared to clay soils, resulting in increased water demands for crops cultivated in sandy soil conditions. The reason for this phenomenon is that sandy soils have a higher rate of water drainage compared to clay soils.

- **Crop Type**

Various crops have varying water requirements. As an illustration, it can be observed that alfalfa exhibits a greater water need in comparison to wheat. The reason for this phenomenon is attributed to the fact that alfalfa possesses a more extensive root system and a greater number of leaves in comparison to wheat.

- **Growth Stage**

Various crops exhibit varying water requirements during different stages of their growth. As an illustration, crops during the vegetative stage generally exhibit reduced water demands compared to crops in the reproductive stage. This phenomenon can be attributed to the fact that crops in the reproductive phase undergo the process of flower and fruit production, necessitating an increased water demand.

3.2.2 Quantifying Irrigation Water Needs

Quantifying irrigation water needs is essential for efficient and sustainable water management in modern agriculture. This entails careful measurements and assessments of crop type, growth stage, soil parameters, climate, and evapotranspiration rates. Weather stations, soil moisture sensors, and satellite imaging provide real-time data for irrigation scheduling. Farmers may optimize water use, and crop productivity, and avoid over- or under-irrigation by assessing crop water needs at different growth stages and modifying irrigation. Quantifying irrigation water needs helps conserve water, reduce water waste, and reduce agriculture's environmental effect, making it crucial for contemporary, sustainable farming.

3.2.2.1 Determining Water Deficits and Irrigation Frequency

The determination of water shortages and irrigation frequency is a crucial aspect of effective and sustainable irrigation management. Water deficits can manifest when plants experience insufficient water supply to fulfill their physiological requirements. This phenomenon can arise due to several circumstances, such as the occurrence of drought, elevated temperatures, and the presence of strong winds. There are various methodologies for assessing water shortages in plants, which encompass:

Plants undergoing water deficiencies frequently manifest signs such as wilting, leaf curling, and leaf abscission. Soil moisture sensors are utilized for quantifying the moisture content present inside the soil. This data can be utilized to ascertain the occurrence of water deficiencies in plants. The utilization of plant stress sensors enables the quantification of plant stress levels. This data can be utilized to ascertain the occurrence of water deficiencies in plants.

After the identification of a water deficit, it becomes crucial to ascertain the appropriate irrigation frequency required to rectify the shortage. The frequency of irrigation will be contingent upon various aspects, such as the prevailing climate, soil composition, crop variety, and developmental phase. Below are several recommendations for determining the frequency of irrigation:

- Utilize a crop water requirements model to assess and approximate the irrigation demands of your agricultural plants.
- When selecting the frequency of irrigation, it is important to consider many factors such as the climate, soil type, crop type, and development stage.
- Soil moisture sensors are employed to monitor soil moisture levels, hence facilitating the maintenance of optimal water supply to crops, ensuring their adequate hydration.
- It is important to calibrate the irrigation system to ascertain its accurate dispensation of water.

3.2.2.2 Estimating Water Losses Due to Evaporation and Runoff

There are various methodologies for assessing water losses resulting from evaporation and runoff (Gibson et al. 2017; Bodner et al. 2015), which encompass:

Evaporation

Pan Evaporation The measurement of pan evaporation is conducted by the utilization of a standardized evaporation pan. The quantification of water evaporation from the pan is employed in the computation of the reference crop evapotranspiration (ETo). The reference evapotranspiration (ETo) represents the quantity of water that a reference crop, such as grass, would consume under ideal growth circumstances. The evapotranspiration (ETo) can be modified by using crop coefficients (Kc) to accommodate the varying water needs of different crops and growth phases.

Physical-Based Models Physical-based models, such as the Penman-Monteith equation, use several parameters, such as temperature, humidity, wind speed, and solar radiation, to approximate the process of evaporation.

Runoff

Rational method The rational method is a straightforward approach utilized to estimate the runoff from smaller catchment areas. The analysis revolves around the interplay of rainfall intensity, catchment area, and runoff coefficient.

Hydrologic Models Hydrological models, such as the Soil and Water Assessment Tool (SWAT), are employed to replicate the hydrological processes occurring within a catchment and make estimations regarding runoff (Tomar et al. 2022). After estimating the water losses resulting from evaporation and runoff, it is possible to deduct these losses from the overall rainfall to approximate the quantity of water that can be infiltrated and stored in the soil.

3.2.2.3 Real-Time Monitoring and Adjustment of Irrigation Practices

The practice of real-time monitoring and adjustment of irrigation methods involves the utilization of sensors and various technological tools to gather data about soil moisture, plant stress, and other variables that impact the water requirements of crops (Adeyemi et al. 2017). The data is subsequently employed to dynamically modify irrigation schedules to optimize water allocation for crops, hence minimizing water wastage. Real-time monitoring and adjustment of irrigation practices offer several advantages, such as:

- **Improved Water Efficiency**

The implementation of real-time monitoring and adjustment of irrigation methods has the potential to mitigate water wastage by ensuring that crops receive watering only when necessary. This has the potential to result in substantial conservation of water, particularly in regions where water availability is limited.

- **Increased Crop Yields**

The implementation of real-time monitoring and adjustment of irrigation methods has the potential to enhance agricultural yields through the prevention of both excessive and insufficient irrigation. Excessive irrigation has the potential to induce root rot and various other complications, whilst inadequate irrigation can result in plant stress and diminished crop production.

- **Reduced Costs**

The implementation of real-time monitoring and adjustment of irrigation methods has the potential to mitigate irrigation costs through the reduction of water wastage and labor expenses.

There exists a variety of technological solutions that can be employed for real-time monitoring and adjustment of irrigation techniques. These technologies encompass:

- **Soil Moisture Sensors**

Soil moisture sensors are utilized for quantifying the moisture content within the soil. This data can be utilized to ascertain the optimal timing and quantity of irrigation necessary to fulfill the water needs of the crop.

- **Plant Stress Sensors**

The utilization of plant stress sensors enables the quantification of plant stress levels. This data can be utilized to ascertain the occurrence of water deficiencies in plants and the subsequent requirement for irrigation.

- **Weather Stations**

Weather stations are utilized to gather data about several meteorological parameters, including temperature, humidity, wind speed, and sun radiation. The provided data can be utilized to estimate the water requirements of crops and then modify irrigation schedules by these estimations. The implementation of real-time monitoring and adjustment of irrigation practices holds significant potential in enhancing water efficiency, augmenting crop yields, and mitigating expenses. With the ongoing advancements in technology and its increasing affordability, it is anticipated that a growing number of farmers worldwide will embrace its adoption.

3.3 Efficient Irrigation Techniques

3.3.1 Drip Irrigation Systems

The system functions by delivering water directly to the root zone via a network of tubes, pipelines, and emitters, thereby minimizing water loss caused by evaporation or runoff as in Fig. 3.1. Drip irrigation presents a multitude of benefits, notably encompassing water saving. This is achieved through the precise delivery of moisture directly to the plants, resulting in a reduction in overall water consumption when compared to conventional irrigation techniques. Additionally, it facilitates enhanced plant health and yields through the maintenance of regular moisture levels in the root zone. Moreover, this system has the potential to be automated and integrated with sensors to provide real-time monitoring. This capability facilitates accurate regulation and fine-tuning of irrigation processes, hence enhancing resource efficiency. Consequently, this technology holds significant value in the context of sustainable agricultural practices.

Fig. 3.1 Drip irrigation system

3.3.1.1 Principles and Benefits of Drip Irrigation

Drip irrigation refers to a micro-irrigation technology that facilitates the targeted delivery of water to the root zone of plants using a network of emitters. This practice aids in the mitigation of water wastage and evaporation, as well as the enhancement of water distribution efficiency to the vegetation. Drip irrigation systems function by dispensing water in a gradual and targeted manner, namely to the area surrounding the root system of plants. This process is achieved using a network of emitters, which are compact devices designed to dispense water at a regulated pace. Emitters have the option of being positioned either on the soil surface or buried slightly beneath it.

- **Benefits of Drip Irrigation**

Drip irrigation systems offer numerous advantages (Taylor and Zilberman 2017), such as:

Water Efficiency: Drip irrigation systems exhibit a high degree of water efficiency by effectively transporting water straight to the root zone of plants, hence minimizing the occurrence of evaporation.

Uniform Watering: Drip irrigation systems are designed to provide water evenly to all plants, irrespective of their varying sizes or shapes.

Reduced Weed Growth: Drip irrigation systems mitigate weed proliferation by delivering water directly to the plant roots rather than the foliage.

Improved Crop Yields: Drip irrigation systems have the potential to enhance crop productivity by delivering water to plants promptly, ensuring their water requirements are met.

Reduced Soil Erosion: Drip irrigation systems mitigate soil erosion by employing a methodical and targeted approach to water delivery, ensuring a gradual and precise supply of water straight to the root zone of plants.

3.3.1.2 Drip System Components and Design Considerations

Drip irrigation systems consist of various components as shown in Fig. 3.2, which encompass:

Water Source: The potential water sources include a municipal water supply, a well, or a cistern.

Filter: The utilization of a filter facilitates the elimination of pollutants present in the water, which have the potential to obstruct the emitters.

Pump: The function of the pump is to enhance the water pressure to facilitate its distribution to the emitters.

Mainline: The primary conduit is a cylindrical structure responsible for transporting water from the pumping mechanism to the outlets known as emitters.

Lateral lines: Lateral lines refer to minor conduits that diverge from the primary pipeline, facilitating the transportation of water to separate emitters.

Emitters: Emitters refer to devices that facilitate the gradual discharge of water into the soil.

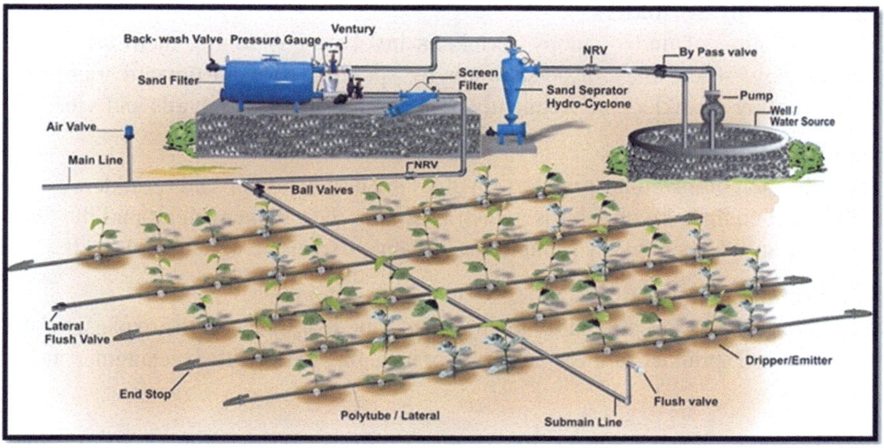

Fig. 3.2 Drip irrigation system components

When formulating a drip irrigation system, several aspects necessitate careful consideration, including: The selection of the appropriate filtration and pumping system is contingent upon the characteristics of the water source. The spacing of the emitters and the flow velocity of the water are contingent upon the specific crop being irrigated. The selection of the emitter type and flow rate of water is contingent upon the soil type. The configuration of the terrain will impact the arrangement of the drip irrigation system. The selection of materials and components is contingent upon budgetary constraints. After the completion of the design phase, the installation process of the drip irrigation system can commence. Drip irrigation systems have the potential to be installed either by manual efforts or with the assistance of a professional irrigation contractor.

3.3.2 *Sprinkler Irrigation Methods*

Sprinkler irrigation is a form of irrigation system that distributes water over the soil surface in a manner that closely resembles natural rainfall. Sprinkler irrigation systems can provide irrigation for a diverse range of crops, encompassing fruits, vegetables, flowers, trees, and shrubs. Additionally, they are highly suitable for irrigating landscapes and gardens. There exist various sprinkler irrigation methods, encompassing:

Overhead Sprinklers

The installation of overhead sprinklers involves the placement of pipes above the crop canopy, from which water is emitted in a downward trajectory. The most prevalent form of sprinkler irrigation system is the overhead sprinkler.

Under-Canopy Sprinklers
The installation of under-canopy sprinklers involves positioning them within the crop canopy, where they disperse water by spraying it in an upward trajectory. Under-canopy sprinklers are frequently employed to irrigate orchards and vines.

Traveling Sprinklers
The utilization of traveling sprinklers involves the installation of said sprinklers onto a mobile cart equipped with wheels, which facilitates their movement around the field. Traveling sprinklers are very suitable for irrigating expansive agricultural fields.

Side-Roll Sprinklers
Side-roll sprinklers are affixed to an elongated pipe that traverses the expanse of the field. Side-roll sprinklers are very suitable for irrigating fields with a rectangular shape.

3.3.2.1 Water Distribution Patterns and Uniformity

The water dispersion characteristics of a sprinkler irrigation system refer to how water is emitted from the sprinklers. The uniformity of water distribution refers to the extent to which water is evenly distributed across the designated area of irrigation. Several factors influence the water distribution pattern and regularity of a sprinkler irrigation system. These factors include:

Various types of sprinklers exhibit distinct water distribution patterns. As an illustration, impact sprinklers generate a revolving water stream that effectively encompasses a circular region. Rotor sprinklers are designed to provide a revolving water stream that encompasses a rectangular region, providing irrigation coverage across various paths. The spatial arrangement of the sprinklers has an impact on the evenness of water dispersion. Increased separation between objects leads to a less consistent distribution of water.

The distance at which water is sprayed from sprinklers is influenced by water pressure. An increase in water pressure leads to an extended range of projectile motion. The distribution pattern and homogeneity of water can be influenced by the presence of wind. Efforts should be made to optimize the design of sprinkler irrigation systems to mitigate the potential impacts of wind.

3.3.2.2 Improving Water Distribution Uniformity

Several measures can be undertaken to enhance the uniformity of water distribution in a sprinkler irrigation system, encompassing the following:

- It is recommended to employ sprinklers that possess a regular water distribution pattern.
- Distribute the sprinklers uniformly across the watered region.
- Ensure the maintenance of sufficient water pressure.
- Develop a system to mitigate the impact of wind.

The consistency of water distribution plays a significant role in assessing the effectiveness of a sprinkler irrigation system. Achieving a high level of water distribution uniformity is crucial in ensuring that adequate water supply is received by all plants within the irrigated area. This phenomenon has the potential to result in enhanced agricultural productivity and decreased water consumption. Below are many further suggestions for enhancing water distribution uniformity:

- Utilize a sprinkler irrigation design calculator as a tool to assist in the determination of
- the most suitable sprinkler type, spacing, and water pressure for effective irrigation.
- It is advisable to conduct frequent inspections of the sprinkler system to ascertain the optimal functionality of each sprinkler and to ensure uniform water pressure throughout the system.
- It is recommended to periodically perform system flushing to effectively eliminate any potential obstructions.

3.4 Soil Moisture Management

3.4.1 Importance of Soil Moisture Monitoring

Soil moisture monitoring refers to the systematic measurement and continuous tracking of the quantity of water present in the soil. Ensuring adequate water supply for crops and plants is a crucial undertaking for farmers and land managers, as it serves as a pivotal element in their agricultural practices (Umesha et al. 2018). Soil moisture monitoring holds significant importance for various reasons, which encompass:

The implementation of soil moisture monitoring systems can be advantageous for farmers as it enables them to optimize crop yields by effectively managing irrigation practices, hence avoiding both excessive and insufficient watering. Excessive irrigation has the potential to induce root rot and various other complications, whilst insufficient irrigation can result in plant stress and diminished crop production. The implementation of soil moisture monitoring techniques enables farmers to mitigate water wastage by ensuring that irrigation is conducted solely when crops necessitate it. This has the potential to yield cost savings and mitigate the environmental repercussions associated with agricultural practices.

Implementing soil moisture monitoring techniques can assist farmers in enhancing soil health by effectively managing soil moisture levels and avoiding both excessive wetness and dryness. These factors can potentially result in enhanced soil structure, heightened nutrient accessibility, and less erosion. There are several ways available for monitoring soil moisture, which encompasses a range of techniques. These approaches enable the assessment and measurement of soil moisture levels. Soil moisture monitoring can be conducted by manual techniques, such as employing a soil probe or collecting soil samples. The procedures are characterized by a high degree of manual labor and have the potential to yield imprecise results.

Sensor-based techniques for soil moisture monitoring involve the utilization of specialized sensors to quantify the moisture levels present within the soil. The approaches exhibit higher levels of accuracy and require less labor compared to manual methods. Sensor-based systems for monitoring soil moisture can be employed to observe and measure the levels of moisture present in the soil in real time. This data can be utilized to autonomously modify irrigation schedules and guarantee the consistent provision of adequate water to crops. The monitoring of soil moisture is a crucial instrument for promoting sustainable practices in the field of agriculture. Through the monitoring of soil moisture levels, farmers have the potential to enhance crop yields, minimize water wastage, and enhance soil health.

3.4.1.1 Relationship Between Soil Moisture and Plant Health

Soil moisture refers to the quantity of water contained inside the soil. Water is a crucial determinant of plant well-being, as it is needed for the survival and growth of plants. The correlation between soil moisture and plant health is intricate and contingent upon many aspects, encompassing plant species, climatic conditions, and soil composition. Plants require water to facilitate several physiological activities (Zhao et al. 2021), such as:

- **Photosynthesis**

Photosynthesis is a vital biological process in which plants use solar energy, together with water and carbon dioxide, to synthesize oxygen and glucose. The presence of water is essential for the process of photosynthesis to take place.

- **Transpiration**

Transpiration is a physiological phenomenon in which plants emit water vapor into the surrounding atmosphere via their foliage. Transpiration plays a crucial role in plant physiology by facilitating temperature regulation and nutrient distribution inside the plant.

- **Cell Growth and Development**

Plants need water for cellular growth and development. The amount of water that a plant need depends on the plant species, climate, soil type, and soil moisture levels. Some plants, such as cacti and succulents, have evolved to thrive in arid environments and can survive with minimal water. Other plants, such as water lilies and rice, need a lot of water because they evolved to thrive in environments with high moisture levels. Plants in hot, dry climates need more water to offset the high rate of evaporation. Plants in cold, wet climates need less water. Sandy soils drain water more quickly than clay soils, so plants in sandy soils need to be watered more often. Clay soils hold water better, so plants in clay soils may need less water. If the soil moisture levels are too low, plants can experience water stress, which can lead to wilting, leaf drop, and stunted growth. If the soil moisture levels are too high, plants can experience root rot, which is a fungal infection that can damage or kill the plant's root system. The amount of water that a plant need depends on several factors, including the plant species, climate, soil type, and soil moisture levels. It is important to water plants regularly, but be careful not to overwater them, as this can lead to root rot.

3.4.1.2 Techniques for Measuring Soil Moisture Levels

There exists a diverse range of methodologies for quantifying soil moisture content (Dobriyal et al. 2012), each exhibiting distinct merits and demerits. Several often-used approaches include:

- **Gravimetric Method**

The technique is often regarded as the most precise approach to quantifying soil moisture content, albeit it is accompanied by the drawbacks of being labor-intensive and potentially causing damage to the soil. The gravimetric technique encompasses the collection of a soil specimen, subsequent desiccation using oven drying, and subsequent quantification of its mass to ascertain the moisture content.

- **Volumetric Method**

The technique exhibits lower levels of accuracy compared to the gravimetric method; however, it compensates for this drawback by offering advantages in terms of speed and reduced destructiveness. The volumetric approach entails the utilization of a device known as a tensiometer to quantify the soil water tension. Soil water tension refers to the quantification of the cohesive force responsible for retaining water within the soil.

- **Electrical Conductivity Method**

The underlying premise of this method is that the electrical conductivity of soil exhibits an increase in direct correlation with the rise in water content. The electrical conductivity method exhibits a reasonably expedient and user-friendly approach, albeit with diminished precision compared to the gravimetric or volumetric techniques.

- **Neutron Probe Method**

The neutron probe method entails the utilization of a neutron probe to quantify the moisture content present in the soil. The neutron probe method is characterized by its relative accuracy and non-destructive nature; nonetheless, it is important to note that this technology is also associated with higher costs compared to alternative approaches.

The optimal approach for assessing soil moisture levels is contingent upon the requirements of the given application. The gravimetric approach may be the most suitable choice when a high level of precision is necessary. If a more expeditious and less detrimental approach is required, the utilization of the electrical conductivity or neutron probe method may present itself as a more favorable alternative. In addition to the methodologies, a plethora of novel technologies are now in development to quantify soil moisture levels. The technologies encompass:

- **Soil Moisture Sensors**

Soil moisture sensors can effectively monitor and assess soil moisture levels in a real-time manner. Soil moisture sensors are commonly deployed within the soil matrix to quantify the water content by diverse methodologies, including electrical conductivity or neutron scattering.

- **Satellite Remote Sensing**

The utilization of satellite remote sensing enables the assessment of soil moisture levels across extensive geographical regions. Satellite remote sensing operates through the quantification of microwave radiation emanating from the soil. The influence of soil moisture content on microwave radiation emitted from the soil renders satellite remote sensing a viable approach for estimating soil moisture levels.

The emergence of these novel technologies presents the opportunity to deliver enhanced precision and promptness in the assessment of soil moisture levels. This information possesses potential utility for farmers and land managers in enhancing water utilization efficiency and augmenting crop yields.

3.4.2 Irrigation Scheduling

The practice of irrigation scheduling involves the determination of optimal timing and quantity of water application for agricultural purposes. Efficient water management is a crucial instrument employed by farmers and land managers to optimize crop irrigation, minimizing water wastage while ensuring adequate hydration for agricultural purposes. Several elements must be considered during the development of an irrigation schedule, which include:

Various crops have varying water requirements. As an illustration, it is worth noting that veggies necessitate a greater amount of water compared to grains. The climate exerts an influence on the rate at which water undergoes evaporation from both the soil and plants. In arid and warm regions, agricultural plants require a greater amount of water compared to regions characterized by cooler and more humid temperatures. Sandy soils have a higher rate of water drainage compared to clay soils, necessitating increased water supply for crops cultivated in sandy soil conditions. Various crops exhibit varying water requirements throughout different phases of growth. Seedlings require a greater amount of water in comparison to fully developed plants.

After considering these elements, it is possible to formulate an irrigation schedule. Various methodologies exist for the formulation of irrigation schedules. A frequently employed approach involves the utilization of a crop water requirement model. Crop water requirement models include several factors such as crop type, climate conditions, soil characteristics, and growth stage to estimate the water demands of the crop. Soil moisture monitoring is a frequently employed technique for the development of irrigation schedules. The process of soil moisture monitoring entails the quantification of water content inside the soil. This data can be utilized to ascertain the appropriate timing and quantity of water application for crops, to sustain the targeted soil moisture level. The process of irrigation scheduling can present a multifaceted undertaking, although it holds significant value as a tool employed by farmers and land managers to effectively allocate water resources to crops, hence minimizing water wastage.

3.4.2.1 Developing Irrigation Schedules Based on Soil Moisture

The utilization of soil moisture as a basis for irrigation scheduling is seen as a more accurate methodology compared to conventional approaches, such as time-based or calendar-based scheduling methods. Soil moisture monitoring enables the determination of optimal irrigation timing and quantity for crops, depending on their specific water requirements, as opposed to following a predetermined irrigation schedule. To design an irrigation plan that is contingent upon soil moisture levels, it is imperative to undertake the following steps:

- **Choose a Soil Moisture Monitoring Method**

There exists a variety of soil moisture monitoring techniques, encompassing both manual approaches (e.g., employing a soil probe or collecting soil samples) and sensor-based methodologies. Sensor-based approaches have been found to exhibit higher levels of accuracy and reduced labor requirements compared to manual methods. However, it is worth noting that the utilization of sensor-based methods may entail higher costs.

- **Determine the Desired Soil Moisture Level for Your Crops**

The optimal soil moisture level for crops is contingent upon factors such as the specific crop species, prevailing climatic conditions, and the type of soil in question. Most plant species exhibit a preference for soil moisture levels ranging from 40% to 60% of the field capacity.

- **Set Up Your Soil Moisture Monitoring System**

To implement a sensor-based soil moisture monitoring system, it is necessary to properly install the sensors inside the designated field. It is imperative to adhere to the installation guidelines provided by the manufacturer.

- **Start Monitoring Soil Moisture Levels**

After the installation of the soil moisture monitoring system, the user can commence the monitoring of soil moisture levels. It is imperative to consistently check soil moisture levels, particularly in times of elevated temperatures and arid conditions.

- **Irrigate When the Soil Moisture Level Drops Below the Desired Level**

When the soil moisture level decreases below the predetermined threshold, it becomes necessary to initiate irrigation. Sufficiently irrigate the soil to restore the moisture content to the desired level.

3.5 Sustainable Crop Selection and Management

3.5.1 Drought-Tolerant and Resistant Crops

Drought-tolerant and drought-resistant crops are highly significant resources in the context of climate change and the issues posed by water scarcity. Drought-tolerant crops possess the ability to endure periods of water scarcity and adjust to situations of restricted moisture, whereas drought-resistant types demonstrate robust resistance to water-related stress, hence sustaining their productivity even in arid settings. These crops frequently exhibit traits such as extensive root systems, optimal water utilization, and decreased rates of transpiration. In recent years, there have been notable breakthroughs in the field of biotechnology and breeding techniques, resulting in the emergence of genetically modified and hybrid crops that demonstrate improved resistance to drought conditions. Resilient crops are of utmost importance in ensuring global food security as they enable farmers to sustain yields in locations with limited water resources, hence decreasing dependence on irrigation and alleviating the detrimental effects of extended droughts on agricultural activities. Additionally, the entities actively advocate for the use of sustainable and responsible farming techniques, with a particular emphasis on the conservation of water resources. These efforts aim to enhance the resilience of the agricultural sector.

3.5.1.1 Improving Water-Use Efficiency at the Plant Level

Enhancing water-use efficiency at the plant level is a crucial factor in promoting sustainable agriculture and ensuring food security. The availability of water is constrained, and the phenomenon of climate change exacerbates its scarcity. Enhancing water-use efficiency at the plant level has the potential to diminish water consumption, enhance crop yields, and mitigate the environmental repercussions associated with agricultural practices. There exist various strategies to enhance water-use efficiency at the plant level, encompassing:

Drought-tolerant crops exhibit a reduced water requirement for growth in comparison to conventional crops. This intervention has the potential to mitigate water usage and enhance agricultural productivity in regions susceptible to arid conditions. The implementation of sound agricultural practices, such as the adoption of crop rotation, cover cropping, and reduced tillage techniques, has the potential to enhance soil health and bolster the resilience of crops in the face of drought conditions.

Various irrigation systems can be employed to enhance water-use efficiency. An illustration of this may be seen in the implementation of drip irrigation and subsurface irrigation techniques, whereby water is supplied directly to the root zone of plants. This targeted approach effectively mitigates evaporation and minimizes water loss.

The application of mulch is effective in mitigating evaporation and minimizing water loss from the soil. In addition to its primary function, mulch can also contribute to the enhancement of soil health and the mitigation of erosion. Plant breeders are currently engaged in efforts to enhance the water-use efficiency of crop varieties through the development of novel cultivars. These cultivars have the potential to mitigate water usage and enhance agricultural productivity in regions susceptible to drought conditions. In conjunction with the approaches, there exists a range of nascent technologies that can be employed to enhance water utilization efficacy within the context of industrial facilities. The technologies encompass:

Precision agricultural technologies like GPS and remote sensing can monitor crop health and detect water stress, enabling tailored irrigation and management strategies. Nanotechnology is being used to develop novel materials and devices to improve water-use efficiency at the plant level. Nanoparticles can be used to produce coatings to mitigate evaporation from foliage and to create sensors to monitor soil moisture levels. These technologies can help agricultural practitioners and land stewards improve water utilization efficiency at individual plant levels, contributing to a more sustainable agricultural framework.

3.5.2 Conservation Tillage and Cover Crops

Conservation tillage as shown in Fig. 3.3 below refers to a soil management approach that aims to minimize the extent of soil disturbance. This is achieved by the practice of minimizing the frequency of soil ploughing or tilling. The implementation of conservation tillage practices has been shown to have positive effects on soil health, erosion control, and water conservation.

Cover crops as shown in Fig. 3.4 refer to crops that are intentionally planted to provide coverage to the soil surface. Cover crops are commonly utilized in agricultural practices to occupy the spaces between rows of crops, following the harvest of a primary crop, or during periods of inactivity in the growing season. Cover crops have a key role in enhancing soil health, mitigating soil erosion, and exerting weed suppression.

The integration of conservation tillage with cover crops can synergistically contribute to the development of a more sustainable agricultural system. Conservation tillage practices contribute to the preservation of soil integrity, while the incorporation of cover crops serves to enhance soil quality. By employing the simultaneous implementation of these two methodologies, agricultural practitioners can enhance their agricultural productivity, mitigate their ecological footprint, and achieve economic savings.

Fig. 3.3 Conservation tillage

Fig. 3.4 Cover crops

3.5.2.1 Benefits of Using Conservation Tillage and Cover Crops

The implementation of conservation tillage practices and the incorporation of cover crops have been found to contribute to the enhancement of soil health through the augmentation of organic matter content, the improvement of soil structure, and the stimulation of microbial activity. The implementation of conservation tillage practices and the utilization of cover crops have been found to effectively mitigate soil erosion by providing a protective barrier against the erosive forces of wind and rain.

The utilization of conservation tillage practices and the incorporation of cover crops have been shown to enhance water infiltration and retention through the augmentation of soil organic matter content and the improvement of soil structure. This practice aids in mitigating runoff and enhancing water accessibility for agricultural purposes. Cover crops play a crucial role in weed suppression by engaging in competitive interactions for essential resources like light, water, and nutrients. This approach has the potential to decrease the utilization of herbicides and yield cost savings. The implementation of conservation tillage practices and the incorporation of cover crops have been found to contribute positively to the enhancement of crop yields through the amelioration of soil health, mitigation of soil erosion, and suppression of weed growth.

3.5.2.2 Minimizing Soil Erosion and Water Runoff

Conservation tillage and cover crops are two important sustainable agricultural practices that can help to reduce soil erosion and water runoff. Conservation tillage maintains soil surface coverage with plant residues, increases soil organic matter content, and reduces soil disturbance frequency. This helps to protect the soil from the erosive forces of wind and rain and improve its structure, making it more resilient to erosion. Cover crops provide a protective layer of vegetation on the soil surface, slow down water movement, and absorb excess nutrients. This helps to shield the soil from the erosive effects of wind and rain, reduce both runoff and erosion, and mitigate nutrient pollution from runoff. Both conservation tillage and cover crops have a number of other benefits, including improved soil health, increased water infiltration, reduced greenhouse gas emissions, and increased crop yields. Conservation tillage and cover crops are important tools for farmers who want to protect their soil and water resources while also producing sustainable and profitable crops.

3.5.2.3 Cover Crops to Enhance Soil Water Retention

Cover crops have the potential to augment soil water retention through various mechanisms, which include:

Cover crops, sometimes known as green manure, are plants that are grown primarily to protect and improve the soil. They mostly add organic matter to the soil when they decay. The presence of organic matter contributes to the enhancement of soil structure and the augmentation of water-holding capacity. Cover crops have a pivotal role in enhancing soil structure through the creation of channels that facilitate water infiltration and the mitigation of compaction. Cover crops play a significant role in mitigating evaporation by effectively covering the soil surface and providing shade to the underlying soil. Cover crops facilitate and enhance biological processes within the soil, hence contributing to the enhancement of soil health and the augmentation of its water retention capacity.

3.6 Integration of Technology and Data

3.6.1 Precision Agriculture Techniques

Precision agriculture is an evidence-based methodology in agriculture that employs advanced technology to effectively monitor and regulate crops and livestock with enhanced accuracy (Shaikh et al. 2022). This phenomenon has the potential to result in higher crop yields, decreased expenses, and enhanced ecological sustainability. Precision agriculture encompasses a diverse array of techniques, among which several prominent ones can be identified.

Global Positioning System (GPS) And Mapping
The utilization of Global Positioning System (GPS) technology enables the generation of comprehensive field maps, facilitating the accurate navigation of tractors and other machinery for precise application of inputs, including fertilizers and pesticides.

Remote Sensing
Remote sensing technology, such as satellite imaging and drones, can effectively monitor the health of crops, detect and identify pests and diseases, as well as track the utilization of water resources.

Sensors
Sensors can quantify diverse parameters within many domains, including but not limited to soil moisture, nutrient concentrations, and crop vitality. The data can thereafter be utilized to facilitate well-informed decision-making about irrigation, fertilization, and several other practices related to crop management.

3.6.1.1 Variable Rate Application (VRA)

The utilization of Variable Rate Application (VRA) technology enables the application of inputs to fields at varying rates, contingent upon the specific requirements of the crop. This has the potential to mitigate waste and enhance operational efficiency.

Precision agriculture techniques have the potential to enhance several agricultural practices, including:

- **Crop Production**

Precision agriculture is a valuable approach that can enhance both crop yields and quality through the efficient management of agricultural inputs by farmers. An instance of utilizing Variable Rate Application (VRA) in agriculture involves farmers applying fertilizer at an exact rate tailored to the specific requirements of each plant. Alternatively, farmers can employ remote sensing techniques to detect regions within a field that exhibit signs of stress and want additional water supply.

- **Pest And Disease Management**

Precision agriculture has the potential to enhance farmers' ability to diagnose and manage pests and illnesses with greater efficacy. One use of remote sensing in agriculture involves the utilization of this technology by farmers to monitor the spatial distribution and movement patterns of pests and illnesses within a given field. Additionally, sensors can be employed to assess the overall health of crops and identify specific regions that are susceptible to potential risks.

- **Water Management**

Precision agriculture has the potential to enhance water efficiency for farmers. One potential use involves the utilization of sensors by farmers to monitor soil moisture levels, enabling them to implement irrigation practices alone when deemed essential.

- **Livestock Management**

The application of precision agriculture techniques can be extended to enhance the management of animals as well. As an illustration, agricultural practitioners can employ Global Positioning System (GPS) technology to ascertain the precise whereabouts of their livestock, while also utilizing sensors to effectively monitor their well-being.

The sector of precision agriculture is characterized by its dynamic nature, with a continuous influx of newer technologies. With the rising affordability and accessibility of these technologies, precision agriculture is positioned to assume a progressively significant role in the realm of sustainable agriculture.

3.6.1.2 Role Of Precision Agriculture in Water Management

The implementation of precision agricultural techniques holds significant potential in the realm of water management, as it enables farmers to optimize their water usage with enhanced efficiency and effectiveness. This holds particular significance in areas characterized by limited availability or heightened demand for water resources. There are several methods by which precision agriculture can contribute to enhancing water management:

Variable Rate Irrigation (VRI)
Vegetation response irrigation (VRI) systems employ sensor technology to quantify soil moisture levels and various other parameters within a given agricultural area. Subsequently, water is administered at varying rates based on the specific requirements of the cultivated crop. This approach has the potential to mitigate water wastage and optimize water allocation for crops, thereby striking a balance between providing adequate hydration and avoiding excessive irrigation.

Precision Irrigation Scheduling
Precision irrigation scheduling technologies utilize meteorological data and crop-specific water demands to assist agricultural practitioners in determining the optimal

timing and quantity of irrigation. This approach has the potential to mitigate water wastage and optimize water distribution for crops, hence ensuring a timely and adequate water supply.

Crop Monitoring
Precision agricultural technologies, including remote sensing and unmanned aerial vehicles (UAVs), have the potential to effectively monitor the health of crops and detect regions under physiological strain. This data can be utilized to strategically allocate irrigation and other resources to the regions that require them the most.

Leak Detection
The application of precision agriculture technologies can help facilitate the identification of leaks in irrigation systems. This practice has the potential to mitigate water wastage and yield financial savings for farmers.

Furthermore, precision agriculture has the potential to enhance water management practices by facilitating farmers' comprehension of their water requirements and enabling them to make well-informed choices on water utilization.

3.6.1.3 Variable Rate Irrigation and Fertilization Based on Spatial Variability

Variable rate irrigation (VRI) and fertilization (VRF) are precision agricultural methodologies that employ sensor-based technology to administer water and fertilizer to agricultural fields at varying rates, tailored to the specific requirements of the crop. This approach has the potential to mitigate waste, enhance operational effectiveness, and augment productivity. Spatial variability pertains to the phenomenon wherein the characteristics of a given area, such as soil composition, nutrient concentrations, and crop vitality, exhibit differences across different locations. The observed fluctuation can be attributed to various variables, including geography, drainage patterns, and historical management practices.

Variable rate irrigation (VRI) and variable rate fertilization (VRF) techniques can effectively manage spatial variability within agricultural fields. By strategically administering inputs at varying rates over different sections of the field, these techniques can cater to the specific requirements of the crops being cultivated. As an illustration, a Variable Rate Irrigation (VRI) system might be employed to administer increased water quantities to regions of a field characterized by sandy soil, which is inherently more susceptible to drought. Similarly, a Variable Rate Fertilization (VRF) system could be utilized to distribute higher amounts of fertilizer to places within the field exhibiting lower nutrient levels. Virtual Reality Interfaces (VRI) and Virtual Reality Frameworks (VRF) can be effectively deployed using a diverse range of technological solutions, encompassing:

The Global Positioning System (GPS) can generate very detailed field maps, enabling the exact guidance of tractors and other agricultural machinery for accurate input applications. The utilization of remote sensing technology, including satellite imaging and unmanned aerial vehicles (UAVs), enables the monitoring of crop health and the identification of specific regions within a field that require additional irrigation or fertilization. Sensors can be affixed onto tractors and various types of equipment to quantify soil moisture levels, nutrient concentrations, and more variables inside agricultural areas. The data can thereafter be utilized to implement inputs at varying rates based on the specific requirements of the crop. Virtual Root Infusion (VRI) and Variable Rate Fertilization (VRF) are advanced precision agriculture methodologies that possess the capability to enhance farmers' financial performance while concurrently safeguarding the environment. With the decreasing costs and rising availability of technology, there is a growing potential for its significant contribution to sustainable agriculture.

3.7 Sustainable Drainage and Runoff Management

3.7.1 Managing Agricultural Drainage

The management of agricultural drainage holds significant importance due to various factors, which include:

- To enhance agricultural productivity and optimize the quality of crops, various strategies can be implemented.
- To mitigate environmental degradation
- To safeguard water resources, it is imperative to implement effective measures.
- To preserve the health of soil, it is imperative to implement appropriate measures.

There exist various methods for the management of agricultural drainage, which encompass: Surface drainage systems, such as ditches and furrows, are employed to eliminate surplus water from the soil's surface. Subterranean drainage systems, such as tile drains and perforated pipes, are employed to eliminate surplus water from the soil's root zone. Conservation drainage practices, such as the implementation of cover cropping and decreased tillage techniques, have been shown to effectively mitigate the volume of runoff and drainage originating from agricultural areas.

3.7.1.1 Impacts of Poor Drainage on Soil and Water Quality

Inadequate drainage can have adverse effects on the quality of both soil and water. Insufficient drainage has the potential to cause waterlogging, a condition that restricts the availability of oxygen to plant roots and impedes the absorption of nutrients. Consequently, this can lead to inhibited crop growth and diminished agricultural productivity. Furthermore, the presence of excessive water in soils can result in compaction and heightened susceptibility to erosion, ultimately contributing to the loss of soil quality. Furthermore, the accumulation of stagnant water in places with inadequate drainage can worsen salt problems, resulting in decreased soil fertility. The surplus water has the potential to transport various pollutants, including pesticides and fertilizers, into adjacent water bodies, resulting in pollution and deterioration of water quality. In essence, the intricate relationship among inadequate drainage, soil composition, and water quality highlights the significance of implementing appropriate land management strategies to maintain agricultural output and protect our valuable natural resources.

3.7.1.2 Designing Effective Drainage Systems

The development of efficient drainage systems is a crucial component of land management, with a special emphasis on its significance in the field of agriculture. These systems encompass a well-planned arrangement of pipes, ditches, and channels designed to effectively eliminate surplus water from agricultural fields to alleviate the detrimental impacts of waterlogging and soil erosion, all while ensuring the preservation of water quality. The design of an effective drainage system necessitates a thorough comprehension of the topographical features, soil composition, and climatic conditions prevailing in the region, alongside an evaluation of the hydrological patterns. Thorough evaluation of variables such as the arrangement and dimensions of drainage tiles or ditches is important to achieve maximum efficiency. An effectively planned drainage system not only improves agricultural productivity and mitigates soil degradation but also plays a significant role in water conservation and pollution mitigation, rendering it an indispensable instrument for sustainable and efficient farming practices.

3.7.1.3 Mitigating Water Pollution Through Proper Drainage Management

The management of adequate drainage is of utmost importance in addressing and mitigating the issue of water pollution, which holds significant implications for both the environment and agriculture. The implementation of efficient drainage systems can have a substantial impact on mitigating the transfer of contaminants from agricultural areas to adjacent aquatic environments. These systems play a crucial role in water quality management and the preservation of aquatic ecosystems by regulating water flow and mitigating the transport of excessive fertilizers, pesticides, and other pollutants. In addition, the implementation of strategically engineered drainage systems, such as vegetated buffers and sediment basins, can effectively enhance the filtration process of pollutants, thereby preventing their entry into water bodies. By incorporating appropriate drainage management practices, we not only maintain water quality but also contribute to sustainable agriculture, boosting both the long-term health of our ecosystems and the quality of our food and water supplies.

3.8 Conclusion

Effective water management plays a vital role in ensuring the sustainability of agricultural output, as it seeks to meet the increasing worldwide need for food and fiber, all the while safeguarding the environment. Water conservation practices such as drip irrigation and deficit irrigation have the potential to effectively mitigate water use. Additionally, precision agricultural techniques, exemplified by variable rate irrigation, offer a means to optimize water distribution for crop cultivation. Enhanced drainage systems effectively mitigate the issues of waterlogging and salinization, whilst integrated water resource management comprehensively addresses the diverse requirements of all water stakeholders. The implementation of water price mechanisms and the establishment of water markets have the potential to foster a more sustainable trajectory for agricultural water utilization.

References

Adeyemi O, Grove I, Peets S, Norton T (2017) Advanced monitoring and management systems for improving sustainability in precision irrigation. Sustainability 9(3):353

Bodner G, Nakhforoosh A, Kaul HP (2015) Management of crop water under drought: a review. Agron Sustain Dev 35:401–442

Dobriyal P, Qureshi A, Badola R, Hussain SA (2012) A review of the methods available for estimating soil moisture and its implications for water resource management. J Hydrol 458:110–117

Gibson JJ, Birks SJ, Jeffries D, Yi Y (2017) Regional trends in evaporation loss and water yield based on stable isotope mass balance of lakes: the Ontario Precambrian Shield surveys. J Hydrol 544:500–510

Irfeey AMM, Alotaibi BA, Najim MM, Shah AA (2023) Water valuation in urban settings for sustainable water management. Water 15(17):3105

Kim SH, Hejazi M, Liu L, Calvin K, Clarke L, Edmonds J et al (2016) Balancing global water availability and use at basin scale in an integrated assessment model. Clim Chang 136(2):217–231

Manzoor A, Dar AH, Pandey VK, Shams R, Khan S, Panesar PS et al (2022) Recent insights into polysaccharide-based hydrogels and their potential applications in food sector: a review. Int J Biol Macromol 213:987–1006

Mishra RK (2023) Fresh water availability and its global challenge. Br J Multidiscip Adv Stud 4(3):1–78

Rasul G (2016) Managing the food, water, and energy nexus for achieving the Sustainable Development Goals in South Asia. Environ Dev 18:14–25

Shaikh TA, Rasool T, Lone FR (2022) Towards leveraging the role of machine learning and artificial intelligence in precision agriculture and smart farming. Comput Electron Agric 198(107):119

Singh SK, Kanga S, Gulati B, Raiˇc M, Sajan B, Đurin B, Singh S (2023a) Spatial and temporal analysis of hydrological modelling in the Beas Basin using SWAT+ Model. Water 15:3338. https://doi.org/10.3390/w1519333

Singh SK, Kotecha MJ, Kanga S, Pankhaniya SK, Agrawal S, Meraj G (2023b) Urban Heat Island (UHI) Resilience plan in varying climatic conditions using geospatial approach: a case study of Rajkot City. In: Climate change, agriculture and society, vol 1, pp 307–338

Singh S, Meraj G, Kumar P, Singh SK, Kanga S, Johnson BA (2023c) Decoding Chambal River shoreline transformations: a comprehensive analysis using remote sensing, GIS, and DSAS. Water 15(9):1793

Sqib N, Rai PK, Kanga S, Kumar D, Đurin B, Singh SK (2023) Assessment of ground water quality of Lucknow City under GIS framework using Water Quality Index (WQI). Water 15(17):3048

Sud A, Kanga R, Singh SK, Meraj G, Kanga S, Kumar P, Bhardwaj V (2023) Simulating groundwater potential zones in mountainous Indian Himalayas—a case study of Himachal Pradesh. Hydrol 10(3):65

Taylor R, Zilberman D (2017) Diffusion of drip irrigation: the case of California. Appl Econ Perspect Policy 39(1):16–40

Tomar JMS, Kar SK, Ahmed A, Sankar M, Jayaprakash J, Kaushal R, Chaturvedi OP (2022) Hydrological fluxes from Sal (Shorea robusta) forests under various disturbance gradients of Himalayan foothills. Ecohydrol Hydrobiol 22(3):541–552

Tuninetti M, Tamea S, D'Odorico P, Laio F, Ridolfi L (2015) Global sensitivity of high-resolution estimates of crop water footprint. Water Resour Res 51(10):8257–8272

Umesha S, Manukumar HM, Chandrasekhar B (2018) Sustainable agriculture and food security. In: Biotechnology for sustainable agriculture. Woodhead Publishing, pp 67–92

Upadhyay K, Elangovan R, Subudhi S (2023) Establishing thermal comfort baseline in a sub-tropical region through a controlled climate chamber study. Advances in Building Energy Research, pp 1–23

Wang D, Saleh NB, Byro A, Zepp R, Sahle-Demessie E, Luxton TP et al (2022) Nano-enabled pesticides for sustainable agriculture and global food security. Nat Nanotechnol 17(4):347–360

Zhao H, Di L, Sun Z, Hao P, Yu E, Zhang C, Lin L (2021, July) Impacts of soil moisture on crop health: a remote sensing perspective. In: 2021 ninth international conference on agro-geoinformatics (agro-geoinformatics). IEEE, pp 1–4

Chapter 4
Remote Sensing and Precision Agriculture: A Sustainable Future

Rehan Mehmood Sabir, Kashif Mehmood, Abid Sarwar, Muhammad Safdar, Nalain E. Muhammad, Nimra Gul, Fareeha Athar, Muhammad Danish Majeed, Jaffar Sattar, Zafran Khan, Mariam Fatima, Muhammad Adnan Hussain, and Hafiz Muhammad Bilawal Akram

Abstract To address challenges like resource scarcity, destruction of the environment, and food security, the chapter examines the transformative impact of remote sensing and precision agriculture. In addition to its potential for soil mapping, crop monitoring, health assessment, and early pest and disease identification, it draws attention to the value of adopting technology into agricultural sustainability. Using data integration and analysis, the chapter additionally highlights the relationship between remote sensing and precision agriculture, with Geographic Information

R. M. Sabir (✉) · M. Safdar
Department of Irrigation and Drainage, University of Agriculture, Faisalabad, Pakistan

Agricultural Remote Sensing Lab (ARSL), University of Agriculture, Faisalabad, Pakistan

K. Mehmood
College of Economics and Management, China Agriculture University, Beijing, China

A. Sarwar
Department of Irrigation and Drainage, University of Agriculture, Faisalabad, Pakistan

N. E. Muhammad
Department of Irrigation and Drainage, University of Agriculture, Faisalabad, Pakistan

National University of Science and Technology, Islamabad (NUST), Islamabad, Pakistan

N. Gul · Z. Khan
Department of Plant Breeding and Genetics, University of Agriculture, Faisalabad, Pakistan

F. Athar · H. M. B. Akram
Department of Agronomy, University of Agriculture, Faisalabad, Pakistan

M. D. Majeed
Department of Irrigation and Drainage, University of Agriculture, Faisalabad, Pakistan

Barani Agriculture Research Institute (BARI) Chakwal, Chakwal, Pakistan

J. Sattar
Department of Irrigation and Drainage, University of Agriculture, Faisalabad, Pakistan

Department of Agricultural Engineering, Khwaja Fareed University of Engineering and Information Technology, Rahim Yar Khan, Pakistan

S. Kanga et al. (eds.), *Transforming Agricultural Management for a Sustainable Future*, World Sustainability Series, https://doi.org/10.1007/978-3-031-63430-7_4

Systems (GIS) functioning as a fundamental component for incorporating data. The foundation for comprehensive analysis is the combination of field data and remote sensing, which is backed by machine learning algorithms and data processing methods. The application of remote sensing and precision agriculture has multiple sustainability advantages, such as resource optimization, environmental awareness, enhanced production, financial success, and resistance to market changes. In addition, technological challenges place an extreme value on technology adoption and education and identify new trends including the combination of remote sensing with the Internet of Things (IoT) and artificial intelligence. The chapter's conclusion draws together essential ideas, highlighting the interdependence of precision agriculture, sustainable agricultural methods, and remote sensing. It serves as a preview of a day when agricultural areas will be advanced in technology and ecologically harmonious.

Keywords Remote Sensing · Precision Agriculture · Geographic Information Systems (GIS) · Sustainability · Machine Learning · Internet of Things (IoT)

4.1 Introduction

4.1.1 The Role of Technology in Revolutionizing Agriculture

Many aspects of agriculture are affected by agricultural technology, which includes fertilizers, pesticides, seed technology, etc. Biotechnology and genetic modification have produced increased crop production and resistance to pests. Mechanization has made tilling and agricultural harvesting more efficient and reduced the amount of manual labor required. Irrigation techniques and transport networks have been enhanced, processing apparatus has decreased waste and the outcomes can be observed in all areas. Robotics, precision agriculture, artificial intelligence, blockchain technology, and more are currently on the leading edge of current technological advancements. In recent decades, technology has played an extremely important part in revolutionizing agriculture, resulting in what is commonly referred to as "precision agriculture". This transformation has significantly impacted many facets of agriculture, including increased productivity and sustainability and enhanced resource management.

M. Fatima
Department of Entomology, University of Agriculture, Faisalabad, Pakistan

M. A. Hussain
Institute of Plant Breeding and Biotechnology, MNS-University of Agriculture, Multan, Pakistan

4.1.2 Overview of Remote Sensing and GIS and Precision Agriculture

Precision agriculture is central to providing solutions to the industry's most immediate problems. Precision agriculture is defined as "the application of technologies that integrate sensors, information systems, improved machinery, and experienced management to maximize agricultural productivity within sustainable systems (Kumar et al. 2023)." Precision agriculture focuses on financial viability, safeguarding the environment, and sustainability. It constitutes a system that facilitates the reorganization of the overall agricultural system towards sustainable, low-input, and high-efficiency agriculture (Ahmad et al. 2023). Despite this, continues to remain difficult to figure out how much water must be available for diverse crop irrigation management systems. Sensors are used to select, construct, and manage irrigation systems to meet the specific irrigation demands for each crop field on a farm. In a period of limited water availability, we are required to better control our water resources to produce enough sustenance to meet future needs. For the administration and planning of irrigation systems, accurate knowledge of how to utilize water resources is presently difficult to come out of. Providing accurate information at scales ranging from farm fields to entire river basins, including millions of hectares of irrigated land, is not a straightforward task. Nevertheless, general information regarding the agricultural and hydrological characteristics of the land surface can be derived for large areas using remote sensing techniques (Gulati et al. 2023). There have been substantial developments in the effectiveness of applications of remote sensing to identify and monitor crop growth and other biophysical characteristics over the past two decades; however, there exist a few issues that need to be resolved. Depending on the data collection and analysis capabilities, methods using remote sensing are currently regarded as efficient and effective agricultural water management measurements (Kanga et al. 2020). The advancement of technologies such as remote sensing, mobile computation, telemetry, and satellite monitoring has significantly contributed to the solution to the problem of water management. It is crucial to maintain an established relationship between farm water availability and crop production for a variety of commodities. The application of remote sensing techniques has substantially improved the characterization of water surface sources, the prediction of precipitation and temperature, the calculation of moisture content of soil, soil topography, and evapotranspiration (Tomar et al. 2023). It is now possible to monitor flood, drought, and irrigation management conditions in near real-time using high-resolution satellite data (Liaghat and Balasundram 2010.

4.2 Understanding Remote Sensing

Remote sensing is the procedure of detecting and monitoring the physical properties of a region through measurement of both its reflected and emitted radiation (typically from a satellite or aircraft). Specialized cameras acquire remotely sensed pictures, which researchers use to "sense" aspects of the Earth. Examples include:

- Satellites and aircraft equipped with cameras acquire images of vast portions of the Earth's surface, enabling us to view much more than is possible from the ground.
- Sonar technology on ships is capable of producing images of the seafloor without requiring the necessity to travel to the ocean floor.
- Satellite cameras are capable of being utilized to capture images of ocean temperature fluctuations (Fig. 4.1).

4.2.1 Principles of Remote Sensing

In Principles of Remote Sensing (RS), a relationship between radiation and the target of interest is required. There are eight components involved in the RS procedure.

1. Energy Source (A)
2. Radiation and the Atmosphere (B)
3. Interaction with the Target (C)
4. Recording of Energy by the Sensor (D)
5. Transmission and Reception (E)
6. Data Processing (F)
7. Interpretation and Analysis (G)
8. Application (H) (Fig. 4.2)

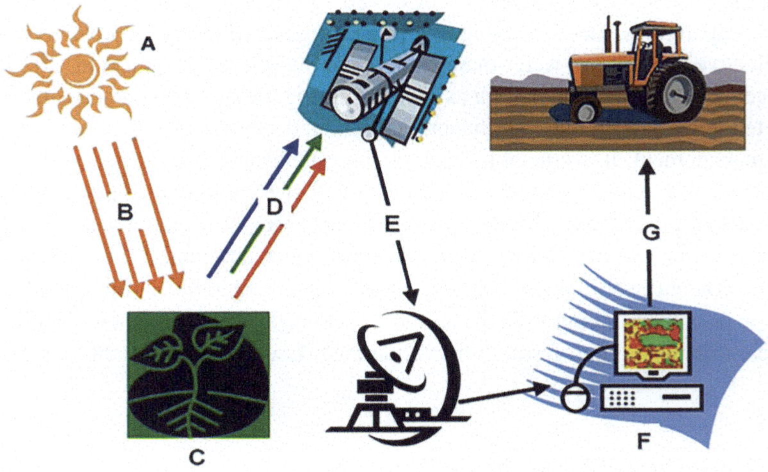

Fig. 4.1 Application of remote sensing and GIS in precision agriculture

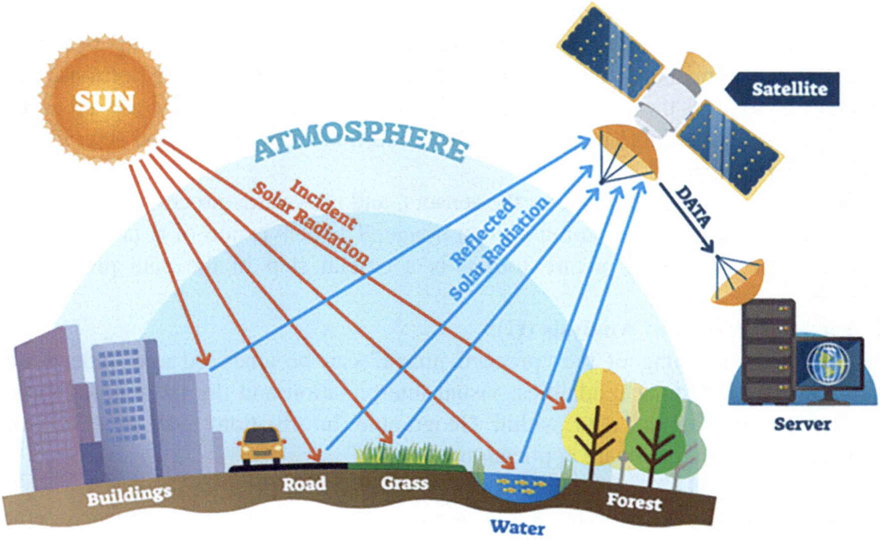

Fig. 4.2 Remote sensing image acquiring process

4.2.2 *Principles Components of Remote Sensing*

1. **Energy Source (A):**
   The primary prerequisite for remote sensing involves the presence of an energy source that either illuminates or supplies electromagnetic energy to the specific target of investigation. The main source of energy is the sun.
2. **Radiation and the Atmosphere (B):**
   During the dispersion of energy emitted by the object being targeted, it becomes necessary to employ a sensor to collect and record the electromagnetic radiation. The process by which energy is transferred from a source to an item can be understood through the phenomena of absorption and scattering of electromagnetic radiation during transmission.
3. **Interaction with the Target (C):**
   Upon traversing the atmosphere, the energy engages with the target, its interaction contingent upon the respective characteristics of both the target and the radiation. The phenomenon under consideration involves the interaction of electromagnetic radiation (EMR) with an object, followed by the processes of reflection and emission.
4. **Recording of Energy by the Sensor (D):**
   During the dispersion or emission of energy from the target, it becomes necessary to use sensors to gather and record the electromagnetic radiation. The sensor's assessment of energy. Photographic or non-photographic sensors are utilized in various applications.

5. **Transmission and Reception (E):**
   The movement of energy essential to the operation of sensors frequently occurs in electronic form and is directed toward the receiving and processing portions. The process involves the transfer of energy from the object to the sensor, as well as the subsequent recording and transmission of information to the ground station.
6. **Data Processing (F):**
   The data has been obtained by the sensors, and afterwards processed in the part related to data transformation into an image. The conversion of data into either digital or hard copy picture format is a crucial step in the data processing workflow.
7. **Interpretation and Analysis (G):**
   The literal meaning of the "proceed image" can be analyzed and understood through two distinct modalities: visual interpretation and digital interpretation. The examination of data within Geographic Information Systems (GIS) or Remote Sensing software applications.
8. **Application (H)**
   Ultimately, utilize the extracted information from the images to enhance comprehension of the target.

   - **Explanation of Remote Sensing as a Data Collection Methods:**
     Remote sensing is an approach utilized for the aim of gathering data because knowledge about objects or occurrences present on the Earth's surface is acquired without the need for direct physical contact. Remote sensing is a methodology that encompasses the utilization of a diverse array of sensors, tools, and technologies, which are affixed to aircraft, satellites, drones, or other platforms, to acquire data from an extensive distance.
   - **Types of Remote Sensing Platforms:**
     Remote sensing platforms refer to the apparatus or instruments that are equipped with sensors for data collection. These entities can be classified into three or four primary classifications.

     1. Sensors Technologies
     2. Satellites,
     3. Drones,
     4. Aircraft.

1. **Sensor Technologies:**

Remote sensing incorporates a diverse range of sensor technologies, such as cameras, multispectral and hyperspectral imagers, radar, lidar, and other advanced instruments. The sensors can acquire many forms of data, including visible light, infrared radiation, microwave signals, or laser pulses, contingent upon the application (Fig. 4.3).

Fig. 4.3 Sensor technologies

Fig. 4.4 Various satellites

2. **Satellites:**
   Earth-observing satellites are deployed in orbit around the Earth to acquire data throughout their traversal of various geographic areas. These systems offer comprehensive worldwide coverage and are well-suited for the surveillance of significant environmental transformations on a big scale. Various types of satellites are used such as Landsat, Sentinel, and GSAT-15 etc. (Fig. 4.4)
3. **Drones:**
   The utilization of remote sensing technology empowers operators of drones to develop a flight plan by offering a wider range of information coverage compared to conventional survey methods. Remote sensing plays a crucial role in furnishing pertinent data regarding the positioning of structures and other obstructions that may impede and jeopardize drone operations inside a specified region (Fig. 4.5).

Fig. 4.5 Drone

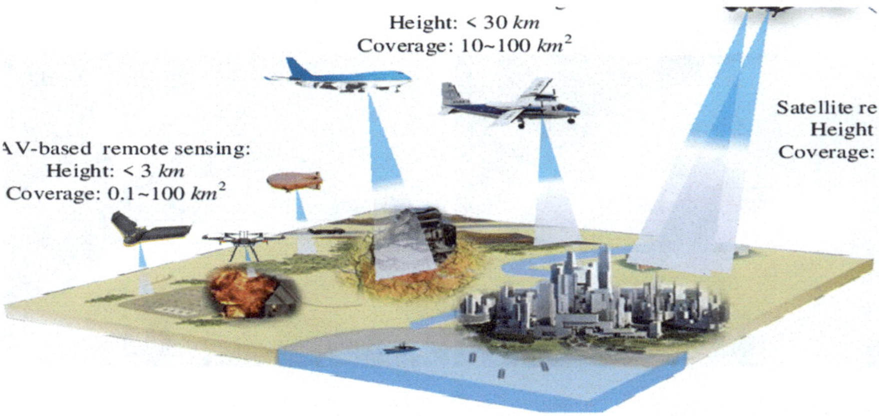

Fig. 4.6 Aircraft

4. **Aircraft:**
   Both manned and unmanned aircraft, encompassing airplanes and drones, possess the capability to transport sensors in closer proximity to the Earth's surface. This proximity enables the acquisition of imagery with enhanced resolution and facilitates greater adaptability in data-gathering processes (Xiang et al. 2018) (Fig. 4.6).

4.2.3 Remote Sensing Technologies

- **Satellite-Based Remote Sensing and Its Applications:**
  Satellite-based remote sensing is a technological approach that encompasses the utilization of man-made satellites in Earth's orbit to gather data about the Earth's

surface, atmosphere, and outer space from a distant vantage point. The satellites are outfitted with a diverse array of sensors and devices that enable the acquisition of data in the form of images, spectra, and measurements. Subsequently, this data is transferred to Earth for analysis. The utilization of this technology facilitates advancements in scientific research, aids in the preservation of the environment, enhances disaster management capabilities, and informs decision-making processes across a wide range of industries and sectors.

- **Applications:**

  1. **Environmental Monitoring:**

     - **Land Use and Land Cover Mapping:**
       Monitoring alterations in land cover, encompassing phenomena such as the expansion of urban areas, the depletion of forests, and modifications in agricultural land utilization.
     - **Forest Monitoring:**
       The evaluation of forest well-being, the surveillance of deforestation rates, and the identification of illicit logging practices.
     - **Wetland Monitoring:**
       The objective of this is to identify alterations in wetland areas and assess their subsequent effects on surrounding ecosystems.
     - **Coastal Zone Management:**
       The monitoring of coastal erosion, sea-level rise, and alterations in coastal ecosystems is of paramount importance in understanding and addressing the dynamic nature of coastal environments.

  2. **Agriculture:**

     - **Crop Monitoring:**
       The objective is to evaluate the condition of crops, and their developmental stages, and estimate their output to enhance agricultural operations.
     - **Drought Monitoring:**
       The purpose is to monitor and analyze soil moisture levels to identify areas that are susceptible to drought conditions.
     - **Precision Agriculture:**
       The strategic utilization of fertilizers and insecticides in accordance with specific crop circumstances.

  3. **Weather and Climate:**

     - **Weather Forecasting:**
       The provision of essential data for weather prediction models encompasses several parameters, such as temperature, humidity, and cloud cover.
     - **Climate Change Analysis:**
       The investigation of climate change implications includes the monitoring of greenhouse gas concentrations, ice sheet melting, and temperature patterns.

4. **Natural Disaster Management:**
   - **Hurricane and Cyclone Tracking:**
     The objective is to observe and track the development and trajectory of storms to offer timely alerts.
   - **Flood Detection:**
     The aim is to identify regions that are prone to flooding and to evaluate the amount and severity of flood events, as well as the resulting damage.
   - **Wildfire Monitoring:**
     The identification and monitoring of wildfires, as well as the evaluation of their propagation and severity.
   - **Earthquake and Volcano Monitoring:**
     The monitoring of the earth's movements and volcanic eruptions plays a crucial role in the assessment of hazards.
5. **Water Resources Management:**
   - **Water Quality Assessment:**
     The surveillance of aquatic ecosystems to detect and assess the presence of contaminants and the process of eutrophication.
   - **Reservoir and Lake Monitoring:**
     Evaluating water levels and water storage in reservoirs and lakes.
   - **River and Watershed Management:**
     The aim focuses on the examination of river flow patterns and the assessment of watershed conditions to facilitate the implementation of sustainable water resource management strategies.
6. **Urban Planning and Development:**
   - **Urban Growth Analysis:**
     This mainly aims to monitor the process of urban expansion, infrastructural development, and population growth inside urban areas.
   - **Transportation Planning:**
     The evaluation of traffic patterns and road networks plays a crucial role in the field of transportation planning.
   - **Land Suitability Assessment:**
     The task at hand involves the identification of appropriate sites for the implementation of building and infrastructure endeavors.
7. **Archaeology and Cultural Heritage:**
   - **Site Detection:**
     The identification and documentation of archaeological sites and cultural heritage aspects through the utilization of satellite imagery.
   - **Site Conservation:**
     The monitoring and safeguarding of cultural sites to prevent looting and mitigate any damage.

8. **Remote Sensing of Space:**
    - **Astronomy:**
    The act of observing celestial entities, including stars, planets, and galaxies, through the utilization of space-based telescopes, such as the Hubble Space Telescope.
    - **Space Debris Monitoring:**
    The objective is to monitor and forecast the trajectory of space debris with the aim of safeguarding satellites and space stations.

- **Unmanned Aerial Vehicles (UAVs) and Their Role in Data Acquisition:**

In recent times, there has been a growing interest among scientific academics and industry professionals in the field of UAV remote sensing, owing to its wide-ranging potential for application. Unmanned Aerial Vehicles (UAVs) possess the capability to be equipped with a diverse range of sensors, including optical, infrared, and LIDAR. Consequently, they have emerged as a crucial platform for remote sensing observations. Researchers can acquire several high-resolution photographs through the utilization of UAV remote sensing, wherein each pixel corresponds to a measurement of either a centimeter or a millimeter. This includes an analysis of the aircraft platforms, data types, and elements employed in various application categories.

There are two main areas within the field of remote sensing: precision agriculture and forestry. (1) Precision agriculture involves the observation of crop diseases, estimation of crop yields, and monitoring of crop environments. (2) On the other hand, forestry remote sensing focuses on identifying forest diseases and observing forest disasters, among other related activities. (3) The application of remote sensing techniques in the monitoring and analysis of power systems; (4) The impact of artificial infrastructure on the natural environment. In recent years, it has been observed that the processing of image data, including RGB, multi-spectral, and hyper-spectral data, has predominantly employed neural network methods. Notably, in the context of crop disease monitoring, multi-spectral data has emerged as the most extensively studied type of data. However, it is worth noting that the current applications of LIDAR data still lack a comprehensive end-to-end neural network processing approach. Considering these observations, this review comprehensively examines the various aspects of UAV platforms, sensors, and data processing methods. Furthermore, based on the developmental trajectory of specific application fields and the existing limitations in implementation.

4.3 Remote Sensing in Precision Agriculture

Precision agriculture (PA) encompasses a range of techniques and technology that enable the identification of soil and crop variability within a field. This capability allows for the enhancement of farming operations at a sub-field level, leading to the optimization of agronomic inputs. Variable-rate technology (VRT) enables the adjustment of soil and crop-applied inputs to be tailored for specific locations, allowing for site-specific application. Currently, there is ongoing advancement in sensing technologies, encompassing both ground-based and remote methods, which have become more affordable for the purpose of acquiring field-level data. To achieve operational success in the context of Variable Rate Technology (VRT), it is important to possess maps that encompass several aspects of crop cultivation such as crop growth, crop diseases, weed distribution, crop nutrient deficits, and other pertinent crop and soil variables. Consequently, the utilization of remote sensing technology, specifically through the acquisition of images by sensors installed on satellites, aircraft, or ground-based equipment, has become an essential component of Variable Rate Technology (VRT) (Fig. 4.7).

Remote sensing imaging has the potential to be utilized in various applications, including the mapping of soil parameters, classification of different crop species, identification of crop water stress, monitoring of weeds and crop diseases, and mapping of crop yield. The utilization of remote sensing in the field of precision agriculture is subject to various factors, including the selection of data collection platforms (such as satellites, aerial vehicles, or ground-based systems), the characteristics of the sensor employed (including the number and spectral width of bands captured, distinguishing between multi and hyperspectral capabilities), as well as the spatial (high, medium, or low), temporal (hourly, daily, or weekly), and radiometric (8-, 12-, or 16-bit) resolutions at which data is acquired. When utilizing remote sensing imagery for agricultural decision-making, it is imperative to thoroughly

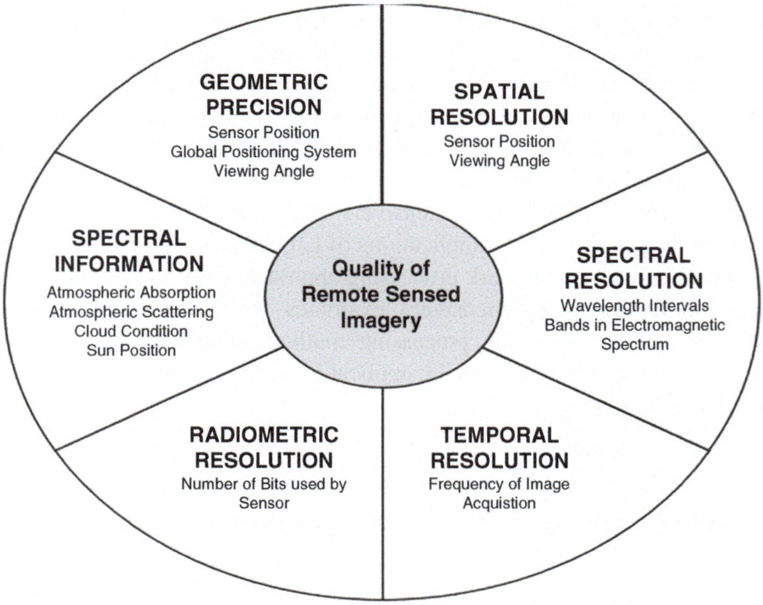

Fig. 4.7 Factors influencing the quality of remotely sensed images (ohioline.osu.edu)

assess various factors. These factors include (1) the degree of accuracy with which the image aligns with the corresponding ground location, commonly referred to as geometric precision; (2) the extent to which the image accurately represents ground features, encompassing spatial and spectral resolutions; and (3) the overall quality of spectral information conveyed in the acquired images.

4.3.1 Crop Monitoring and Health Assessment

- **Using Remote Sensing to Monitor Crop Development and Health**
  The utilization of remote sensing technology for the purpose of monitoring crop development and health represents a significant and advantageous application within the field of agriculture. Remote sensing is a method employed to get data pertaining to an object or region without the need for direct physical interaction. In the field of agriculture, the practice is commonly conducted through the utilization of satellites, drones, or other aerial platforms that are outfitted with a range of sensors. Remote sensing is a valuable tool for monitoring crop development and health (Mutanga et al. 2017).

  1. **Vegetation Indices:**
     Vegetation indices, such as the Normalized Difference Vegetation Index (NDVI) and Enhanced Vegetation Index (EVI), are frequently employed in the evaluation of crop health and growth. These indices quantify the extent of green vegetation and can serve as indicators of crop health, either indicating stress or vigor. Typically, crops that are in a state of good health exhibit elevated values of the Normalized Difference Vegetation Index (NDVI) or Enhanced Vegetation Index (EVI), whereas crops experiencing stress or unhealthiness tend to display lower values of these indices (Bannari et al. 1995).
  2. **Satellite Imagery:**
     Satellites outfitted with multispectral and hyperspectral sensors are utilized to get imagery of expansive agricultural regions at consistent intervals. These images provide the capability to monitor alterations in crop vitality, development, and land utilization. Both free and commercial satellite data sources, such as Landsat, Sentinel, and Planet Labs, provide significant and valuable information to farmers and academics.
  3. **Aerial Imagery:**
     Drones, also known as unmanned aerial vehicles (UAVs), provide the capability to acquire high-resolution photos of crops from several perspectives through the utilization of cameras and multispectral sensors. Aerial imagery offers a higher level of detail compared to satellite images and can be tailored to suit certain domains or areas.

4. **Thermal Imaging:**
   Thermal cameras provide the capability to identify temperature fluctuations within crops, serving as potential indicators of stress, irrigation complications, or disease. The practice of monitoring temperature variations aids in the identification of regions that may require attention.
5. **LiDAR (Light Detection and Ranging):**
   The utilization of LiDAR technology involves the utilization of laser beams to generate intricate three-dimensional representations of the topography and flora. Remote sensing technology has the potential to facilitate the evaluation of crop height, canopy structure, and the comprehensive topography of agricultural fields, thereby contributing to the advancement of precision agriculture and irrigation management.
6. **Time-Series Analysis:**
   Over a period, the utilization of remote sensing data enables the generation of time-series analyses, which facilitate the monitoring of agricultural growth and development. Through the process of comparing data collected at various time intervals, it becomes possible to discern patterns, deviations from the norm, and noteworthy aspects that warrant attention.
7. **Disease and Pest Detection:**
   The utilization of remote sensing techniques facilitates the identification of initial indications of agricultural illnesses and insect infestations through the analysis of alterations in plant reflectance and other relevant indicators. The implementation of this early warning system facilitates prompt action and mitigates agricultural yield losses.
8. **Yield Prediction:**
   The integration of remote sensing data with machine learning models enables farmers and researchers to enhance the precision of agricultural output predictions. These projections can aid with marketing decisions and budget allocation.
9. **Irrigation Management:**
   The utilization of remote sensing techniques enables the acquisition of valuable information regarding soil moisture levels, hence facilitating the implementation of accurate irrigation management strategies. This practice aids in the preservation of water resources and mitigates the risks associated with excessive or insufficient irrigation, which can have detrimental effects on agricultural yields.
10. **Crop Health Maps:**
    The utilization of remote sensing data enables the generation of maps that visually depict the overall condition and vitality of crops within a certain agricultural area. These maps possess the capacity to provide guidance to farmers in their decision-making processes regarding fertilizer application, pesticide utilization, and other interventions.

4.3.2 *Soil Mapping and Analysis*

- **Remote Sensing for Soil Property Mapping**
  The utilization of remote sensing in soil property mapping is an esteemed methodology employed to evaluate and delineate diverse soil properties over expansive regions, obviating the necessity for extensive on-site investigations. The process entails the utilization of remote sensing technology and satellite imaging to acquire data pertaining to soil attributes, encompassing texture, moisture content, organic matter content, nutrient levels, pH, and other relevant factors. The soil attributes mentioned are of utmost importance in the fields of agriculture, land management, and environmental monitoring. This chapter provides a comprehensive review of the utilization of remote sensing techniques for the purpose of soil property mapping.

  1. **Data Collection:**
     The practice of remote sensing involves the utilization of a diverse array of sensors that are affixed to satellites, airplanes, or drones. These sensors are responsible for capturing and detecting electromagnetic radiation, encompassing wavelengths within the visible, infrared, and microwave spectra, which are either reflected or emitted by the Earth's surface. Various wavelengths of light exhibit distinct interactions with soil properties, thereby enabling the extraction of specific information.
  2. **Image Acquisition:**
     Remote sensing platforms are utilized to acquire remote sensing imagery, which consists of photos or data. Soil property data collection can be influenced by the spatial resolutions, spectral bands, and temporal frequencies of different sensors.
  3. **Pre-processing:**
     Preprocessing is frequently necessary for raw remote sensing data to address atmospheric influences, geometric distortions, and radiometric fluctuations. This process guarantees that the collected data effectively and faithfully depicts the topography of the Earth's surface.
  4. **Feature Extraction:**
     After undergoing pre-processing, remote sensing data is utilized to extract pertinent information associated with soil qualities. The features encompass a range of characteristics, such as vegetation indices, surface temperature, and spectral reflectance patterns, among other relevant factors.
  5. **Modeling and Analysis:**
     The collected features are subjected to statistical and machine learning algorithms to make estimations about soil qualities. The variety of models encompasses both basic regression models and more sophisticated algorithms, including Random Forests, Support Vector Machines, and neural networks.
  6. **Validation:**
     The evaluation of the precision of soil property maps derived from remote sensing data is conducted by means of ground truthing and validation. To

validate and calibrate the models, particular locations are chosen for the collection of field data and soil samples.

7. **Mapping:**
   After undergoing validation, the models are employed to generate spatial representations of soil characteristics within the designated research region. The maps offer significant insights into the characteristics, disparities, and patterns pertaining to soil conditions.
8. **Decision Support:**
   The integration of soil property maps into decision support systems enables their utilization in diverse applications such as precision agriculture, land-use planning, environmental monitoring, and natural resource management. These maps can be utilized by farmers, land managers, and policymakers to facilitate informed decision-making about land use and resource allocation (Forkuor et al. 2017).

- **Soil Compaction Assessment Through Satellite Imagery**
  - **How Satellite Technology is Helping to Monitor and Mitigate Soil Compaction:**
    The utilization of satellite technology is progressively assuming a significant role in the monitoring and mitigation of soil compaction. The phenomenon of soil compaction is a significant concern for agricultural practitioners, as it diminishes the capacity of water and air to permeate the soil and access plant roots, thus resulting in diminished agricultural productivity.

    The utilization of satellite photography enables the identification of temporal variations in soil compaction. Researchers can detect areas of compaction and monitor their evolution by observing alterations in soil color, texture, and structure. This data can subsequently be utilized to guide decisions on soil management, including the timing of tilling operations and the application of fertilizers.
  - **How Satellites are Used to Monitor and Reduce Soil Compaction:**
    The utilization of satellites for the purpose of monitoring and mitigating soil compaction, a phenomenon associated with diminished agricultural productivity and related concerns, is seeing a growing trend. The issue of soil compaction is significant to farmers due to its adverse effects on the soil's capacity to retain water and nutrients, ultimately resulting in potential crop failure. Compaction may arise because of the utilization of heavy machinery, overgrazing, and various other anthropogenic activities. Satellites are employed for the purpose of monitoring soil compaction with the aim of mitigating its adverse effects. Through the utilization of satellite imagery, researchers have the capability to identify alterations in soil texture, moisture content, and various other parameters that may serve as indicators of compaction. Upon the detection of soil compaction, farmers can implement measures aimed at mitigating its effects. One potential approach is the utilization of tillage techniques to ameliorate soil structure by loosening its composition.

- **The Impact of Satellites on Soil Compaction Mitigation and Monitoring** The utilization of satellites has significantly influenced the mitigation and monitoring of soil compaction. Satellite photography has emerged as a highly helpful instrument in contemporary times, aiding farmers, academics, and land managers in the evaluation of soil compaction and its consequential impact on crop productivity. The phenomenon of soil compaction poses a significant challenge in the realm of agricultural production, as it has the potential to diminish crop productivity and inflict enduring harm upon the structural integrity of the soil. Satellites have the capability to capture high-resolution photos of the earth's surface, enabling them to discern minute alterations in soil composition that serve as indicators of compaction. The utilization of high-resolution satellite imagery has significantly enhanced the ability of researchers and land managers to effectively detect and monitor soil compaction on the earth's surface. This has enabled them to devise more efficient ways for minimizing compaction and assessing their success (Al-Dousari et al. 2000).

4.3.3 Pest And Disease Detection

- **Early Identification of Pests and Diseases Using Remote Sensing:**

Remote sensing offers a comprehensive perspective of agricultural areas by recording data pertaining to vegetation indices, temperature fluctuations, and spectral characteristics. The provided data functions as a foundation for identifying abnormalities and potential occurrences of pests or diseases. The utilization of remote sensing data enables the identification of patterns and correlations between environmental parameters and occurrences of pests and diseases. Timely identification of agricultural pests and illnesses enables prompt responses and focused control strategies. Farmers can employ suitable pest management tactics, such as accurate application of pesticides, utilization of biological control methods, or cultivation of crop varieties that are resistant to pests. By doing so, they may effectively decrease the likelihood of crop losses and mitigate the adverse effects on the environment (Fig. 4.8).

- **Hyperspectral Imaging for Spectral Signatures of Pathogens**

Hyperspectral imaging (HSI) exhibits considerable promise as a non-invasive diagnostic modality for the detection of diseases. This section examines the relationship between leaf features and spectral reflectance of leaves affected by Cercospora leaf spot, powdery mildew, and leaf rust at various phases of development. The utilization of light microscopy was employed to elucidate the alterations in the host tissue's morphology resulting from the colonization of pathogens. A hyperspectral imaging line scanning spectrometer, namely the ImSpector V10E model, was employed under controlled settings to conduct continuous screening and monitoring of disease symptoms throughout the pathogenesis process. This spectrometer offers a spectral resolution of 2.8 nm within the wavelength range of 400 to 1000 nm, and a spatial resolution of 0.19 mm. The utilization of a pixel-wise mapping technique to analyze

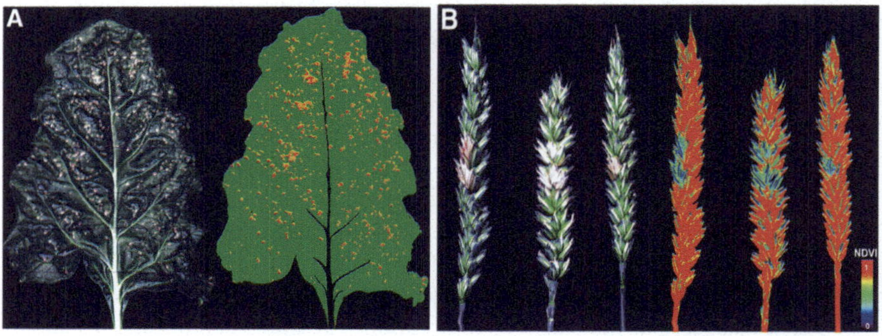

Fig. 4.8 Plant disease detection based on hyperspectral images

spectral reflectance throughout the visible and near-infrared spectrum facilitated the identification and comprehensive characterization of sick tissue at the individual leaf level. The relationship between leaf structure and leaf spectral reflectance patterns was established. The spectral signatures shown by pathogens are particular to the disease and are influenced by their interaction with the host tissue. The impact of infections on leaf reflectance was contingent upon both the developmental stage of the disease and the specific subarea of the symptoms. The utilization of spectral reflectance, in conjunction with the Spectral Angle Mapper classification technique, facilitated the distinction of mature symptoms into distinct zones that encompass all stages of ontogenetic development, ranging from early to advanced symptoms.

4.4 Precision Agriculture: Data Integration and Analysis

Precision agriculture, alternatively referred to as precision farming or smart farming, is a progressive methodology in the field of agriculture that uses the integration and analysis of data to enhance the efficiency and effectiveness of crop and livestock production. This methodology is dependent on cutting-edge technologies, including sensors, GPS guidance systems, drones, and data analytics, to enhance the decision-making process in agricultural practices, resulting in improved efficiency and informed choices.

4.4.1 Geographic Information Systems (GIS)

Geographic Information Systems (GIS) is a powerful set of tools and technologies used for capturing, storing, analyzing, and visualizing spatial or geographic data. GIS technology enables individuals, organizations, and governments to better understand and manage various aspects of the physical world by linking data to specific locations on the Earth's surface.

- **Integration Of Remote Sensing Data into GIS Platforms**

The incorporation of remote sensing data into Geographic Information Systems (GIS) platforms is a robust methodology that amplifies the spatial analysis functionalities of GIS and yields valuable insights across diverse domains, such as environmental monitoring, land use planning, agriculture, disaster management, and other relevant areas. This paper provides a comprehensive overview of the integration of remote sensing data into Geographic Information Systems (GIS) platforms:

1. **Understanding Remote Sensing Data:**
   The process of remote sensing entails the acquisition of information pertaining to the Earth's surface by non-contact means, often employing satellites, airplanes, unmanned aerial vehicles (UAVs), or ground-based sensors. Remote sensing data encompasses several types of information, such as images derived from satellite imaging and aerial photographs, with non-image data including temperature, radiation, and spectral data.
2. **Types of Remote Sensing Data:**
   Remote sensing data can be classified into different categories, including optical (visible and near-infrared), thermal infrared, radar, and hyperspectral data.
3. **Data Acquisition:**
   Remote sensing data is obtained by means of sensors and platforms, and it is imperative to consider parameters such as spatial resolution, spectral resolution, and temporal frequency when choosing data sources.
4. **Data Preprocessing:**
   Preprocessing is a crucial step in the utilization of remote sensing data, typically conducted prior to integration. This preparatory phase encompasses several activities, such as radiometric calibration, geometry correction, atmospheric correction, and image enhancement.
5. **GIS Data Preparation:**
   It is necessary to establish and prepare geographic information system (GIS) data layers and databases to effectively accommodate remote sensing data. This process entails the development of suitable layers for the integration of data.
6. **Georeferencing and Projection:**
   It is imperative to ensure the proper georeferencing and projection of remote sensing data and GIS layers to a consistent coordinate system to provide precise spatial analysis.
7. **Analysis and Visualization:**
   Geographic Information System (GIS) tools can be utilized to conduct many spatial analyses, including but not limited to land cover classification, change detection, and suitability analysis. Utilize various visual representations such as maps, charts, and graphs to effectively communicate information derived from the integrated data.

8. **Decision Support and Reporting:**
   Utilize the amalgamated data to make well-informed decisions and generate comprehensive reports for diverse applications, including but not limited to urban planning, disaster response, and natural resource management.
9. **Continuous Monitoring:**
   Remote sensing data possesses the capability to offer a consistent and uninterrupted flow of information, rendering it well-suited for the purpose of continuously monitoring and evaluating dynamic events.
10. **Data Management and Storage:**
    To guarantee the enduring accessibility and usability of integrated data, it is imperative to implement appropriate procedures for data management and storage.
11. **Training and Expertise:**
    It is imperative to ensure that experts in the field of Geographic Information Systems (GIS) possess the requisite skills and knowledge to proficiently handle remote sensing data.

- **Creating Spatial Databases for Informed Decision-Making**

The development of geographical databases for the purpose of making informed decisions is an essential component of contemporary data management and analysis, particularly in disciplines such as urban planning, environmental research, and corporate intelligence, among others. Spatial databases enable organizations to effectively store, manage, retrieve, and analyze data that is geographically and location-based in nature. These databases serve as a fundamental resource for making well-informed decisions since they allow users to extract valuable insights from the spatial correlations and patterns included in their data.

1. **Define Your Objectives:**
   Begin by establishing a concise and precise articulation of your goals and comprehending the distinct criteria and demands of your undertaking. One should contemplate the nature of the data that requires management and the specific forms of judgments that necessitate support through spatial information.
2. **Choose the Right Database Management System (DBMS):**
   Choose a database management system that provides support for spatial data. Some commonly used alternatives for spatial databases are PostgreSQL with the PostGIS addition, Oracle Spatial, Microsoft SQL Server with spatial extensions, and SQLite with spatialite.
3. **Data Collection and Preparation:**
   Gather the requisite spatial data for your project. The data can be derived from a multitude of sources, including but not limited to GPS devices, satellite imaging, surveys, and publicly available databases. It is imperative to ensure that the data being utilized is both correct and full, while also adhering to proper formatting guidelines.
4. **Database Design:**
   Developing a database structure that can effectively handle spatial data is essential. The process entails the creation of tables with the purpose of storing

both spatial and non-spatial attributes. It is imperative to provide due consideration to the many aspects of data types, indexes, and restrictions.

5. **Spatial Data Import:**
   The spatial data should be imported into the database. The selection of a certain database management system (DBMS) will determine the range of tools and approaches that can be employed for this undertaking. It is imperative to ensure the preservation of data integrity during the importation procedure.
6. **Spatial Indexing:**
   To optimize spatial queries, it is recommended to establish spatial indexes on the appropriate columns. Indexing plays a crucial role in enhancing the efficiency of spatial data queries by minimizing the volume of data that must be scanned.
7. **Data Quality Assurance:**
   Perform data quality assessments to detect and resolve any inaccuracies or discrepancies within your spatial data. It is imperative to verify that the data being utilized is both current and precise.
8. **Spatial Analysis Tools:**
   The integration of spatial analytic tools or libraries may be necessary depending on the specific objectives at hand. One can employ Python libraries such as GeoPandas, QGIS, or ArcGIS for the purpose of conducting spatial analysis and visualization.
9. **Query and Analysis:**
   Utilize SQL or alternative spatial query languages to derive significant insights from the geographic database. A range of geographical analysis techniques can be employed, including buffering, proximity analysis, clustering, and more methods.
10. **Visualization:**
    Utilize visual representations of the outcomes derived from your geographical analysis to facilitate informed decision-making processes. Maps, charts, and dashboards possess significant potential as instruments for effectively communicating spatial information to relevant parties.
11. **Regular Maintenance:**
    It is imperative to consistently update and maintain the spatial database to ensure its alignment with the most current information. This encompasses the incorporation of fresh data, elimination of outdated data, and enhancement of database efficiency as required.

4.4.2 Data Processing and Analysis

Data processing and analysis are fundamental elements of contemporary decision-making processes that rely on data. The processing of data serves as the primary stage in leveraging the potential of the data. The process entails the acquisition and systematic arrangement of data to render it amenable for analytical purposes. This can involve a wide range of activities, including the collection and storage of financial records in a database, as well as the compilation of user interactions on a website. The achievement of effective data processing is crucial in guaranteeing the correctness and consistency of data, thereby establishing a solid foundation for conducting relevant analysis. Upon completion of data processing, the true insights are revealed through the process of data analysis. Data analysis involves the exploration and examination of patterns, trends, and correlations that are inherent in the dataset. This procedure utilizes a range of statistical, mathematical, and computational methodologies to extract significant information.

- **Preprocessing Of Remote Sensing Imagery for Accurate Results**

The preprocessing of remote sensing imagery is crucial in guaranteeing the precision and dependability of the outcomes derived from these potent data sources. The crucial stage encompasses a sequence of procedures designed to improve the caliber of unprocessed satellite or aerial photographs prior to their utilization in analysis. Radiometric adjustment is considered a crucial preprocessing procedure that aims to mitigate the impact of several factors, such as air conditions, sensor sensitivity, and sun illumination angle, on the brightness changes observed in remote sensing data. This correction assures the maintenance of constant pixel values throughout the whole image, enabling reliable comparison and analysis of various components within the scene. In brief, the preprocessing of remote sensing imagery is an essential requirement for attaining precise and significant outcomes in several applications, including environmental monitoring, land use classification, disaster management, and urban planning.

- **Machine Learning and Data Analytics for Pattern Recognition**

Machine Learning and Data Analytics are essential components in the field of Pattern Recognition, facilitating the ability of computers to identify significant patterns and structures within extensive and intricate datasets. Pattern recognition refers to the cognitive process of discerning regularities or unique characteristics within data. This concept encompasses a diverse array of applications, such as the identification of patterns in images and voice, as well as the detection of anomalies in financial transactions. One of the primary advantages of Machine Learning in the field of Pattern Recognition lies in its ability to effectively extract pertinent information from unorganized or unstructured data. Data analytics is a valuable tool that enhances the process by which data is prepared and examined, revealing concealed insights that might inform the choice and refinement of machine learning models.

4.5 Sustainability Benefits of Remote Sensing and Precision Agriculture

The integration of Remote Sensing technology and Precision Agriculture techniques has introduced a novel phase of sustainable farming methods, presenting a wide range of advantages for both the environment and agricultural productivity. Remote sensing, a technology commonly employed using satellites and drones, offers important real-time information regarding soil conditions, crop vitality, and meteorological patterns within extensive agricultural regions. Precision Agriculture, a farming approach that heavily relies on this data, facilitates the precise and focused utilization of water, fertilizers, and pesticides. This practice effectively minimizes wastage and environmental contamination, while concurrently enhancing crop productivity.

4.5.1 Resource Optimization

Resource optimization refers to the strategic procedure of effectively managing and utilizing accessible resources to attain the highest possible output, while simultaneously eliminating both waste and unnecessary expenses. Its primary objective is to achieve equilibrium between the availability and utilization of resources, such as time, money, energy, and materials, to efficiently fulfill predetermined goals and objectives.

- **Water Management Through Efficient Irrigation**

Efficient irrigation plays a vital role in the implementation of sustainable water management strategies, especially in areas characterized by limited water supplies or susceptibility to depletion. Conventional irrigation practices frequently give rise to the overconsumption of water, hence causing water inefficiency and environmental deterioration. Nevertheless, by implementing contemporary irrigation technologies and adopting efficient methods, it is possible to maximize water utilization, improve agricultural productivity, and mitigate adverse ecological consequences. Furthermore, the implementation of sustainable irrigation practices has the potential to minimize the adverse effects of climate change through the reduction of greenhouse gas emissions that are typically linked with conventional irrigation techniques.

- **Precision Application of Fertilizers and Pesticides**

The precise application of fertilizers and pesticides has emerged as a crucial development in contemporary agriculture, fundamentally transforming the approach employed by farmers in crop management. This methodology entails the accurate and focused administration of fertilizers and insecticides to designated regions within a field, hence maximizing the efficiency of resource allocation and reducing the potential adverse effects on the environment. Consequently, this approach has the potential to enhance crop yields, minimize production expenses, and mitigate environmental impact. One of the primary advantages of precision application is its capacity to mitigate excessive use of fertilizers and insecticides. Rather of employing a blanket application of these chemicals throughout a whole agricultural area, farmers now have the capability to precisely detect and solve certain localized concerns, such as deficits in nutrients or infestations of pests.

- **Reducing Resource Wastage and Environmental Impact**

In the contemporary day, it is of utmost importance to prioritize the reduction of resource wastage and the minimization of our environmental impact. This urge arises from the heightened focus on sustainability and the long-term well-being of our planet. The promotion of a circular economy is considered a vital strategy in attaining this objective. The notion places significant emphasis on the optimization of resource utilization through the implementation of strategies such as material and product reduction, reuse, and recycling. The change in perspective not only results in the preservation of resources, but also serves to alleviate pollution, minimize landfill trash, and restrain the release of greenhouse gases, fostering a future that is more sustainable and ecologically conscious. In general, via the collective adoption of sustainable practices and the conservation of resources, we can offset the negative consequences of resource depletion and minimize our ecological footprint, thus ensuring the preservation of the Earth for future generations.

4.5.2 Economic Advantages

Economic advantages cover a wide array of benefits that emerge from a proficient and resilient economic system. One of the key economic benefits is the augmentation of prosperity and enhancement of living standards among the populace of a given nation. When an economy has consistent growth, it results in the creation of additional employment possibilities, increased incomes, and enhanced prospects for both individuals and enterprises. Fundamentally, a prosperous economy cultivates the development of novel ideas, the establishment of new businesses, and a positive cycle of advancement that has the potential to move a country ahead in the international sphere.

- **Increased Yield and Profitability with Optimized Inputs**

In contemporary agricultural practices, the persistent endeavor to attain increased yields and profitability necessitates the implementation of tactics that optimize inputs, hence proving to be highly effective in reaching this objective. Inputs in agriculture encompass a range of resources utilized by farmers in their agricultural activities, such as seeds, fertilizers, water, and pesticides. The prudent and meticulous administration of these inputs can have a substantial influence on the overall productivity and financial viability of an agricultural enterprise. The utilization of soil testing, remote sensing techniques, and weather data can yield significant insights pertaining to soil health, moisture content, and the likelihood of pest or disease infestations. Equipped with this data, agricultural practitioners can make well-informed judgments regarding the specific nature and quantity of resources to employ. Through the optimization of inputs in this manner, farmers have the potential to enhance both their output and profitability, while also fostering sustainable and responsible farming methods.

- **Cost Savings Through Targeted Interventions and Reduced Inputs**

The implementation of focused interventions and decreased resource utilization have emerged as basic ideas within contemporary precision agriculture, leading to significant cost reductions. Through the application of new technology such as GPS, remote sensing, and data analytics, farmers can enhance their resource allocation and crop management tactics, resulting in significant reductions in costs. One of the primary methods by which this objective is accomplished is the implementation of accurate variable-rate application techniques for inputs, including fertilizers, herbicides, and irrigation water. Precision agriculture enables farmers to customize the allocation of resources by utilizing real-time data and considering the distinct requirements of various sections within a field, as opposed to applying resources uniformly. In general, the implementation of focused interventions and decreased resource usage in precision agriculture yields cost savings that not only enhance the financial performance of farmers but also promote sustainable and ecologically conscious farming methods.

4.6 Challenges And Future Directions

The challenges and future directions in precision agriculture are many, including the complex and ever-evolving character of this chapter. The primary obstacles encompass the necessity for enhanced data accuracy and integration, guaranteeing data privacy and security, and facilitating the accessibility of precision agriculture to small-scale farmers in developing areas. Furthermore, there is a continued emphasis on the advancement of sustainable methods that aim to reduce environmental effects while maximizing productivity. The incorporation of unmanned aerial vehicles, robotic systems, and self-driving vehicles will contribute to the advancement of precision agriculture techniques. Simultaneously, the ongoing development of data analysis and remote sensing technologies will facilitate the acquisition of more comprehensive knowledge regarding crop conditions and the optimization of crop productivity. The resolution of these difficulties and the utilization of emerging technology provide the potential to transform global agriculture to effectively address the needs of a burgeoning population, while simultaneously mitigating resource inefficiencies and environmental degradation.

4.6.1 Technical Challenges

The field of precision agriculture presents a range of technical issues that necessitate resolution to effectively leverage technological advancements in the agricultural sector. The challenges encompass the requirement for dependable and cost-effective sensor technologies to facilitate the collection of data, guaranteeing uninterrupted connectivity and network coverage in rural regions for the real-time transmission of data, and the creation of user-friendly software and interfaces to enable farmers to interpret and effectively utilize the data. Furthermore, the incorporation of intricate data analytics and machine learning algorithms for the purpose of handling the extensive volumes of data produced by precision farming systems continues to provide a substantial obstacle. It is imperative to overcome these technical challenges to fully unlock the potential of precision agriculture in enhancing efficiency, sustainability, and productivity in agricultural operations.

- **Overcoming Limitations in Remote Sensing Data Quality**

Addressing constraints in the quality of remote sensing data is a critical issue in the field of precision agriculture, as the accuracy and dependability of such data play a crucial role in making well-informed decisions regarding farming operations. An effective strategy for improving data quality involves the utilization of sophisticated sensor technologies capable of capturing more accurate and comprehensive data pertaining to crop health, soil conditions, and weather patterns. Advancements in multispectral and hyperspectral sensors have facilitated enhanced spectral resolution, hence permitting the identification of nuanced fluctuations in crop vitality and nutrient concentrations. Furthermore, the incorporation of data derived from a variety of sensors, encompassing satellites, drones, and ground-based sensors, has the potential to yield an all-encompassing and mutually corroborated dataset. This approach mitigates the influence of constraints such as cloud cover or discrepancies in sensor calibration. The utilization of these techniques facilitates the rectification of mistakes and artifacts within the data, amplification of image resolution, and extraction of significant information from datasets that are noisy or incomplete. By conducting training exercises on extensive and reliable datasets, it becomes feasible to automate the preprocessing of data and improve the precision of crop monitoring and predictive modeling in the field of precision agriculture.

- **Ensuring Data Accuracy and Consistency**

The maintenance of data accuracy and consistency holds the utmost importance in the field of precision agriculture since the quality of data has a direct impact on the efficacy of decision-making procedures. Ensuring data accuracy necessitates the appropriate calibration and upkeep of sensors and measurement devices employed during the process of data collecting. The implementation of regular calibration procedures is essential to ensure that these devices consistently and accurately produce reliable readings throughout their operational lifespan. In addition, the process of validating data through ground-truthing involves comparing the acquired data with physical observations conducted in the field. This has resulted in improved accessibility to consistent and current information for stakeholders involved in the agricultural value chain. Through the prioritization of data accuracy and consistency, precision agriculture has the potential to optimize the allocation of resources, increase crop management practices, and ultimately improve agricultural sustainability and productivity.

4.6.2 Emerging Trends

The current trajectory of precision agriculture indicates a forthcoming transformation in farming practices, characterized by enhanced sustainability, efficiency, and reliance on data-driven methodologies. A notable phenomenon in the field of precision agriculture is the growing incorporation of artificial intelligence (AI) and machine learning (ML) techniques. These technologies provide the automated examination of extensive quantities of data, encompassing crop images, weather patterns, and soil data, with the aim of delivering instantaneous observations and anticipatory analytics. Drones outfitted with multispectral and hyperspectral cameras possess the capability to acquire high-resolution imagery of crops, thereby facilitating meticulous surveillance of crop health and timely identification of potential concerns. The growing trends highlight the industry's transition towards automation, data-centric insights, and sustainability within the field of agriculture.

- **Fusion Of Remote Sensing with Internet of Things (IoT)**

The integration of remote sensing and the Internet of Things (IoT) signifies a pioneering development in precision agriculture, with the potential to significantly transform data acquisition and decision-making procedures. The integration of remote sensing technologies, such as satellites and aerial images, with Internet of Things (IoT) devices is becoming more prevalent. These IoT devices encompass a range of technologies, including soil moisture sensors, weather stations, and automated drones. The ongoing development of this trend shows significant potential in advancing sustainability and efficiency within the field of precision agriculture, ultimately making valuable contributions to food security and environmental conservation.

- **Integration of Artificial Intelligence for Real-Time Decision Support**

The incorporation of artificial intelligence (AI) into the realm of precision agriculture has emerged as a significant and disruptive trend, particularly in the context of real-time decision support. Artificial intelligence (AI) algorithms are being progressively utilized for the purpose of processing and analyzing large volumes of data obtained from diverse sources, including sensors, drones, satellites, and weather stations. AI systems provide the capability to swiftly discern patterns, trends, and anomalies within datasets, so furnishing farmers with immediate and up-to-date observations regarding the well-being of crops, soil quality, and meteorological predictions. The ultimate goal of these recommendations is to maximize crop yields while simultaneously preserving valuable resources.

4.7 Conclusion

In the final analysis of this book chapter pertaining to remote sensing and precision agriculture, we arrive at the intersection of a paradigm-shifting movement in contemporary agricultural methodologies. The incorporation of remote sensing technologies has initiated a paradigm shift, wherein the utilization of data-driven approaches is augmenting agricultural sustainability, efficiency, and productivity. The utilization of satellite images to detect even the most minor alterations in crop health, along with the implementation of ground-based sensors for precise monitoring of soil conditions, has provided farmers with unparalleled knowledge regarding their agricultural fields. The capacity to promptly respond to dynamic environmental elements and crop requirements is a defining characteristic of this technological revolution, offering the potential to sustainably nourish an expanding global population while mitigating the ecological consequences of agricultural practices. The imperative for fostering collaboration among scientists, engineers, policymakers, and farmers cannot be overstated, since it is crucial for the advancement and seamless integration of these technologies into conventional agricultural operations. To ensure that small-scale farmers and individuals residing in resource-constrained locations may derive benefits from these instruments, it is imperative to prioritize accessibility and affordability in the ongoing improvements. Furthermore, it is imperative that our pursuit of a more sustainable future in agriculture is guided by ethical considerations, data security, and environmental stewardship. Remote sensing and precision agriculture have the potential to significantly contribute to the resolution of global food security and sustainability issues in the twenty-first century, given their continuous advancements and dedication to responsible use.

References

Ahmad T, Gupta SK, Singh SK, Meraj G, Kumar P, Kanga S (2023) Unveiling nature's resilience: Exploring vegetation dynamics during the COVID-19 Era in Jharkhand, India, with the Google Earth Engine. Climate 11(9):187

Al-Dousari AM, Misak R, Shahid S (2000) Soil compaction and sealing in Al-Salmi area, western Kuwait. Land Degrad Dev 11(5):401–418

Bannari A, Morin D, Bonn F, Huete A (1995) A review of vegetation indices. Remote Sens Rev 13(1–2):95–120

Forkuor G, Hounkpatin OK, Welp G, Thiel M (2017) High resolution mapping of soil properties using remote sensing variables in south-western Burkina Faso: a comparison of machine learning and multiple linear regression models. PLoS One 12(1):e0170478

Gulati B, Sharma R, Kanga S, Singh SK, Sajan B, Meraj G, Kumar P, Ramanathan AL (2023) Unraveling the relationship between stubble burning and air quality degradation in Punjab: a temporal and spatial analysis (2019–2022). J Clim Change 9(2):43–53. IOS Press. https://ohioline.osu.edu/factsheet/fabe-5541

Kanga S, Singh PK, Ohri A (2020) Evaluating the environmental sustainability of smart cities in India: the design and application of the Indian smart city environmental sustainability index. Sustain For 13(1):327

Kumar P, Singh SK, Gupta SK, Kanga S, Meraj G (2023) Uncovering the hydro-meteorological drivers responsible for forest fires utilizing geospatial techniques. Theor Appl Climatol. https://doi.org/10.1007/s00704-023-04497. Springer Nature

Liaghat S, Balasundram SK (2010) A review: the role of remote sensing in precision agriculture. Am J Agric Biol Sci 5(1):50–55

Mutanga O, Dube T, Galal O (2017) Remote sensing of crop health for food security in Africa: potentials and constraints. Remote Sensing Applications: Society and Environment 8:231–239
Tomar P, Singh SK, Kanga S, Pattanaik A (2023) Water bodies mapping and monitoring using high-resolution satellite images. Sustain Agri Food Environ Res 11:1–18
Xiang TZ, Xia GS, Zhang L (2018) Mini-UAV-based remote sensing: techniques, applications and prospectives. arXiv 2018. arXiv preprint arXiv:1812.07770

Chapter 5
Agroforestry Systems and Its Impact on Tribal Livelihoods: A Case study of Sungra Village of District Kinnaur, Himachal Pradesh

Khub Ram and Seema Choudhary

Abstract Agroforestry is the general term for land use techniques, technologies, or systems in which woody perennials are integrated with agricultural crops or animals in the same system of land management through spatial or temporal arrangements. Agroforestry has its roots in the concept of Agri-silviculture. A sustainable and diverse strategy to land usage, agroforestry systems include trees, crops, and livestock on the same plot of land. The study tries to encompass both primary and secondary sources of the data. The analysis comprises of the use of the simple random sampling technique for the selection the households. Therefore, the present study tries to analyse the impact of agroforestry on the tribal people's livelihoods. The collected data has been tabulated, processed, analysed and interpreted by applying suitable statistical techniques. The present study has applied sustainable livelihood framework (SLF) (Department for International Development (DFID) 1999). In which improvement in different livelihood capitals/assets such as natural, human, physical, social and financial have been analysed. In the present investigation of agroforestry systems, the farmers of the present study area have mainly practised Agro-horticulture. The major agriculture crops grown by the farmers are wheat, maize, rajmah ogla and fafra (buckwheat), potato and peas. As per the findings of the study the social and financial capitals have significantly improved in terms of information about farming and helps from the Government and Non-government organisations. Positive improvement has been witnessed in the study area due to agroforestry practices in the past few decades.

Keywords Agroforestry · Livestock · Agricultural crop · Agri-horticulture · Livelihoods · DFID · Sustainable livelihood framework

---

K. Ram (✉) · S. Choudhary (✉)
Department of Geography, HP University, Shimla, India

S. Kanga et al. (eds.), *Transforming Agricultural Management for a Sustainable Future*, World Sustainability Series, https://doi.org/10.1007/978-3-031-63430-7_5

5.1 Introduction

Agroforestry is a general term applied for land use techniques, technologies, or systems in which woody perennial plant are integrated with agricultural crops or animals in the same system of land management through spatial or temporal arrangements. Agroforestry has its roots in the concept of Agri-silviculture which refers to sustainable and diverse strategy for land usage, comprising of trees, crops, and livestock on the same plot of land (Atangana et al. 2013). Agroforestry is a term used for the use of available arable land in spatial and temporal arrangement in which perennial trees are grown along with the crops and livestock (Kumar et al. 2023). This fulfils economic and ecological requirements from the agroforestry practices in terms of needs for food, fodder for livestock, and fuel wood (Kanga et al. 2020). The ecologically benefits also encompass the exchange of biological elements, such as the addition of leaf litter to the soil or the feeding of cattle with such litter as browse before returning it to the soil as manure (Young 1988). In the literature related to agroforestry, the words "systems" and "practices" are usually used as synonyms. In practical terms, an agroforestry system is a particular local instance of a technique, determined by the environment, plant species, and how they are arranged, managed, and integrated into the socio-economic system (Dagar et al. 2013). However, the integration of trees on farms and throughout the agricultural landscape, agroforestry is a multifaceted, ecologically oriented natural resource management method which is believed to diversify and maintain output for improved social, economic, and environmental advantages. Through the integration of trees on farms and throughout the agricultural landscape, agroforestry believed to be a multifaceted, ecologically oriented natural resource management method. It diversifies and maintain output for improved social, economic, and environmental advantages (Quandt et al. 2017). The farmers in tribal communities are limited by harsh and desolate terrain and depend mostly on their traditional knowledge of agricultural practices (Ahmad et al. 2023). It is necessary to thoroughly research the requirements, goals, and cultural norms of these hill people in order to create a specific action plan (Pandey et al. 2019).

In India, agroforestry has played a key role in supplying the country's needs for a living since ancient times. India's tribal have been practising agroforestry from time immemorial. However, an organized agroforestry research was primarily started in India in the early 1970s through industrial involvement in the planting of marketable tree species through Indian Council of Agricultural Research (ICAR) institutes. Consequently, the ICAR established the All India Coordinated Research Project (AICRP) on Agroforestry in April 1983 (Dagar et al. 2013). Therefore, in spite of harsh climate and difficult terrain in the northwestern Himalayan region of India the people practice intensive agriculture for their livelihoods (Farooq et al. 2022). The different agro-climatic zones' hill farming practises for subsistence are mostly determined by the climate, soil, and ecology. The farmer considers all of these factors and practice various agroforestry systems in which fruit-based agroforestry along with crops are common (Tomar et al. 2023).

The traditional agroforestry systems and practices consist of the intentional cultivation of trees with various crops and livestock for multiple benefits such as firewood, fibre, food, fruit, etc. and are time-tested and well-suited to a variety of situations. These systems developed over the years and were found suitable for conservation of natural resources namely soil, water and vegetation. (Kashyap et al. 2014).

5.2 Types of Agroforestry Systems

Nair (1985) broadly classified the agroforestry systems on the basis of structure, function, socio-economic base and ecology. These approaches are not independent rather these are interrelated. In this the major types are Agri silviculture (crops and trees includes shrubs/trees and trees) Silvo pastoral (pasture/animals and trees) Agro-silvopastoral (crops, pasture/animal: and trees) and Others (multipurpose tree lots, apiculture with trees, aquaculture with trees, etc.) For food, wood, and fodder as well as for agroforestry, farming, and raising animals, mountain people have long relied on the forest. It is one of the finest methods to preserve land, water, and increase fertility, it also serves as a fantastic supplier of non-timber wood (Gulati et al. 2023). In rural areas, people either go to the plains to find work or rely entirely on the forest, agriculture and horticulture for all of their needs, including farming and raising animals (Kumar et al. 2018).

In the western Himalayas, there are different agroforestry especially systems which are totally based on the temperature, altitude, and topography (Kumar et al. 2018). The conventional agroforestry systems in the highlands, however, are extremely similar to natural ecosystems as they offer both concrete and abstract functions that are comparable to those provided by forests for farmers' sustainable livelihood (Arunachalam et al. 2019). The types of agroforestry have been varied for Himalayan region. For this various agroforestry systems have been identified for different zones of western Himalayas. Particularly for Himachal Pradesh agroforestry systems are as follow: Low hills (<1000 m) subtropical zone comprising of Agri-silviculture, Agri-horticulture, Agri-horti-silviculture, and Silvi-pasture (horticulture trees, forest trees, crops and grasses), the mid hills (1000–1500 m) Sub-temperate comprises Agri-silviculture, Agri-horticulture, Agri-horti-silviculture, Silvi-pasture (horticulture trees, forest trees, crops and grasses), the high hills (1500–2500 m) Temperate zone comprises of Agri-silviculture, Horti-agriculture, Horti-silviculture, and silvi-pasture (horticulture trees, forest trees, crops and grasses) and Dry temperate zone (>2500 m) Agri-silviculture, Horti-silviculture, Silvi-pasture (horticulture trees, forest trees, crops and grasses). Kumari et al. (2008)

identified Agri silviculture, Agri-silvipastoral, Agri-silivihorticulture, Agri-hortisilviculture, Horti-pastoral and Silvi-pastoral agroforestry system in the Kandaghat developmental block of district Solan. Similarly, Kumari et al. (2008) carried out a study traditional agroforestry system practised in Lahul (Lahul and Spiti) and Kinnaur district of Himachal Pradesh. The study identified tree based various agroforestry systems practise in Lahul and district Kinnaur such as Agri-horticulture, Agri-silviculture, Silvi-pastoral and pastoral -horticulture. The agri-horticulture was prevalent type of agroforestry system practised in district Kinnaur and agri-silviculture in Lahul region.

5.3 Objective of the Study

The present research was undertaken

1. to address the impact of agroforestry system on livelihoods of tribal population of district Kinnaur.
2. to assess the improvement in livelihood capitals such as Human, Social, Physical, Natural and Financial owing to agroforestry system.

5.4 Study Area

The study area is village Sungra located in the Nichar developmental block of the Kinnaur district of Himachal Pradesh. Geographically, Kinnaur district is located between 31° 05′ 55″ and 32° 05′ 20″ north latitude and 77° 45′ 00″ and 79° 00′ 50″ east longitude. Three mountain ranges viz. the Zanskar, the Great Himalaya and the Dhauladhar. The average altitude of the study area is between approximately 1600 to 1800 metre above sea level. The total geographical area of the study area is 219.57 hectare (Census of India 2011) The study area touches its boundary with villages namely, Kachrang in the north, Gharshu in north-west, Kangosh in south-west and Punaspa in west direction. The study area is bounded by demarcated protected forest village in the south-west direction (Census of India 2011).

There are 785 households living in Sungra, a sizeable community. According to the 2011 census, the village comprises of 1233 female and 1436 male. This village out of total population, about 1310 were engaged in work activities. The further classification shows that 73.6% of workers are classified as Main Worker. while 26.4% as Marginal worker who work for their livelihood for less than 6 months. Of 1310 workers engaged in Main Work, 189 were cultivators (owner or co-owner) while 31 were Agricultural labourers (Census of India 2011).

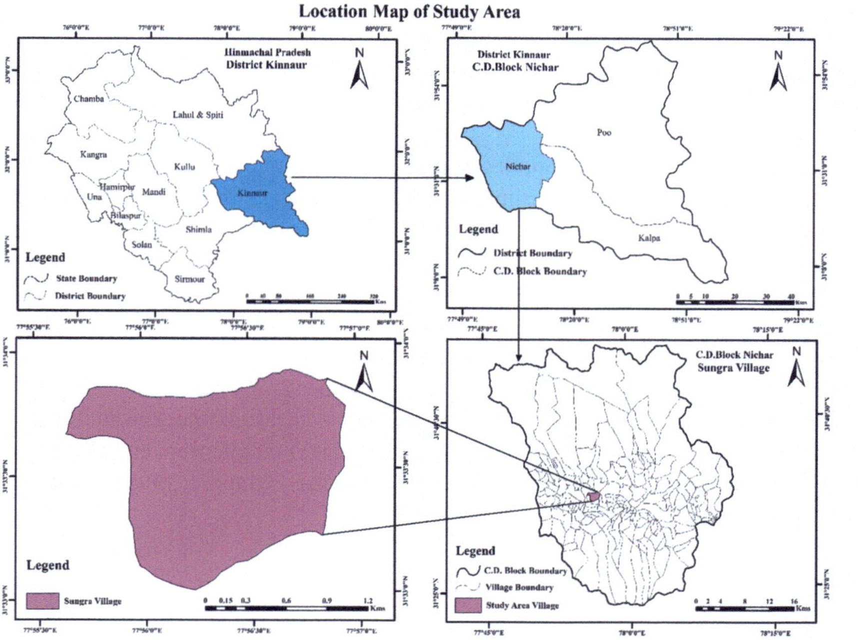

Source: District Census Handbook, Census of India 2011

5.5 Database and Methodology

The present study has been based mainly on primary sources of data collected by the use of prepared schedule and observation method. The present study has aimed to analyse pattern of agroforestry carried in the district Kinnaur. As such Nichar development block has been selected as the unit for the analysis of the agroforestry systems and its impact on tribal people livelihood.

The collected data has been tabulated, processed, analysed and interpreted by applying suitable statistical techniques. The Arc GIS software 10.3 has been used for the composing the maps. The present study has applied sustainable livelihood framework (SLF) (Department for International Development (DFID) 1999). The Sustainable Livelihood (SL) framework (DFID 1999) illustrates the key dynamics of a collection of assets that constitute the basis for people's livelihoods. The key influences on people's livelihoods are included in the sustainable livelihood's framework, along with the typical relationships between them. It may be used to both plan new development initiatives and evaluate how current initiatives are affecting the viability of human livelihood. (DFID 1999). The capitals or assets are divided into five groups using this framework in which Human assets (farming knowledge, management skills, traditional knowledge); Social (relationship, participation, information); Physical (soil productivity, household infrastructure, use of machinery or tools); Natural (tree coverage, soil conservation of properties); and Financial (household income, savings, expenditure, credits). The Livelihood Improvement Index (LII) was subsequently determined by considering the individual and combined responses provided by respondents in relation to the assets related to their livelihoods.

The present study has been applied simple random sampling technique to select the household. District Kinnaur comprises of the three-development block namely Kalpa, Pooh and Nichar of these three Nichar holds highest population and thus has been selected for the study. Thereafter, in Nichar block Sungra village records highest population of 785 households. In order to conduct the study about 10% sampled households have been selected for the study from this village. Therefore, total number of 79 households have been surveyed. The survey for the collection of primary data was carried out in the month of September 2023. Each respondent was asked throughout the household survey about their educational background, property ownership, agroforestry system, income, crops and tree planting, etc. However, the question regarding the different livelihood capitals (Human, Social, Natural, Physical and Financial capitals) have been asked.

5.6 Limitation of the Study

The sample could be collected only from one block as there was massive landslide during the field visit in August and September 2023. The study could have brought more variation if sample from other blocks were collected.

5.7 Results and Discussion

The livelihood activities in any area are the outcome of interaction between its environment and availability of technical know-how. It also depends of the capabilities of the population. The present study has diverse population, social and economic characteristics. The study area has recorded majority of the farmers between the age group 45 to 55 which is more than 40% of the total number of respondents. When combined with age group of 25–35 it shows that more than 65% of people with respect working age group are involved in agriculture activities. So, it shows that agriculture activities are the main source of earning livelihood in the study area. The lowest proportion of the farmers has been recorded between the age group 35–45 which is only 3.8% (Table 5.1). The educational qualification level has categorised into the five groups. The largest proportion of the farmers have senior secondary level education (34.2%) and there are few farmers which have middle level of education (5.1%) (Table 5.1). The second highest proportion (21.5%) of farmers have no formal education and 20.3% farmers are graduated.

Table 5.1 Demographic and socio-economic characteristics

	Categories	Respondents	
Age group		No.	Percent
	above 55	24	30.4
	45–55	32	40.5
	35–45	3	3.8
	25–35	20	25.3
	below 25	0	0.0
Educational Qualification			
	No formal education	17	21.5
	Middle	4	5.1
	Secondary	15	19.0
	Sr. Secondary	27	34.2
	Graduate	16	20.3
Land Holding			
	2-5 h	34	43.0
	1-2 h	41	51.9
	<1	4	5.1

5.8 Land Holding

The evaluation of the farmer's physical assets, particularly their land holding, is of significant value. The sampled households bring out the fact that, the majority (52%) of the farmers are holding of 1 to 2 hectares of land followed by 43.0% of small land holding with 2–5-hectare land. There are only 5.1% farmers who have less than one hectare of land and fall under marginal category. The cultural practices of polyandry can be attributed the fact that majority of the farmers have large farms as the division of land have been vary minimum (Table 5.1).

5.9 Educational Qualification

Attainment education is a positive indicator that a farmer is an aware person who know the use technology or the new innovations in the field of agriculture. As per Table 5.1 majority (34%) of the farmers are educated up to senior secondary followed by graduation (20%), secondary (15%) and middle class (4%). A very small proportion of farmers (17%) are not having formal education.

5.10 Agroforestry System

In district Kinnaur four types of agroforestry systems have been identified such as Agri silviculture, Agri-silvipastoral, Agri-silivihorticulture, Agri-hortisilviculture, Horti-pastoral and Silvi-pastoral (Kumari et al. 2008). Agri-silviculture – the

Table 5.2 Crops/trees and livestock inventory of the study area

Village (Study Area)	Agri-horticulture		
Sungra	Crops/trees	Medicinal Plants	Livestock raised
	Wheat+ogla+peas+rajmah+shimla-mirch+apple +green chillies+apricot+potato+maize+fafra +rajmah+tomato+plum	Tulsi +aloevera	cow+sheep +goat +poultry

conscious and deliberate use of land for the concurrent production of agricultural crops (including tree crops) and forest crops. (King 1979). Agri-silvipastoral is a combination of Agri silviculture and silvi-pastoral system (pasture/livestock and trees) (Carne 2008). Agri-hortisilviculture made up with the agriculture crops and fruit trees such as apple, plum, apricot etc. (Bijalwan 2013). The present study area has observed mainly Agri-horticulture system of agroforestry practise by the farmers. In which the farmers most prevalent crops and trees are grown such as potato, peas, maize, ogla, fafra, rajmah, Shimla-mirch, green chillies, apricot and apple. The medicinal plants grown by the farmers are Tulsi and aloe vera. The farmers in the study area also raised livestock commercially for their livelihood support along with the agriculture and horticulture. The livestock rearing is an integral part of the farming system which is equally important as agriculture and horticulture. The main animals and avian species are cow, sheep, goat and poultry respectively raised for their livelihoods (Table 5.2).

5.11 Analysis of Impact of Agroforestry on Livelihood Improvement of Tribal People

The sustainable livelihoods approach is a method of considering the aims, scope, and order of importance of development efforts. It facilitates the creation of development initiatives that are based on people, flexible, and participative, Multilevel, Partnerships between the public and private sectors. This approach makes it easier to identify realistic action priorities (Serrat 2017). People are the primary focus of the livelihoods approach. It seeks to comprehend people's livelihood assets and how they apply them to achieve successful livelihood results. It aims at understanding people's livelihood assets/capitals and successful life outcomes. (DFID 1999).

The present study has applied the sustainable livelihood framework to assess the impact on livelihoods capitals of the tribal people of Sungra village. A Sustainable Livelihood Framework has been applied to assess the five capitals for the sustainable livelihood in the Sungra village. The Livelihood Improvement Index has been used in this study. It shows how respondents evaluate changes to the capitals or assets under consideration. The agroforestry system was traditionally practised by the rural people in which farmers grew various types of trees with other crops. These trees provide the necessary wood, fuel, fodder and other materials. While the crops provided food for survival. There has been changes in the Agro-forestry system as well. The older ones were more sustainable while the recent ones are good from financial point of view.

In order to analyse the impact of agroforestry on livelihood of the people in study area. A livelihood improvement index has been framed. This index takes into consideration five variable or livelihood capitals namely, human, physical, natural, social and financial. As a result, the farmers have significantly improved. Their social capital is reflected in the culture and interpersonal interactions of groups and individuals (Akter et al. 2022). About 74.68% respondents of the study area witnessed that there is increase in the access to information about the farming and related agroforestry in the present scenario. The tribal denotes to an area which remains under tough condition and cut off from rest of the world for a long period of time. However, the scenario has changed today. After coming tribal sub development plan (1974–75) the tribal areas have shown significant improvement in all sectors. The tribal people are able to get new innovative ideas of the farming. The government and non-government organisation are quite active in producing knowledge about farming and practising agroforestry to the local in the study area. This is validated from the fact that majority of the respondents of the study have registered an increase in the social relationship within the society. The strengthened social relationship in the outcome of showing knowledge about traditional ways of mixed farming. About 62.03% of the respondents stated that the government and non-government organisation lends financial help to improve the traditional agroforestry system. As a result, the financial capitals have shown the significant improvement. This shows improvement in access to credit from Government and Non-Government organisations as well as improvement in people's household income, savings and household expenditure.

Human capital includes the skills, knowledge and physical capability to pursue the livelihood strategies (Scoones 1998). In the findings of present study, majority of the respondent's human capital has registered a slight increase in terms of sustainable knowledge of farming, farmland management (about 78.48%) and enhance in the traditional knowledge of farming. A very less proportion of farmers (1.27%) stated that their knowledge about sustainable farming has highly improved due to practising of traditional agroforestry. About 46.84% respondents reported that there is no change or modifications in the traditional knowledge of farming. It is so because farmers believe that crop selection are well-suited to their specific environment. There is rich knowledge among the farmers about seed saving, and crop rotation with seasons.

Infrastructure and producer goods necessary to support livelihoods are referred to as physical capital (Zhang and Fang 2020) which also includes competent workers and equipment needed to sustain livelihoods (Pandeya et al. 2017). Agroforestry raises productivity by harnessing the economic benefits of both crops and trees, producing agricultural and forestry product yields that improve physical capital (Hanif et al. 2018). In the present study, in the physical capitals have shown satisfactory improvement. About 73.42% respondents perceived that due to practising agroforestry there is slight increase in the forest and crops productivity. In terms of timber harvested from trees planted in the farmland about 70.89% stated that there has improvement in the physical infrastructure of household. However, 67.09% farmers have used machinery and there is improvement in the use machinery and modern tools to maintain the farmland.

Natural capital includes all of the components of the environment that people depend on for their livelihood, such as topography, soil, water, climate, minerals, forests, and biodiversity. (Fahad et al. 2022). For example, the capacity of certain farmers to sustain themselves might be severely harmed by the lack of access to productive land. Due to their precarious living conditions, poor households may unintentionally misuse the natural environment, which would deplete soil nutrients (Mazibuko 2013). However, in the context of present study natural capital of the farmers shown slight increased. About 89.97% of the respondent perceived that there is slight increase in the tree coverage and 8.86% reported significant increase due to practising the agroforestry. Although, in terms of soil conservation and increase in physical properties of the soils majority of the respondents (58.23%) did not perceive any changes. However, nearly 62.03% farmers perceived slight increase in the agriculture productivity and 36.71% respondents have registered significant increase (Table 5.3).

5.12 Conclusion

In the present investigation of agroforestry systems, the farmers of the present study area have mainly practised Agro-horticulture. The major agriculture crops grown by the farmers are wheat, maize, rajmah ogla and fafra (buckwheat), potato and peas. The leaves of ogla are employed in the creation of various vegetable dishes, after ripened while the seeds are utilized to produce flour. In the study area fruit trees are majorly planted in the farmland in which apple (Malus domestica) is common fruit tree. In spite of apple fruit tree plum (genus prunus) is also a major fruit tree grown by the farmers of the study area. These fruit trees are native to the temperate zone are common in the study area which lies in the temperate region of Himachal Pradesh. In the study area there are a smaller number of framers who grown medicinal plants in their farmland such as Tulsi (*Ocimum sanctum* Linn) and Aloe vera (genus aloe). With agriculture and horticulture, the livestock commercially raised by the farmers are cow, sheep, goat and poultry.

As per the findings of the study the social and financial capitals have significantly improved in terms of information about farming and helps from the Government and Non-government organisations. This has in turn also improved traditional agroforestry practises. The respondents of the study area that feel there has been a noticeable improvement in both social and financial capitals as well. Improved financial assets enable the people to strengthen the household's physical infrastructure and also use modern machinery to maintain the farmland. The agroforestry system practised by the farmers has also improve the natural capital in terms of positive change in tree

Table 5.3 Respondents observed improvement in their livelihood capitals

Livelihood Capitals	Highly Increased	Increased	Slightly Increased	No Change	LII[a]	Rank Order
Human						
1. Increase the knowledge of sustainability of farming	1.27	18.99	65.82	13.92	107.59	11
2. Increase management skills of farmland	0.00	18.99	78.48	2.53	116.46	8
3. Enhance traditional knowledge of farming	0.00	6.33	46.84	46.84	59.49	12
Social						
1. Increase relationship with social organization (in terms of share knowledge about traditional way of mixed farming)	2.53	27.85	51.90	17.72	115.19	9
2. Increase participation in the government organization and NGOs to enhance traditional agroforestry farming	2.53	36.71	44.30	16.46	125.32	6
3. Increase in the access to information about farming	3.80	74.68	20.25	1.27	181.01	1
Physical						
1. Increase Forest and crops productivity	1.27	24.05	73.42	1.27	125.32	6
2. Improved household physical infrastructure (in terms of timber harvested from trees planted in farmland)	0.00	25.32	70.89	3.80	121.52	7
3. Improvement in the use of machinery/tools to maintain farmland	1.27	67.09	29.11	2.53	167.09	2
Natural						
1. Increase tree coverage	1.27	8.86	89.87	0.00	111.39	10
2. Increase soil conservation and physical properties of soils	0.00	10.13	31.65	58.23	51.90	13
3. Increase agricultural productivity	1.27	36.71	62.03	0.00	139.24	4
Financial						
1. Improve household income	2.53	29.11	67.09	1.27	132.91	5
2. Increase savings	2.53	24.05	70.89	2.53	126.58	6
3. Increase household expenditure	0.00	39.24	60.76	0.00	139.24	4
4. Access to credit from GOs and NGOs to improve traditional agroforestry practices	1.27	62.03	34.18	2.53	162.03	3

[a]Livelihood Improvement Index (Compiled by authors)

coverage, soil conservation, physical properties of soils and agriculture productivity. The respondents have been perceived positive overall improvement in the all the livelihood capitals due to practising agroforestry. Overall, it can be concluded that a positive improvement has been witnessed in the Sungra village of Nichar Block of Kinnaur due to agroforestry practices in the past few decades.

References

Ahmad T, Gupta SK, Singh SK, Meraj G, Kumar P, Kanga S (2023) Unveiling nature's resilience: exploring vegetation dynamics during the COVID-19 Era in Jharkhand, India, with the Google Earth Engine. Climate 11(9):187

Akter R, Hasan MK, Kabir KH, Dar D, Roshni NA (2022) Agroforestry systems and their impact on livelihood improvement of tribal farmers in a tropical moist deciduous forest in Bangladesh. Trees, Forest and People 9:100315

Arunachalam A, Chavan SB, Handa AK, Anil Kumar S, Bhaskar KA, Mohapatra T (2019) Agroforestry systems for the Indian Himalayan Region. Indian Council of Agricultural Research, New Delhi, p 19

Atangana A, Khasa D, Chang S, Degrande A (2013) Tropical agroforestry. Springer, Dordrecht, p 35

Bijalwan A (2013) Forest biodiversity: conservation and management. In: Bijalwan A, Kala CP (eds) Agroforestry in Garhwal Himalaya, 1st edn. Pointer Publisher, pp 28–51

Carne RJ (2008) Agroforestry land use: the concept and practice. Aust Geogr Stud 31(1):79–90

Dagar JC, Singh AK, Arunachalam A, Dagar JC et al (eds) (2013) Agroforestry systems in India: livelihood security & ecosystem services, advances in agroforestry, vol. 10. Springer

Department for International Development (DFID) (1999)

District Census Handbook Census of India (2011)

Fahad S, Lan HNT, Manh DN, Duc HT, Nguyen To-The (2022) Analyzing the status of multidimensional poverty of rural households by using sustainable livelihood framework: policy implications for economic growth. Environ Sci Pollut Res 30:16106–16119. 16113

Farooq M, Mushtaq F, Meraj G, Singh SK, Kanga S, Gupta A, Kumar P, Singh D, Avtar R (2022) Strategic slum upgrading and redevelopment action plan for Jammu City. Resources 11(12):120

Gulati B, Sharma R, Kanga S, Singh SK, Sajan B, Meraj G, Kumar P, Ramanathan AL (2023) Unraveling the relationship between stubble burning and air quality degradation in Punjab: a temporal and spatial analysis (2019–2022). J Clim Change 9(2):43–53

Hanif MA, Roy RM, Bari MS, Ray PC, Rahman MS, Hasan MF (2018) Livelihood improvements through agroforestry: evidence from Northern Bangladesh. Small-scale For 17:505–522. https://doi.org/10.1007/s11842-018-9400-y519

Kanga S, Singh PK, Ohri A (2020) Evaluating the environmental sustainability of smart cities in India: the design and application of the Indian smart city environmental sustainability index. Sustainability 13(1):327

Kashyap SD, Dagar JC, Pant KS, Yewale AG et al. (2014) (eds) Agroforestry systems in India: livelihood security & ecosystem services, advances in agroforestry 10. Springer, India, pp 117–154

King KFS (1979) Concepts of agroforestry. In: Chandler T, Spurgeon D (eds) International cooperation in agroforestry by: proceedings of an international conference. ICRAF, Nairobi, pp 1–13

Kumar BM, Handa AK, Dhyani SK, Arunachalam A (2018) Agroforestry in the Indian Himalayan region: overview. In: Gordon AM, Newman SM, Coleman BRW (eds) Temperate agroforestry systems, 2nd edn. CAB International, Germany, pp 153–172

Kumar P, Singh SK, Gupta SK, Kanga S, Meraj G (2023) Uncovering the hydro-meteorological drivers responsible for forest fires utilizing geospatial techniques. Theor Appl Climatol. https://doi.org/10.1007/s00704-023-04497. Springer Nature

Kumari A, Sehgal RN, Kumar S (2008) Traditional agroforestry systems practiced in Lahaul (Lahaul. & Spiti) and Kinnaur district of Himachal Pradesh. Indian For 134(8):1003–1010

Mazibuko S (2013) Understanding underdevelopment through the sustainable livelihoods approach. Community Dev 44(2):173–187. https://doi.org/10.1080/15575330.2012.683798

Nair PKR (1985) Classification of agroforestry systems. Agrofor Syst 3:97–128

Pandey DK, De HK, Geetarani L, Hazarika BN (2019) Problems and prospects of agricultural diversification in shifting cultivation area of north east India: an empirical study. J Pharmacogn Phytochem 8(3):69–73

Pandeya R, Jha SK, Alataloc JM, Archie KM, Gupta AK (2017) Sustainable livelihood framework-based indicators for assessing climate change vulnerability and adaptation for. Himalayan Commun Ecol Indic 79(2017):338–346. 344

Quandt A, Neufeldt H, McCabe JT (2017) The role of agroforestry in building livelihood resilience to floods and drought in semiarid Kenya. Ecol Soc 22(3):10

Scoones I (1998) Sustainable rural livelihoods: a framework for analysis., IDS Working Paper 72. IDS, Brighton

Serrat O (2017) The sustainable livelihoods approach. In: Knowledge solutions. Springer, Singapore. https://doi.org/10.1007/978-981-10-0983-9_5

Tomar P, Singh SK, Kanga S, Pattanaik A (2023) Water bodies mapping and monitoring using high-resolution satellite images. Sustain Agric Food Environ Res 11

Young A (1988) Biogeography and development in the humid tropics. J Biogeogr 15(1):19–30

Zhang C, Fang Y (2020) Application of capital-based approach in the measurement of livelihood sustainability: a case study from the Koshi River basin community in Nepal. Ecol Indic 116(3):106474. https://doi.org/10.1016/j.ecolind.2020.1064748

Chapter 6
Climate Change Impact on Crop Production: Insights from Smallholder Farmers in Flood-prone Areas of North Bengal, India

Subham Dey Sarkar, Tuhin Dey Roy, Pranoy Dey, Sushmita Singh, Anuusua Poddar, Riya Roy, and Arindam Basak

Abstract The agriculture sector is highly susceptible to the impacts of climate change, as it frequently encounters various extreme climatic events, including droughts, floods, natural disasters, and salinity intrusion. In order to effectively address the challenges posed by climate change and minimise its negative effects on crop production, it is crucial to evaluate the perceptions of farmers. The objective of this study was to evaluate the perception of smallholder farmers regarding the impact of climate change on crop production in flood-prone regions of North Bengal, India. A combination of quantitative and qualitative data collection methods was utilised in this study. The collection of primary data was conducted by administering a structured questionnaire through a household survey conducted in June and July of 2023. The study participants consisted of 225 smallholder farmers residing in the flood-prone regions of North Bengal. The data were analysed utilising descriptive statistics, the coefficient of correlation, and stepwise regression. The primary findings indicate that the perceived effects of climate change on crop production were classified as severe, with a subsequent classification of moderate impact. Out of the 15 statements concerning the impact of climate change on crop production, it was widely recognised that the most significant consequence was the escalation of pest infestations. The elevated frequency and intensity of flooding events have been identified as the second most significant consequence of climate change. The

S. D. Sarkar (✉) · S. Singh · A. Poddar · R. Roy · A. Basak
Department of Geography and Applied Geography, University of North Bengal, Siliguri, West Bengal, India
e-mail: rs_subhamd@nbu.ac.in

T. D. Roy
Department of Geography, Siliguri Mahila Mahabidyalaya, Siliguri, West Bengal, India

P. Dey
Department of Geography and Applied Geography, University of North Bengal, Siliguri, West Bengal, India

Department of Geography, Birpara College, Birpara, West Bengal, India

S. Kanga et al. (eds.), *Transforming Agricultural Management for a Sustainable Future*, World Sustainability Series, https://doi.org/10.1007/978-3-031-63430-7_6

socio-economic characteristics of the respondents that had the most significant impact on smallholder farmers' perception were education, farming experience, and training experience. The study has identified significant issues that policymakers and development practitioners should address. The methodologies utilised and the outcomes obtained in this study have the potential to be applied in other flood-prone regions of India that share comparable socio-economic and regional characteristics.

Keywords Climate change · Crop production · Smallholder farmer · Flood · North Bengal

6.1 Introduction

Climate change poses a significant threat to agricultural productivity, making it the most perilous natural hazard with unfavourable consequences (Enete and Amusa 2010). Climate change refers to the observed alterations in the average values of climate variables, including but not limited to precipitation and temperature, over an extended duration, generally spanning 50 years or more (Roy et al. 2018; IPCC 2007, 2012). According to the findings of the Intergovernmental Panel on Climate Change (IPCC 2007), climate change can be attributed to various factors, including natural variations in the environment as well as human-induced activities (Praveen and Sharma 2019). In recent times, human activities such as deforestation for agricultural and industrial purposes, excessive utilization of unprocessed petrochemicals, and heightened reliance on air conditioning systems have led to the accelerated emission of greenhouse gases (GHGs) (Morton 2007; Praveen and Sharma 2019). The phenomenon in question has exerted a substantial influence on the altering composition of the Earth's atmosphere, resulting in the occurrence of global warming and subsequent climate change (Lobell et al. 2008). The agricultural sectors and climate change are intricately connected on a worldwide scale, and their relationship has significant relevance due to the growing disparity between the world's population and food production (Praveen and Sharma 2019). Agriculture is an economic endeavour that heavily relies on weather and climate conditions to cultivate essential resources such as food, fibre, and other necessities vital for human sustenance (Praveen and Sharma 2019). The impacts of climate change on agriculture exhibit spatial disparities on a global scale, as indicated by the Intergovernmental Panel on Climate Change (IPCC 2014). It is commonly recognized that the influence of climate change plays a crucial role in exacerbating the vulnerability of agricultural production (Hossain et al. 2013; Ruane et al. 2013; Yu et al. 2010; IFPRI 2013). The potential consequences of this phenomenon may have an impact on agricultural productivity, specifically with regard to the overall amount and quality of crops. Furthermore, individuals from many regions across the globe are currently experiencing food insecurity as a consequence of climate change (Pretty et al. 2003). Additionally, projections indicate that there would be a gradual decrease in global agricultural production, estimated to reduce by 2–6% every decade (Smith et al. 2014). Agricultural production and food prices are influenced by climatic conditions, encompassing factors such as temperature and precipitation (Hertel et al. 2010; Ahmed et al. 2009; Ahmed et al. 2011; UNDP 2009; Porter et al. 2014; Kobayashi and Furuya 2011; Salam et al. 2016; Alamgir et al. 2018; Kumar and Sharma 2022; Alamgir et al. 2021).

Climatic variability is a significant determinant, accounting for over 60% of production variability, and consequently, it plays a critical role in shaping the income of farmers (Osborne and Wheeler 2013; Ray et al. 2015; Matiu et al. 2017). Environmental stress has been making it harder and harder for farmers to make a living from farming and to grow crops (Kabir et al. 2017). Moreover, Climate change has a significant impact on the initiation and duration of growing seasons (Fiwa et al. 2014; Zhao et al. 2015; Lemma et al. 2016), as well as the intensity and duration of heat and water stress experienced in agricultural production systems (Lobell et al. 2015; Saadi et al. 2015; Schauberger et al. 2017). The repercussions, however, exhibit unequal distribution characterized by significant geographical variations. Certain researchers contend that agricultural practices in countries situated in temperate and polar locations are anticipated to experience advantageous outcomes due to rising average temperatures and prolonged periods of cultivation. On the contrary, a majority of nations placed in tropical and subtropical regions are projected to encounter a reduction in agricultural productivity (Dinar et al. 2008; Lybbert and Sumner 2010). A new study shows how climate change affects the production and yield of four of the most important crops in the world i.e., maize, rice, wheat, and soybeans. It is noteworthy that according to estimates, the cultivation of present maize varieties under the influence of climate change is expected to result in a decrease in rain-fed maize production by an average range of 3.3–6.4% in the year 2030 and 5.2–12.2% in the year 2050. Similarly, the yield of irrigated maize is anticipated to decline by 3–8% in 2030 and 5–14% in 2050 (Tesfaye et al. 2017). In addition to its influence on crop yields and output, climate change also exerts a significant influence on the essential natural resources of land and water that underpin agricultural production (Aryal et al. 2020; Cramer et al. 2018). The anticipated consequences of climate change include a decrease in water supply, juxtaposed with a projected 19% surge in agricultural water usage by the year 2050 (UN-Water 2013). The escalating dependence of Indian farmers on groundwater as a response to drought driven by climate change has resulted in a significant depletion of the groundwater level. This situation is anticipated to exacerbate in the future due to heightened climatic variability (Fishman 2018). According to projections, a significant portion of the Indo-Gangetic Plains (IGP), which serves as a crucial food source, is anticipated to experience unsuitability for wheat cultivation due to the adverse effects of heat stress by the year 2050 (Ortiz et al. 2008). The effects of climate change on agricultural output and natural resources exacerbate food insecurity and poverty in Asian countries, negatively affecting the livelihoods of millions of people in the region (Schmidhuber and Tubiello 2007; Bandara and Cai 2014; Shankar et al. 2015; Wang et al. 2017; Aryal et al. 2019; Edame et al. 2011). According to estimations, there is a probable spike in food price fluctuations for major food crops (such as rice, wheat, maize, and soybean) by a factor of 2.5 between the years 2000 and 2050 (Nelson et al. 2009). Additionally, livestock products (including cattle, hog, lamb, and chicken) are expected to experience a price increase of 1.5 times due to the influence of climate change (Nelson et al. 2009).

Agriculture has a prominent part in the economic development of countries such as India, where approximately half of the population relies on agricultural activities for their sustenance and livelihood (Ministry of Agriculture and Farmers' Welfare 2020). Despite the significant reliance on the agricultural sector, it is evident that agriculture in India continues to be very susceptible to many threats and difficulties associated with climate variability and change (Edame et al. 2011). In addition to its significance in ensuring food security and generating employment opportunities, the agricultural sector also plays a pivotal role in stimulating the development of interconnected industries, mitigating rural poverty, and facilitating the expansion of agricultural exports and foreign exchange (Tripathi and Prasad 2009; Vyas 2003). Despite the positive impacts observed, the agricultural industry has experienced an average growth rate of 3.64% from 1980 to 2019 (World Bank 2020). During the period preceding independence, the growth rate was recorded at 1%, however, in the post-independence era, it had an annual growth rate of 2% (Tripathi and Prasad 2009). In the Indian context, it is noteworthy that a significant proportion of agricultural land, exceeding 50 %, as well as a substantial portion, ranging from 70 to 80%, of the overall land area, is reliant on groundwater for irrigation purposes (Falkenmark and Molden 2008). Research conducted on crop yield in India has revealed that the phenomenon of global warming has resulted in a decrease in wheat production by 5.2% between the years 1981 and 2009, despite the implementation of adaptive strategies (Gupta et al. 2017). Consequently, climate change impacts both wealthy and developing nations. India is widely recognised as a developing country that exhibits a high degree of vulnerability to the impacts of climate change, mostly due to its significant dependence on agriculture as a major source of employment (Bhardwaj et al. 2022).

Several researchers, including Field (2012), Greenough et al. (2001), and Mirza (2003), have conducted estimations indicating that discrepancies in climatic conditions are expected to result in a rise in the occurrence of certain unpredictable climatic events and disasters. Floods, landslides, earthquakes, droughts, and cyclones represent prominent and severe natural hazards that exhibit an increased frequency as a consequence of climate change. Climate-induced disasters, particularly floods, have resulted in significant devastation in agriculture on a global scale (Teo et al. 2018; Eckstein et al. 2019; Ahmad et al. 2019). In the present global context, floods exhibit a higher frequency and greater magnitude compared to other hazards (Niles and Mueller 2016; Ali et al. 2017; Teo et al. 2018; Khan et al. 2021), primarily because to their significant correlation with societal vulnerabilities, economic damages, and loss of human lives (Roco et al. 2017; Miah et al. 2018; Mahmood et al. 2019; Hossain et al. 2020; Ahmad and Afzal 2021). Over the past few decades, there has been a notable rise in the frequency and severity of floods, notably in nations located in South Asia and Southeast Asia (Mendelsohn and Massetti 2017; Quaye et al. 2018; Khan et al. 2019). Specifically, India, Bangladesh, China, and Pakistan have been identified as regions that experience a significant number of devastating flood events (Ahmad et al. 2019; Rizwan et al. 2020). In developing nations, the exacerbation of flood hazards can be attributed to anthropogenic influences, specifically the encroachment of human activities on river systems, in conjunction with environmental and climatic changes (Iqbal et al. 2016; Teo et al. 2018). Floods are the significant factor contributing to economic and social hazards faced by individuals, among the several environmental threats to which people are exposed. These phenomena also lead to an increase in mortality rates. In the context of developing nations, rural populations frequently face heightened vulnerability to flood events due to their limited adaptive capacity and resource endowment (Fahad and Jing 2018; Huong et al. 2019).

The majority of farming groups in India are comprised of smallholder farmers. These individuals are commonly associated with possessing small parcels of land in low-lying areas, typically ranging from 0.21 to 1.0 hectares (FAO 2015). They are characterized by limited economic resources, a high number of dependents within their households and are likely to be the most susceptible to the impacts of climate change. The agricultural populations residing in North Bengal are subjected to recurrent flood events, which can be attributed to the phenomenon of climate change. The decision-making process of farmers under uncertain conditions is significantly influenced by their perception of climate information, environmental choice, and attitude (Ullah et al. 2016; Ali et al. 2017). Small subsistence farmers and rural populations in developing countries are especially impacted by climate variability. This is primarily due to limited access to alternative methods of productivity and a lack of adaptive capacity (IPCC 2007; Skoufias et al. 2011). Perception is the cognitive process involved in the collection and interpretation of information derived from one's surrounding environment. According to Howlader et al. (2015), perception serves as the initial phase in the process of adapting to climate change. In order to enhance coping mechanisms and implement effective measures, it is imperative to ascertain the opinion of smallholder farmers towards climate change in flood-prone regions of North Bengal. Hence, it is essential to deepen our comprehension of the effects of climate change on agriculture and the corresponding adaptation strategies. This knowledge is crucial for promoting the sustainability of agricultural practices and formulating policies that mitigate the vulnerability of smallholder farmers to climate change.

6.2 Materials and Methods

6.2.1 Study Area

The study was carried out in the flood-prone regions of the Alipurduar district located in North Bengal, India. The district of Alipurduar is recognised for its susceptibility to climate change-induced occurrences, including floods. In the Sub-Himalayan foothills, the Alipurduar district is home to many rivers, and it falls under a high rainfall zone, making it one of the prominent districts of high

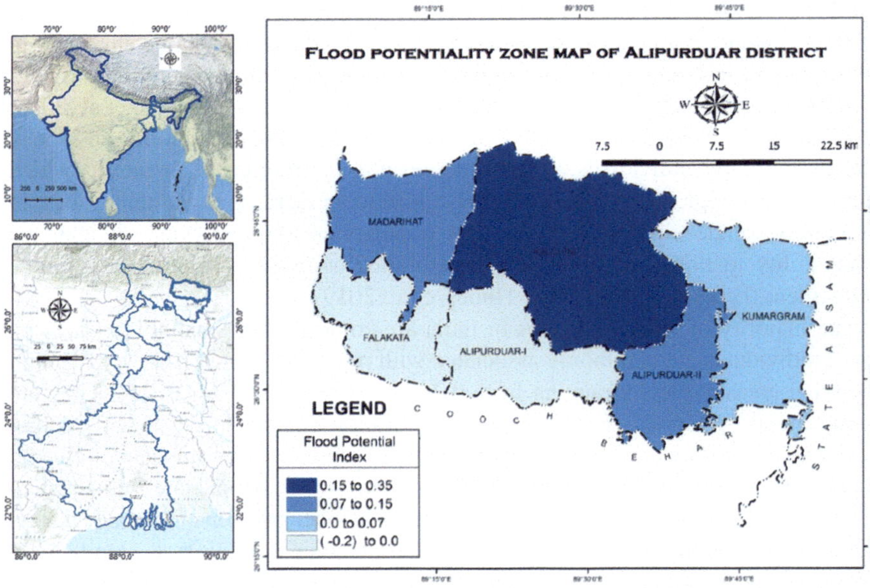

Fig. 6.1 Spatial flood potential mapping of the Alipurduar District, North Bengal

rainfall in West Bengal. According to the *Disaster Management Department, Government of West Bengal* and *Natural Calamity Report, Block Development Office, Government of West Bengal*, Alipurduar district has the highest potentiality of flood. The region is distinguished by its abundant agricultural terrain, whereby small-scale farmers occupy a key position in the cultivation of crops. The study was centred on flood-prone regions since they are prone to catastrophic climatic occurrences that have a substantial influence on agricultural activity and the livelihoods of the local population. The selection of this region was based on its representation of a common climate change scenario encountered by smallholder farmers in India. Therefore, the findings of this study are applicable not only to Alipurduar district but also to other flood-prone regions in India that share similar socio-economic and regional attributes. In the current study, Fig. 6.1 indicates Kalchini followed by Madarihat and Alipurduar II blocks of the Alipurduar district, has a high probability of flooding on the other hand, Alipurduar I and Falakata block has the lowest probability of flooding (Roy et al. 2023).

6.2.2 *Study Design*

The present study is based on the data collected during the cross-sectional survey of smallholder farmer's households located along the flood-prone areas of Alipurduar district, North Bengal. The study makes use of quantitative as well as qualitative methods of research. A household survey was conducted with the help of a comprehensively prepared questionnaire. The questionnaire comprised two sections, the first section focused on the socio-economic attributes of the farmers, while the second section aimed at understanding the perception of smallholder farmers about the impact of climate change on their local crop production and productivity.

6.2.3 Sampling, Data Collection and Analysis

For the present study, a sample size of 225 smallholder farmers has been considered. Smallholder farmers were then randomly selected from the flood-prone areas of Alipurduar district. To ensure the reliability of the survey instrument, a pre-field survey was conducted where 12 smallholder farmers' families were surveyed. Finally, between August and September 2023, data collection (quantitative and qualitative) was done and focused group discussions were also conducted. The present study will incorporate 8 socio-economic attributes, namely, age, education, family size, farm size, farming experience, annual income, access to information sources and training experience of the farmers. These attributes constitute the explanatory variables which influence and shape the perceived notion of farmers regarding the impact of climate change on the local crop production of the study area.

Conversely, to study the farmers' perception of climate change and its impact on crop production, 15 statements covering an array of possible and/or potential impact of climate change on crop production have been framed. Likert Scale was employed here and farmers were made to respond on the basis of the 5-point Likert scale wherein 1 stands for strongly disagree, 2 for disagree, 3 for neutral, 4 for agree and 5 for strongly agree. This way, farmers expressed their perception regarding climate change and its impact on crop production. Since the response of the farmers was restricted to 1 to 5 the scale score ranged from 15 to 75, i.e., if a farmer strongly disagrees with all the statements, then the total score would add up to 15 (15*1) as strongly disagree is given the weightage of only 1 which means there is zero impact of climate change on the local crop production. On the other hand, if another farmer strongly agrees with all the statements, then the total score of all the 15 statements would be 75 (15*5) since a weightage of 5 is given to strongly agree. It would therefore mean that climate change has a maximum impact on the local crop production. Therefore, in the present study, the impact score ranges between 15 to 75, wherein 15 corresponds to no impact while 75 corresponds highest or maximum impact.

Perception index (PI) was calculated to elucidate the most significant impact of climate change on crop production.

$$\text{Perception Index (PI)} = (\text{SA} \times 5 + \text{A} \times 4 + \text{U} \times 3 + \text{D} \times 2 + \text{SD} \times 1)/\text{N} \quad (6.1)$$

With SA denoting the frequency of highly agreeing responses, A denoting agreeing responses, U denoting undetermined responses, D denoting disagreeing responses, SD denoting severely disagreeing responses, and N denoting the overall frequency of responses. The survey data was examined and analyzed using the statistical software for social science, or SPSS, version 16. Pearson The relationship between the explanatory factors and the focus is further evaluated using the Product Moment Correlation Coefficient (r). Additionally, the factors impacting smallholder farmers' perceptions of how climate change will affect agricultural output were determined through the use of stepwise and multiple linear regressions.

6.3 Results

6.3.1 Socio-Economic Characteristics of the Smallholder Farmers

In discussing the perceptions of smallholder farmers regarding climate change's impact on crop production in flood-prone areas of North Bengal, India, it was important to discuss the socio-economic characteristics of smallholder farmers of the study area. Hence, the below table, i.e., Table 6.1 represents the socio-economic characteristics of the smallholder farmers of the study area. From the data analysis, it has been noticed that the majority of the respondents of the study area, i.e., 48.44% were middle-aged. The majority of the respondents, i.e., 61% have studied up to secondary followed by 24.89 studied up to primary, 18.67 with no formal education, 16.44 studied up to higher secondary, and 12.89 studied higher education. In terms of family size small family size was more common than the other two taken into consideration, 58.67 percent out of a total of 100 percent of respondents belonged to small families. The majority of the respondents, i.e., 56% had Less than 0.5 ha of land holdings with an average value of 4.36 ha of land holdings.

The above data has also revealed that most of the respondents, i.e., 48% had moderate farming experience, followed by 36.89% having low experience, and lastly, only 15.11% having high experience. As the data have showed, major share of the respondents, i.e., 52.89%, have earned low income annually, i.e., below Rs. 50,000 and the average annual income of the respondents was Rs. 61577.78. The majority, about 42.67% of the respondents have moderate access to information sources followed by 38.22% having low access and lastly, 19.11% having high access. A lion share of the respondents, i.e., 71.11%, have not been trained in any experience followed by 15.11% having low experience, 11.11% having moderate experience and only 2.67% having high experience.

Table 6.1 Socio-economic characteristics of the smallholder farmers

Characteristics	Category	N	%	Mean	SD
Age (Years)	Young (<35)	48	21.33	**43.68**	**9.62**
	Middle-aged (35–50)	109	48.44		
	Old (>50)	68	30.22		
Education	No formal	42	18.67	**6.29**	**5.14**
	Primary (1–4)	56	24.89		
	Secondary (5–10)	61	27.11		
	Higher secondary (11–12)	37	16.44		
	Higher education (>12)	29	12.89		
Family size	Small (1–4)	132	58.67	**4.36**	**2.41**
	Medium (5–7)	79	35.11		
	Large (>7)	14	6.22		
Farm size	Less than 0.5 ha	126	56.00	**0.60**	**0.36**
	0.5 to 1 ha	44	19.56		
	More than 1 ha	55	24.44		
Farming experience (Year)	Low (<15)	83	36.89	**20.54**	**11.58**
	Moderate (15–30)	108	48.00		
	High (>30)	34	15.11		
Annual Income ('000 Rs.)	Low (<50)	119	52.89	**61577.78**	**27208.85**
	Medium (50–80)	57	25.33		
	High (>80)	49	21.78		
Access to information sources	Low (1)	86	38.22	**2.24**	**1.41**
	Moderate (2–3)	96	42.67		
	High (>3)	43	19.11		
Training experience (Days)	No training	160	71.11	**13.60**	**33.54**
	Low (<30)	34	15.11		
	Moderate (30–60)	25	11.11		
	High (>60)	6	2.67		

6.3.2 Perception of the Smallholder Farmers Regarding the Impact of Climate Change

A Likert scale has been employed to find out the perception of smallholder farmers regarding the climate change impact on crop production. In total 15 statements have been obtained from Focused Group Discussions have been incorporated into the scale. In this below-mentioned Likert scale, the values of strongly agree, agree, neutral, disagree, and strongly disagree have been numbered as 5,4,3,2,1 respectively. As of minimum, a farmer could get 15 (1*15 = 15) by strongly disagreeing with all the below-mentioned statements and as of maximum, a farmer could get 75 (5*15 = 75) by strongly agreeing with all the below-mentioned statements. Thus, a score, ranging from 15 to 75 has been obtained on the scale. The mean score of the perception of respondents regarding individual statement has also been calculated and on the basis of that every individual statement have been ranked according to the perceived importance given by the respondents (Table 6.2).

Being ranked one, escalation in pest infestations on the crops due to changing climate conditions has been found as the most important climate change impact with a mean score of 4.56 as perceived by the respondents. Changing climate has resulted in more frequent and intense flooding events, which harm crops has been found as the second most important climate change impact with a mean score of 4.34 as perceived by the respondents. The farmers have identified climate change has led to more severe and unpredictable weather patterns affecting crop production, with a mean score of 4.28 as the third important impact. However, the Occurrence of droughts has been increased, which negatively impacts crop yields, with a mean score of 2.29 has been perceived as the least important climate change impact on crop production by the respondents in the study area.

Based on the perception score, smallholder farmers have been distributed into four impact categories, i.e. low impact category, ranging below 40 perception score, moderate impact category, ranging from 40 to 50 perception score, high impact category, ranging from 50 to 60 perception score and severe impact category, ranging above 60 perception score, because a farmer, who agreed with the all the statements regarding the impact of climate change, would score 4 in individual statements and 4*15 = 60 in all 15 statements considered, showed in Table 6.3. The above table shows that the majority of the respondents, i.e., 76.44%, have perceived a high climate change impact on crop production followed by 13.78% perceived a moderate impact, 9.33% perceived a severe impact and remaining of the respondents i.e., 0.44%, who has perceived low climate change impact on crop production.

6.3.3 *Correlation Between Explanatory Variables (Socio-economic Characteristics of Smallholder Farmers) and Dependent Variable (Smallholder Farmers' Perception of Climate Change Impact on Crop Production)*

Pearson product-moment coefficient of correlation (r) has been employed to determine the association between explanatory variables, i.e., socio-economic characteristics of smallholder farmers and dependent variable, i.e., smallholder farmers' perception of climate change impact on crop production. A null hypothesis (H_0), there is no association between explanatory variables and dependent variable, has been formulated, where a 5% level of probability and 1% level of probability have been used in order to reject the null hypothesis (H_0). The below Table 6.4 has shown correlation coefficient values (r), and critical correlation values (r_c).

Five of the eight explanatory factors that were taken into consideration showed substantial and positive relationships with smallholder farmers' perceptions of the

Table 6.2 Perception of the smallholder farmers regarding the impact of climate change

Statements	Strongly agree	Agree	Neutral	Disagree	Strongly disagree	Mean score	SD	Rank order
Changes in the quality of soil due to climate-related factors affecting crop health	44	24	111	25	21	**3.20**	**1.16**	12
Changing climate conditions have made it challenging to predict and manage disease outbreaks in crops	45	25	109	27	19	**3.22**	**1.15**	11
Changing climate has resulted in more frequent and intense flooding events, which harm crops	137	42	33	11	2	**4.34**	**0.96**	2
Climate change has caused shifts in the suitability of certain local crops for cultivation	92	53	56	20	4	**3.93**	**1.08**	7
Climate change has led to alterations in the timing of seasons, affecting planting and harvesting schedules	101	40	47	31	6	**3.88**	**1.20**	8
Climate change has led to more severe and unpredictable weather patterns affecting crop production	131	38	49	2	5	**4.28**	**0.98**	3
Climate-induced floods often coincide with crucial planting and harvesting periods leading to reduced crop quality and quantity	108	55	41	12	9	**4.07**	**1.11**	5

(continued)

Table 6.2 (continued)

Statements	Strongly agree	Agree	Neutral	Disagree	Strongly disagree	Mean score	SD	Rank order
Erratic rainfall patterns caused by climate change have resulted in uneven distribution of water for the crops	83	47	40	41	14	**3.64**	**1.31**	10
Escalation in pest infestations on the crops due to changing climate conditions	151	57	10	5	2	**4.56**	**0.76**	1
Higher temperatures have caused stress on the crops, leading to reduced growth and yield	36	8	56	100	25	**2.69**	**1.21**	14
Occurrence of droughts has been increased, which negatively impacts crop yields	0	8	97	72	48	**2.29**	**0.84**	15
Rising temperatures have led to reduced water availability, impacting irrigation and overall crop productivity	37	9	55	99	25	**2.71**	**1.23**	13
Rising water levels during floods have caused crop lodging and physical damage to agricultural fields	114	50	25	24	12	**4.02**	**1.24**	6
The intensity of river bank erosions has been increased	98	43	36	33	15	**3.78**	**1.32**	9

(continued)

Table 6.2 (continued)

Statements	Strongly agree	Agree	Neutral	Disagree	Strongly disagree	Mean score	SD	Rank order
Unseasonal rainfall, exacerbated by climate change, disrupts the growth stages of crops, resulting in lower overall yields	135	39	28	15	8	**4.24**	**1.12**	4

Table 6.3 Distribution of smallholder farmers based on their perception of the impact of climate change

Impact category	Range	N	%	Mean	SD
Low	<40	1	0.44	**54.84**	**4.48**
Moderate	40–50	31	13.78		
High	51–60	172	76.44		
Severe	>60	21	9.33		

Table 6.4 Correlation between explanatory variables (socio-economic characteristics of smallholder farmers) and dependent variable (smallholder farmers' perception of climate change impact on crop production)

Explanatory variables	Correlation coefficient (r)	Critical correlation value (r_c)	
		0.05	0.01
Age	**0.134***	**0.131**	**0.171**
Education	**0.208****		
Family size	0.004		
Farm size	0.043		
Farming experience	**0.186****		
Annual household income	0.016		
Access to information source	**0.174****		
Training experience	**0.185****		

No. of Degrees of freedom (df) = 223; **Correlation significant at the 0.01 level (2-tailed). *Correlation significant at the 0.05 level (2-tailed)

influence of climate change on crop output, according to the data analysis. Farmers' attitudes were found to be significantly influenced by age, education, agricultural experience, access to information sources, and training experience. At a 95% significance level, the correlation coefficient (r value) of 0.134 for age was more than the critical value (rc) of 0.131. As a result, the null hypothesis was rejected, showing a relationship between smallholder farmers' perceptions of the influence of climate change on crop production and their age. In a similar vein, the null hypothesis was rejected for education when the r value (0.208) exceeded the rc value (0.171) at a 99% significance level, so bolstering the alternative hypothesis that education influences farmers' views of the impact of climate change. Significant correlations were also shown by access to information sources, training experience, and farming experience, with corresponding r values of 0.186, 0.174, and 0.185 exceeding the common rc value (0.171) at a 99% significance level. The null hypothesis was thus rejected, confirming that smallholder farmers' perceptions of the influence of climate change on crop output are correlated with their farming experience, access to information sources, and training experience. On the other hand, at a 95% significance level, the r values of family size, farm size, and annual household income (0.004, 0.043, and 0.016) showed minimal correlations with each other. As a result, the null hypothesis was accepted, showing that smallholder farmers' perceptions of the influence of climate change on crop output were not significantly correlated with family size, farm size, or yearly household income.

6.3.4 Stepwise Multiple Regression Analysis

To understand the individual contribution of the explanatory variables, i.e., education, farming experience, and training experience to the variation of smallholder farmers' perception of climate change impact on crop production, multiple regression analysis has been employed. The below Table 6.5 has represents the summary of the multiple regression analysis.

Table 6.5 represented that only three variables, i.e., education, farming experience and training experience have comprised the model. Education (R2 = 0.072) which, was the first variable entered into the model, has contributed the highest of 7.2% in explaining the dependent variable. Farming experience (R2 = 0.042) which, was the second variable entered into the model, has made a contribution of 4.2% in explaining the variation of the dependent variable. The third variable entered into the model was training experience, that has contributed 3.4% in explaining variation in the dependent variable.

Table 6.5 Summary of the stepwise multiple regression analysis

Model	R	R^2	Change in R^2	Variation explained (%)	Significance level
Constant + X2	0.208	0.072	0.072	**7.2**	0.001
Constant + X2 + X5	0.276	0.114	0.042	**4.2**	0.001
Constant + X2 + X5 + X8	0.318	0.148	0.034	**3.4**	0.001

Variables entered: Education (X2), Farming experience (X5), Training experience (X8)

6.4 Discussion

The agriculture industry is vulnerable to the effects of climate change since it's exposed to a lot of different types of extreme weather, from droughts and floods to natural disasters and saltwater intrusion. To tackle climate change and make sure it doesn't affect crop production too much, it's important to look at how farmers feel about it. This study was done to see how smallholder farmers feel about climate change and how it affects crop production in areas of North Bengal that are prone to flooding. It was based on data from a cross-section survey of smallholders' households in the flood-stricken areas of North Bengal. In order to better understand the biggest impacts climate change has on crop production, a perception index (PI) has been calculated. Based on the data collected from smallholder farmers in flood-prone regions of North Bengal, India, the study aimed to evaluate their perception of climate change's impact on crop production. The majority of respondents were middle-aged with secondary education and small family sizes. The current study found that an increase in pest infestations followed by flooding due to changing climate conditions were perceived as the most significant impact on crop production. They also recognize the negative effects of unpredictable weather patterns. However, droughts were perceived as the least important impact.

It is predicted that as a result of changes in an area's climate or weather pattern, a crop may become more susceptible to a variety of pests, diseases, and weeds. Climate change is expected to aggravate pest infestations of different crops since warmer, more humid weather is more conducive to pest growth. However, it will differ according to the pests' ability to adapt to climate change and from place to region (Malhi et al. 2021). The majority of the respondents have perceived a high climate change impact on crop production. Based on the analysis, five variables (age, education, farming experience, access to information sources, and training experience) showed significant and positive associations with smallholder farmers' perception of climate change's impact on crop production. For age and education, the null hypothesis was rejected, indicating an association between these variables and farmers' perceptions. Similarly, farming experience, access to information sources, and training experience also showed significant associations. However, family size, farm size, and annual household income did not show any significant associations with farmers' perceptions.

The perception of farmers is positively and significantly related to education. Additionally, there is evidence of a strong and favourable relationship between education and climate change adaptation. A person can become educated and wise by having access to a variety of information sources. They may encounter and seek advice on various farm-related concerns that assist them in making the necessary adjustments. By taking part in various training programs, individuals can learn more about crop production and how climate change is affecting it. (Akanda and Howlader 2015; Habiba et al. 2014; Islam and Hossen 2016; Mardy et al. 2018; Roy et al. 2018). The multiple regression analysis, employed to understand the individual contribution of the explanatory variables, i.e., education, farming experience, and training experience, to the variation of smallholder farmers' perception of climate change impact on crop production, revealed that education, farming experience, and training experience were the variables considered in the model. Education had the highest contribution in explaining the variation in farmers' perception of climate change's impact on crop production, followed by farming experience and training experience.

Farmers' perceptions of climate change are greatly influenced by their education, years of experience farming, availability of extension services, and farming area (Atchkipa et al. 2017). Education enables farmers to foresee the effects of climate change and comprehend the possible advantages of using various coping mechanisms (Amare and Simane 2017).

6.5 Conclusion

Climate change poses significant challenges to agricultural development and livelihood conditions in developing countries like India. Climate change which refers to long-term changes in climate measures such as rainfall and temperature over a period of 50 years or more can have negative impacts on crop production. Given the size and sensitivity of the agricultural sector, developing countries are particularly affected. In the study area of North Bengal, smallholder farmers have expressed a strong perception of the impacts of climate change on crop production. They have identified pest infestations and flooding as the most significant consequences. Factors such as education, farming experience, access to information sources, and training also influence their perception. This research can help policymakers and practitioners in developing effective strategies to address these challenges and support the resilience of farmers. The smallholder farmers in the study area have indeed recognized the severe impacts of climate change on crop production. It is crucial to implement programs and provide suitable solutions to address their challenges. Immediate actions are needed to tackle issues like pest infestation and water scarcity during peak growing seasons. Research on climate change adaptation by these farmers can yield valuable findings and recommendations. A collaborative approach involving both government and private extension providers is key to addressing these issues effectively.

References

Ahmad D, Afzal M (2021) Impact of climate change on pastoralists' resilience and sustainable mitigation in Punjab, Pakistan. Environ Dev Sustain 23:11406–11426

Ahmad D, Afzal M, Rauf A (2019) Analysis of wheat farmers' risk perceptions and attitudes: evidence from Punjab, Pakistan. Nat Hazards 95(3):845–861

Ahmed SA, Diffenbaugh NS, Hertel TW (2009) Climate volatility deepens poverty vulnerability in developing countries. Environ Res Lett 4(3):034004

Ahmed SA, Diffenbaugh NS, Hertel TW, Lobell DB, Ramankutty N, Rios AR, Rowhani P (2011) Climate volatility and poverty vulnerability in Tanzania. Glob Environ Chang 21(1):46–55

Akanda MGR, Howlader MS (2015) Coastal farmers' perception of climate change effects on agriculture at Galachipa Upazila under Patuakhali District of Bangladesh. Global J Sci Front Res: D Agric Vet 15:31–39

Alamgir MS, Furuya J, Kobayashi S, Binte MR, Salam MA (2018) Farmers' net income distribution and regional vulnerability to climate change: an empirical study of Bangladesh. Climate 6(3):65

Alamgir MS, Furuya J, Kobayashi S, Mostafiz RB, Ahmed MR (2021) Farm income, inequality, and poverty among farm families of a flood-prone area in Bangladesh: climate change vulnerability assessment. GeoJournal 86:2861–2885

Ali S, Liu Y, Ishaq M, Shah T, Abdullah, Ilyas A, Din IU (2017) Climate change and its impact on the yield of major food crops: evidence from Pakistan. Food Secur 6(6):39

Amare A, Simane B (2017) Determinants of smallholder farmers' decision to adopt adaptation options to climate change and variability in the Muger Sub basin of the Upper Blue Nile basin of Ethiopia. Agric Food Secur 6:64

Aryal JP, Sapkota TB, Rahut DB, Jat ML (2019) Agricultural sustainability under emerging climatic variability: role of climate smart agriculture and relevant policies in India. Int J Innovation Sustain Dev

Aryal JP, Sapkota TB, Khurana R, Khatri-Chhetri A, Rahut DB, Jat ML (2020) Climate change and agriculture in South Asia: adaptation options in smallholder production systems. Environ Dev Sustain 22(6):5045–5075

Atchkipa MT, Adzawla W, Yegbemey RN, Diallo M, Aman N (2017) Determinants of farmer's perception and adaptation to climate change: the case of maize and Sorghum farmers in Benin and Nigeria. Int J Sci Res Rev 6:227–243

Bandara JS, Cai Y (2014) The impact of climate change on food crop productivity, food prices and food security in South Asia. Econ Anal Policy 44(4):451–465

Bhardwaj M, Kumar P, Kumar S, Dagar V, Kumar A (2022) A district-level analysis for measuring the effects of climate change on production of agricultural crops, ie, wheat and paddy: evidence from India. Environ Sci Pollut Res 29(21):31861–31885

Cramer W, Guiot J, Fader M, Garrabou J, Gattuso JP, Iglesias A et al (2018) Climate change and interconnected risks to sustainable development in the Mediterranean. Nat Clim Chang 8(11): 972–980

Dinar A, Hassan R, Mendelsohn R, Benhin J (eds) (2008) Climate change and agriculture in Africa: impact assessment and adaptation strategies. Earthscan, London

Eckstein D, Künzel V, Schäfer L, Winges M (2019) Global climate risk index 2020. Germanwatch, Bonn, pp 1–50

Edame GE, Ekpenyong A, Fonta WM, Duru E (2011) Climate change, food security and agricultural productivity in Africa: issues and policy directions. Int J Humanit Soc Sci 1(21):205–223

Enete AA, Amusa TA (2010) Challenges of agricultural adaptation to climate change in Nigeria: a synthesis from the literature. Field actions science reports J Field Actions, 4

Fahad S, Jing W (2018) Evaluation of Pakistani farmers' willingness to pay for crop insurance using contingent valuation method: the case of Khyber Pakhtunkhwa province. Land Use Policy 72: 570–577

Falkenmark M, Molden D (2008) Wake up to realities of river basin closure. Int J Water Resour Dev 24(2):201–215

FAO (2015) The economic lives of smallholder farmers: an analysis based on household data from nine countries. Food and Agriculture Organization of the United Nations, Rome

Field CB (2012) Managing the risks of extreme events and disasters to advance climate change adaptation: special report of the intergovernmental panel on climate change. Cambridge University Press

Fishman R (2018) Groundwater depletion limits the scope for adaptation to increased rainfall variability in India. Clim Chang 147:195–209

Fiwa L, Vanuytrecht E, Wiyo KA, Raes D (2014) Effect of rainfall variability on the length of the crop growing period over the past three decades in central Malawi. Clim Res 62(1):45–58
Greenough G, McGeehin M, Bernard SM, Trtanj J, Riad J, Engelberg D (2001) The potential impacts of climate variability and change on health impacts of extreme weather events in the United States. Environ Health Perspect 109(suppl 2):191–198
Gupta R, Somanathan E, Dey S (2017) Global warming and local air pollution have reduced wheat yields in India. Clim Chang 140(3–4):593–604
Habiba U, Shaw R, Takeuchi Y (2014) Farmers' adaptive practices for drought risk reduction in the Northwest Region of Bangladesh. Nat Hazards 72:337–359
Hertel TW, Burke MB, Lobell DB (2010) The poverty implications of climate-induced crop yield changes by 2030. Glob Environ Chang 20(4):577–585
Hossain M, Jaim WMH, Alam MS, Rahman ANM (2013) Rice biodiversity in Bangladesh: adoption, diffusion and disappearance of varieties: a statistical report from farm survey in 2005. Brac University
Hossain MS, Arshad M, Qian L, Kächele H, Khan I, Islam MDI, Mahboob MG (2020) Climate change impacts on farmland value in Bangladesh. Ecol Indic 112:106181
Howlader MS, Akanda MGR, Zaman AKMM (2015) Adaptation towards climate change effects on coastal agriculture by the farmers of Patuakhali district of Bangladesh. Int J Agric Innov Res 3: 1560–1566
Huong NTL, Yao S, Fahad S (2019) Assessing household livelihood vulnerability to climate change: the case of Northwest Vietnam. Hum Ecol Risk Assess Int J 25(5):1157–1175
IFPRI (2013) Agriculture and adaptation in Bangladesh, current and projected impacts of climate change., Discussion paper 01281. International Food Policy Research Institute, Washington, DC
IPCC (2012) Managing the risk of extreme events and disasters to advance climate change adaptation. A special report of working groups I and II of the intergovernmental panel on climate change. Cambridge University Press, Cambridge, p 594
IPCC (2014) Climate change 2014: impacts, adaptation, and vulnerability. In: Field CB, Barros VR, Dokken DJ, Mach KJ, Mastrandrea MD, Bilir TE et al (eds) Contribution of working group II to the fifth assessment report of the intergovernmental panel on climate change. Cambridge University Press, Cambridge
IPCC, Climate Change (2007) The physical science basis. Contribution of working group I to the fourth assessment report of the Intergovernmental Panel on Climate Change. Cambridge University Press, Cambridge, UK/New York, NY, p 996
Iqbal MA, Ping Q, Abid M, Kazmi SMM, Rizwan M (2016) Assessing risk perceptions and attitude among cotton farmers: a case of Punjab province, Pakistan. Int J Disaster Risk Reduct 16:68–74
Islam MN, Hossen MS (2016) Farmer's perception and agricultural adaptation of climate change in Drought Prone Areas of Bangladesh. J Environ Sci Nat Resour 9:147–150
Kabir MJ, Alauddin M, Crimp S (2017) Farm-level adaptation to climate change in Western Bangladesh: an analysis of adaptation dynamics, profitability and risks. Land Use Policy 64: 212–224
Khan I, Javed T, Khan A, Lei H, Muhammad I, Ali I, Huo X (2019) Impact assessment of land use change on surface temperature and agricultural productivity in Peshawar-Pakistan. Environ Sci Pollut Res 26:33076–33085
Khan I, Lei H, Shah AA, Khan I, Muhammad I (2021) Climate change impact assessment, flood management, and mitigation strategies in Pakistan for sustainable future. Environ Sci Pollut Res 28:29720–29731
Kobayashi S, Furuya J (2011) Comparison of climate change impacts on food security of Bangladesh. Stud Reg Sci 41(2):419–433
Kumar A, Sharma P (2022) Impact of climate variation on agricultural productivity and food security in rural India. Available at SSRN 4144089
Lemma M, Alemie A, Habtu S, Lemma C (2016) Analyzing the impacts of on onset, length of growing period and dry spell length on chickpea production in Adaa District (East Showa Zone) of Ethiopia. J Earth Sci Clim Chang 7(5):349

Lobell DB, Burke MB, Tebaldi C, Mastrandrea MD, Falcon WP, Naylor RL (2008) Prioritizing climate change adaptation needs for food security in 2030. Science 319(5863):607–610

Lobell DB, Hammer GL, Chenu K, Zheng B, McLean G, Chapman SC (2015) The shifting influence of drought and heat stress for crops in northeast Australia. Glob Chang Biol 21(11): 4115–4127

Lybbert T, Sumner D (2010) Agricultural technologies for climate change mitigation and adaptation in developing countries: policy options for innovation and technology diffusion

Mahmood N, Arshad M, Kächele H, Ma H, Ullah A, Müller K (2019) Wheat yield response to input and socioeconomic factors under changing climate: evidence from rainfed environments of Pakistan. Sci Total Environ 688:1275–1285

Malhi GS, Kaur M, Kaushik P (2021) Impact of climate change on agriculture and its mitigation strategies: a review. Sustain For 13(3):1318. https://doi.org/10.3390/su1303131

Mardy T, Uddin MN, Sarker MA, Roy D, Dunn ES (2018) Assessing coping strategies in response to drought: a micro level study in the North-West Region of Bangladesh. Climate 6:23

Matiu M, Ankerst DP, Menzel A (2017) Interactions between temperature and drought in global and regional crop yield variability during 1961–2014. PLoS One 12(5):e0178339

Mendelsohn RO, Massetti E (2017) The use of cross-sectional analysis to measure climate impacts on agriculture: theory and evidence. Rev Environ Econ Policy 11:280–298

Miah JH, Griffiths A, McNeill R, Halvorson S, Schenker U, Espinoza-Orias ND et al (2018) Environmental management of confectionery products: life cycle impacts and improvement strategies. J Clean Prod 177:732–751

Ministry of Agriculture & Farmers' Welfare, (2020) Annual report, 2019–20

Mirza MMQ (2003) Climate change and extreme weather events: can developing countries adapt? Clim Pol 3(3):233–248

Morton JF (2007) The impact of climate change on smallholder and subsistence agriculture. Proc Natl Acad Sci 104(50):19680–19685

Nelson GC, Rosegrant MW, Koo J, Robertson R, Sulser T, Zhu T et al (2009) Climate change: impact on agriculture and costs of adaptation, vol 21. Intl Food Policy Res Inst

Niles MT, Mueller ND (2016) Farmer perceptions of climate change: associations with observed temperature and precipitation trends, irrigation, and climate beliefs. Glob Environ Chang 39: 133–142

Ortiz R, Sayre KD, Govaerts B, Gupta R, Subbarao GV, Ban T et al (2008) Climate change: can wheat beat the heat? Agric Ecosyst Environ 126(1–2):46–58

Osborne TM, Wheeler TR (2013) Evidence for a climate signal in trends of global crop yield variability over the past 50 years. Environ Res Lett 8(2):024001

Porter JR, Xie L, Challinor AJ, Cochrane K, Howden SM, Iqbal MM et al (2014) Food security and food production systems. In: Field CB, Barros VR, Dokken DJ, Mach KJ, Mastrandrea MD, Bilir TE, Chatterjee M, Ebi KL, Estrada YO, Genova RC, Girma B, Kissel ES, Levy AN, MacCracken S, Mastrandrea PR, White LL (eds) Climate change 2014: impacts, adaptation, and vulnerability. Part A: global and sectoral aspects. contribution of working group II to the fifth assessment report of the intergovernmental panel on climate change. Cambridge University Press, Cambridge, UK/New York, NY, pp 485–533

Praveen B, Sharma P (2019) A review of literature on climate change and its impacts on agriculture productivity. J Public Aff 19(4):e1960

Pretty JN, Morison JI, Hine RE (2003) Reducing food poverty by increasing agricultural sustainability in developing countries. Agric Ecosyst Environ 95(1):217–234

Quaye F, Nadolnyak D, Hartarska V (2018) Climate change impacts on farmland values in the Southeast United States. Sustain For 10(10):3426

Ray DK, Gerber JS, MacDonald GK, West PC (2015) Climate variation explains a third of global crop yield variability. Nat Commun 6(1):5989

Rizwan M, Ping Q, Saboor A, Ahmed UI, Zhang D, Deyi Z, Teng L (2020) Measuring rice farmers' risk perceptions and attitude: evidence from Pakistan. Hum Ecol Risk Assess Int J 26(7): 1832–1847

Roco L, Bravo-Ureta B, Engler A, Jara-Rojas R (2017) The impact of climatic change adaptation on agricultural productivity in Central Chile: a stochastic production frontier approach. Sustain For 9(9):1648

Roy D, Kowsari MS, Nath TD, Taiyebi KA, Rashid MM (2018) Smallholder farmers' perception to climate change impact on crop production: case from drought prone areas of Bangladesh. Int J Agric Technol 14(7):1813–1828

Roy D, Das S, Paul S, Paul S (2023) Application of analytical hierarchy process (AHP) method to flood risk assessment at sub-Himalayan region using geospatial data: a case study of Alipurduar District, West Bengal, India. In: Das J, Bhattacharya SK (eds) Monitoring and managing multi-hazards. GIScience and Geoenvironmental modelling. Springer, Cham. https://doi.org/10.1007/978-3-031-15377-8_12

Ruane AC, Major DC, Winston HY, Alam M, Hussain SG, Khan AS et al (2013) Multi-factor impact analysis of agricultural production in Bangladesh with climate change. Glob Environ Chang 23(1):338–350

Saadi S, Todorovic M, Tanasijevic L, Pereira LS, Pizzigalli C, Lionello P (2015) Climate change and Mediterranean agriculture: Impacts on winter wheat and tomato crop evapotranspiration, irrigation requirements and yield. Agric Water Manag 147:103–115

Salam MA, Furuya J, Alamgir MS, Kobayashi S (2016) Policy adaptation cost for mitigation of price variation of rice under climate change in Bangladesh. Centre Environ Inf Sci Tokyo 30 (November):E1–E7

Schauberger B, Archontoulis S, Arneth A, Balkovic J, Ciais P, Deryng D et al (2017) Consistent negative response of US crops to high temperatures in observations and crop models. Nat Commun 8(1):13931

Schmidhuber J, Tubiello FN (2007) Global food security under climate change. Proc Natl Acad Sci 104(50):19703–19708

Shankar KR, Nagasree K, Nirmala G, Prasad M, Venkateswarlu B, Rao CS (2015) Climate change and agricultural adaptation in South Asia. In: Filho L (ed) Handbook of climate change adaptation. Springer, Berlin, pp 1657–1671

Skoufias E, Rabassa M, Olivieri S (2011) The poverty impacts of climate change. Economic Premise, vol 5622. Worldbank

Smith KR, Woodward A, Campbell-Lendrum D, Chadee DD, Honda Y, Liu Q (2014) Human health: impacts, adaptation, and co-benefits. In: Field CB, Barros VR, Dokken DJ, Mach KJ, Mastrandrea MD, Bilir TE et al (eds) Climate change 2014: impacts, adaptation, and vulnerability. Part A: Global and sectoral aspects. Contribution of working group II to the fifth assessment report of the intergovernmental panel on climate change, Cambridge, UK/New York, NY

Teo M, Goonetilleke A, Ahankoob A, Deilami K, Lawie M (2018) Disaster awareness and information seeking behaviour among residents from low socio-economic backgrounds. International journal of disaster risk reduction 31:1121–1131

Tesfaye K, Zaidi PH, Gbegbelegbe S, Boeber C, Rahut DB, Getaneh F, Stirling C (2017) Climate change impacts and potential benefits of heat-tolerant maize in South Asia. Theor Appl Climatol 130:959–970

Tripathi A, Prasad AR (2009) Agricultural development in India since independence: a study on progress, performance, and determinants. J Emerg Knowl Emerg Mark 1:63–92

Ullah R, Shivakoti GP, Kamran A, Zulfiqar F (2016) Farmers versus nature: managing disaster risks at farm level. Nat Hazards 82:1931–1945

UNDP (2009) Policy study on the probable impacts of climate change on poverty and economic growth and the options of coping with adverse effect of climate change in Bangladesh. General Economics Division, Planning Commission, Government of the People's Republic of Bangladesh & UNDP Bangladesh

UN-Water (2013) Climate change (WWW document)

Vyas VS (2003) India's Agrarian structure, economic policies, and sustainable development: variations on a theme. Academic Foundation

Wang SW, Lee W-K, Son Y (2017) An assessment of climate change impacts and adaptation in South Asian agriculture. Int J Clim Change Strateg Manag 9:517–534

World Bank (2020, Sep) Agriculture and food

Yu WH, Alam M, Hossain A, Khan AS, Ruane AC, Rosenzweig C et al (2010) Climate change risks and food security in Bangladesh. World Bank, USA

Zhao J, Yang X, Dai S, Lv S, Wang J (2015) Increased utilization of lengthening growing season and warming temperatures by adjusting sowing dates and cultivar selection for spring maize in Northeast China. Eur J Agron 67:12–19

Chapter 7
Climate Resilient Agroforestry Systems for Sustainable Land Use and Livelihood

Alisha Keprate, D. R. Bhardwaj, Prashant Sharma, Kamlesh Verma, Ghazanfer Abbas, Vaishali Sharma, Kuldipika Sharma, and Shiva Janju

Abstract Agroforestry, an ancient practice, has evolved significantly to provide manifold benefits to humanity. The execution and efficacy of both traditional and improved forms of agroforestry systems have emerged as a potent strategy for ensuring sustainable land use and safeguarding livelihoods despite the formidable challenges posed by escalating climatic variability. This practice, with its myriad ecological advantages such as soil and water conservation, carbon sequestration, enhanced land efficiency, and increased productivity, has proven imperative in empowering countries to confront the challenges posed by climate change. Beyond its ecological impacts, agroforestry plays a crucial role in enhancing the economic, food, and nutritional security of the farming community. This twofold effect not only contributes to environmental replenishment but also enhances the social standing of individual farmers, fostering stronger bonds within the broader community. While numerous agroforestry models implemented globally have shown promise in enhancing adaptive capacity against diverse climate-induced stressors, adopting agroforestry still encounters various challenges including dearth of knowledge, technical skills, motivation, capital, access to qualified seeds, and the absence of a reliable market have emerged as significant obstacles. Addressing these multifaceted challenges is crucial for facilitating the transition from monoculture to agroforestry and unlocking the full potential of this sustainable land-use practice. The ongoing commitment to overcoming these challenges will be pivotal in ensuring the

A. Keprate · D. R. Bhardwaj · P. Sharma (✉) · V. Sharma · S. Janju
Department of Silviculture and Agroforestry, Dr Yashwant Singh Parmar University of Horticulture and Forestry, Solan, Himachal Pradesh, India
e-mail: prashant92749@yspuniversity.ac.in

K. Verma
ICAR-Central Soil Salinity Research Institute, Karnal, Haryana, India

G. Abbas
Forest College and Research Institute, Tamil Nadu Agricultural University, Mettupalayam, Coimbatore, Tamil Nadu, India

K. Sharma
Department of Forest Products, Dr Yashwant Singh Parmar University of Horticulture and Forestry, Solan, Himachal Pradesh, India

S. Kanga et al. (eds.), *Transforming Agricultural Management for a Sustainable Future*, World Sustainability Series, https://doi.org/10.1007/978-3-031-63430-7_7

widespread success and positive impacts of agroforestry on both the environment and communities worldwide. In navigating the complexities of a changing climate, recognizing and investing in these scientifically informed and comprehensive approaches becomes paramount for building resilience and ensuring a sustainable future for both ecosystems and human societies.

Keywords Carbon sequestration · Soil conservation · Biodiversity · Ecosystem services · Adaptive capacity · Climate mitigation

7.1 Introduction

Amid a rapidly changing world, where climate change and unsustainable land use have emerged as critical global challenges, the imperativeness of addressing their far-reaching repercussions cannot be overstated (Bhardwaj et al. 2021, 2022). Climate change, a result of anthropogenic activities, has ushered in an era of increasing temperatures, unpredictable weather patterns, decreased crop production, pest and disease outbreaks and rising sea levels, with far-reaching implications for ecosystems and the global community (Frumkin et al. 2008; Joesidawati et al. 2017; Skendžić et al. 2021). Moreover, land utilization practices, starting with the relentless progression of deforestation, urbanisation and monoculture farming, are actively depleting our soils of nutrients, magnifying greenhouse gas emissions (Verma et al. 2023d), and contributing to biodiversity loss (Gomiero et al. 2011) and resulting in food insecurity (Lal 2013). In light of these pressing issues, it is now of paramount importance to embrace a holistic approach aimed at addressing the root causes of both climate change and unsustainable land use (Ahmad et al. 2023).

Agroforestry, a time-tested land management practice, has quickly gained prominence in the face of these formidable challenges (Sharma et al. 2022, 2023). Its essence is rooted in the purposeful incorporation of woody components alongside agricultural production in the lower stratum (Santiago-Freijanes et al. 2021). This not only provides a concrete route towards climate resilience (Gupta et al. 2023) but also promotes sustainable land use and improves the overall well-being and quality of lives of people by addressing problems like low economic returns, hunger, and nutrition (Montagnini and Metzel 2017; Verma et al. 2021). Through the formation of diverse ecosystems that combine the benefits of agriculture and forestry, it stands as a testament to the potential of harmonizing human needs with ecological stewardship (Jose 2009). While traditional methods often seem to pit the environment against human needs, agroforestry deconstructs this apparent conflict and provides a new way forward (Bhardwaj et al. 2023a, b).

The environment benefits as agroforestry brings back life to depleted lands, combats climate change by sequestering carbon in plants, trees, soil, and nurtures biodiversity within its diversified ecosystems (Sharma et al. 2017a, b). Yet, at the same time, the very people who steward the land—the farmers, the rural communities—experience a transformation in their own livelihoods (Kumar et al. 2023). The equilibrium created by agroforestry fosters a delicate balance that not only conserves our natural heritage but also cultivates human well-being. By thoughtfully weaving the cultivation of crops, trees, and livestock into a coherent tapestry, agroforestry enhances not only the fertility of the soil but also the resilience of local communities. Through this approach, the land transforms into a thriving ecosystem, where the productivity of crops is enhanced by the shade and shelter provided by trees. Not only that, but the livestock also graze on nutritious forage, and their waste, in turn, enriches the soil (Navale et al. 2022). Therefore, we can say that agroforestry exemplifies the beauty of ecological cycles, where nothing goes to waste, and everything finds its place in the intricacies of life on Earth.

7.2 Climate and Agriculture

Climate change is characterized as prominent alterations in the average metrics of meteorological elements over a long period of time (WMO 1992). Traditional agricultural systems face high susceptibility to the repercussions of change in climate which include alterations in weather conditions, water scarcity, changes in growing seasons (Panda et al. 2021, 2022), heightened occurrences of pests and diseases (Altieri et al. 2015), soil degradation, runoff, and erosion (Mall et al. 2017; Meena et al. 2017), along with lack of adaptive capacity. These systems often rely on predictable climates and conventional crop varieties, making them inadequately prepared to confront the challenges brought about by climate change (Singh et al. 2023).

The increase in temperature, coupled with the rise in carbon dioxide and heatwaves rocking the skies (Kumar and Gautam 2014) is now unmistakably manifesting within the agricultural sector, primarily due to its substantial size and responsiveness to atmospheric conditions, upon which the world's food production and economies critically rely (Mendelsohn 2009; Arora 2019). In the realm of climate monitoring, recent scientific advancements have unveiled compelling data, projecting a noteworthy 2 °C temperature increase by 2050 (Tripathi et al. 2023). Studies focusing on crop yields in India have highlighted that, despite adaptation measures, global warming led to a 5.2% decline in wheat yield between 1981 and 2009 (Gupta et al. 2017). Looking ahead, future projections for rain-fed maize show a potential yield decline of 3.3–6.4% in 2030 and a more significant 5.2–12.2% in 2050 due to climate change. Similar projections extend to irrigated crops, suggesting potential yield decreases of 3–8% in 2030 and a more pronounced 5–14% by 2050 if existing varieties of the plant are cultivated (Tesfaye et al. 2017).

Population projections paint a vivid picture, indicating that by 2050, the world's populace is poised to hit 9.7 billion, with a subsequent surge to 10.4 billion in the 2080s, as forecasted by the United Nations (2022). This imminent demographic evolution is poised to amplify the strain on agricultural lands, already grappling with the multifaceted outcomes of climatic alterations, as they strive to meet the intensifying demands for food (Tomar et al. 2023). If the present trajectory of climate change persists, the outlook is ominous, with a significant downturn anticipated in major cereal crop production by 2100—ranging from 5–50% in wheat, 20–45% in maize yields, and 20–30% in rice (Arora 2019). This forewarns of an imminent future where climate induced crop losses escalate at an unparalleled pace, contributing to diminished global crop production (Lobell and Gourdji 2012), price volatility (Haile et al. 2017), and the challenging task of fulfilling the rising needs of an expanding populace. To tackle this challenge, we must embrace climate-smart practices such as agroforestry (Awazi 2022), which is crucial for establishing a sustainable foundation for future agriculture (Altieri et al. 2015).

7.3 Climate Resilience in Agroforestry

Climate resilience pertains to the ability of a system—whether it be an ecosystem, community, infrastructure, or an entire economy—to anticipate, prepare for, respond to, and recover from the inimical effects of climate change. It takes into the account the capacity of the system to withstand shocks, adjust to evolving conditions, and either maintain or swiftly restore functionality when confronted with disturbances related to climate. Resilience goes beyond merely recovering from disruptions; it involves adapting and learning from experiences to improve preparedness for future challenges. It employs a comprehensive approach, taking into account socio-economic, and environmental factors. The aim is to augment the ability of systems to flourish and endure amidst uncertainties and extremes in climate (Kanga et al. 2020).

Agroforestry emerges as a commendable strategy for climate resilience, representing a comprehensive approach to address the complexities arising from a shifting climate (Verma et al. 2021). It is also suggested as a method of generating various ecosystem benefits such as maximizing land productivity and efficiency, soil and water conservation, promoting ecosystem health and biodiversity conservation, and carbon sequestration necessary for alleviating the effects of climate change (Akamani and Holzmueller (2017). Table 7.1 explores diverse global agroforestry systems that showcase practices promoting climate resilience for long-term sustainability. Meanwhile, Fig. 7.1 illustrates both the positive and negative aspects of agroforestry systems in terms of climate resilience.

7.3.1 Enhancing Sustainability and Resilience Through Agroforestry

7.3.1.1 Land Use Management

Agroforestry emerges as a multifaceted and pivotal strategy in the pursuit of optimizing land productivity and efficiency by seamlessly integrating tree cultivation with traditional agriculture and sometimes livestock, harnessing diverse benefits

Table 7.1 Investigating the sustainability impacts of diverse agroforestry systems in existing studies

Agroforestry systems	Geographic region	Main component intercrops	Sustainability benefits	Reference
Hedgerow cropping	India	Leucaena trees	Decreased runoff and soil loss	Narain et al. (1997)
Windbreaks	Central Bohemia region of the Czech Republic.	*Quercus petraea*, *Acer sp*, *Fraxinus excelsior*, *Ulmus glabra*	Wind speed reduction, reduced wind erosion, increased production efficiency, biodiversity, amelioration of microclimate	Vacek et al. (2018)
Alley cropping	Peninsular Malaysia	Oil palm, Pineapple, Bactris, Bamboo, Black pepper, Cacao	Improved vegetation heterogeneity and microclimate regulation	Ashraf et al. (2019)
Homegardens	India	Tomato, Mint, Spinach, Coriander, Basil, Egg-plant, beans, Okra, Pumpkin, Lime, Mango, Papaya, Coco-nut, Pomegranate, Cus-tard apple, Banana, Chiku, Guava, and Fique.	Urban biodiversity con-servation, scenic urban greenscape	Zasada et al. (2020)
Coffee based agroforestry system	Brazil	Coffee	Increased carbon stock	Goncalves et al. (2021)
Parkland system	Ethiopia	*Faidherbia albida*, *Acacia spp*, *Croton macrostachyu*, and *Ziziphus mauritiana*	Carbon sequestration in soil and biomass	Tsedeke et al. (2021)
Riparian Buffer Strips	United Kingdom	Festuloliu grass, Willow, Hornbeam, Sweet chestnut, Hazel, Oak, Lime (*Tilia cordata*), Wych elm	Reduction in total run-off, Reduced sediment loss,	Dunn et al. (2022)
Homegardens	Indonesia	Several native and exotic trees	Biodiversity conserva-tion, Carbon storage	Wiryono et al. (2023)

for sustainable land use. Estimates of global agroforestry cover suggests that 1023 million hectares is under agroforestry (Nair et al. 2009a). The fundamental principle guiding agroforestry is that the union of trees and crops in a piece of land, arranged spatially or temporally, brings about greater structural and functional complexity in contrast to sole cropping (Jose 2009). The integration of trees, with their deep root systems and erosion control measures, not only improves short-term productivity by capturing solar radiation and nutrients but also ensures the long-term sustainability of the land.

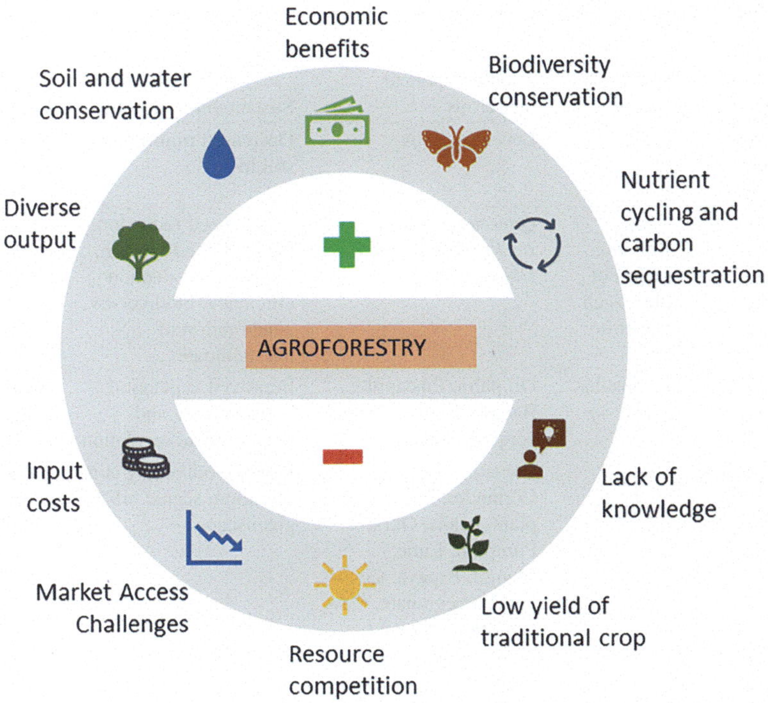

Fig. 7.1 Positive and negative aspects of agroforestry in climate resilience

With tree cover extending to nearly 10% of a billion hectares of agricultural landscapes, projections indicate that approximately 1.6 billion hectares of land will transition to agroforestry management in the imminent future. By diversifying land use, agroforestry minimizes risks associated with monoculture while optimizing spatial and temporal arrangements, resulting in more efficient resource utilization. The resulting structural and functional complexity enhances ecosystem service delivery, rejuvenating degraded land and optimizing resource utilization in farming systems (Rigueiro-Rodríguez et al. 2009; FAO 2014).

In the realm of global land management initiatives, improved agroforestry has gained prominence in recent decades (Plieninger et al. 2020). Dynamic and sustainable, these agroforestry interventions are prevalent in developing countries (Dhyani et al. 2021), covering over 50% of the land (Zomer et al. 2008, 2014, 2016). The recent improved agroforestry systems may exhibit lower diversity, but they often surpass monoculture and forest output in terms of productivity (Nair 2007). For instance, in Europe, advanced form of agroforestry systems has been reported to achieve efficiency levels up to 40% higher compared to traditional farming methods (Graves et al. 2007). The dual advantage, i.e., production and protection, that agroforestry brings to the table makes it one of the most sustainable land-use systems, paving the way for a sustainable future (Dey et al. 2017; Garima et al. 2021; Verma et al. 2023a, b, c).

7.3.1.2 Soil and Water Conservation

The application of agroforestry in soil and water conservation entails the synergistic integration of conservation and production principles, constituting a fundamental requirement for the achievement of sustainable land use practices. Young (1989) defines soil conservation as the maintenance of soil fertility through erosion control, preservation of organic content and nutrients, management of soil properties and avoidance of toxicities. Agroforestry systems offer potential solutions for controlling drainage and erosion, sustaining organic content and physical properties, and optimizing nutrient cycling. Trees in agroforestry systems serve dual ecological purposes: firstly, their substantial surface cover and extensive network of fine-root contribute to controlling erosion, minimizing losses of soil, water and associated nutrients; secondly, the higher biodiversity acts as a "safety net," capturing nutrients (Zhu et al. 2020).

Throughout history, trees, shrubs, bamboos, or palms have been harnessed for soil and water conservation through their deliberate cultivation, wherein they are strategically grown to enhance soil coverage, establish barriers, and uphold or contribute to soil organic matter, thereby mitigating adverse interactions. Widely employed in the tropics, agroforestry practices like compatible crop combinations, windbreaks, alley cropping, multi-storey tree gardens and shelterbelts contribute significantly to soil conservation. These practices operate through the combined effects of foliage cover, detritus, undergrowth, and the strengthening effect of roots (Atangana et al. 2014). A recent study in India specifically demonstrated the effectiveness of agroforestry, incorporating sapota, cowpea, and castor, in significantly reducing total soil loss (37.7%) and runoff (19.1%) in contrast to monoculture (Jinger et al. 2022). Building on this, another noteworthy study by Zhu et al. (2020), through a meta-analysis, demonstrated the effectiveness of agroforestry systems in both temperate and tropical regions. The comparative analysis indicated a substantial reduction in average soil, runoff, nutrient, and pollutant losses by 65%, 58%, 49%, and 50%, respectively. Furthermore, agroforestry exhibits potential for enhancing the physical, chemical, and biological properties of soils, as highlighted by Masebo and Menamo (2016). By embracing a holistic approach, agroforestry not only addresses soil and water conservation challenges but also aids to the overall improvement of soil quality. In essence, it manifests as a versatile and eco-friendly approach, covering water and soil conservation, erosion management, and the refinement soil quality.

7.3.1.3 Ecosystem Health and Biodiversity Conservation

Agroforestry emerges as a powerful catalyst for nurturing biodiversity (McNeely and Schroth (2006) and enhancing ecosystem vitality through its purposeful integration of trees, shrubs, and woody perennials within agricultural landscapes. This intentional blending creates a dynamic and complex patchwork of habitats that serves as a haven for diverse plant and animal species (Bekele 2018). The increased biodiversity in most instances is linked to aspects of agroforestry like well distributed heterogenous vegetation, a range of food sources, favourable microclimate, and enhanced soil health (Udawatta et al. 2021). Research indicates that agroforestry contributes to maintaining the health of ecosystems by enhancing their ability to withstand and recover from various risks and disruptions (Schoeneberger and Bentrup 2017).

The different agroforestry systems play pivotal role in preserving biodiversity at the landscape level, as delineated by Achiso and Masebo (2019):

1. Offering supplementary secondary habitats for species capable of tolerating low level of disturbance.
2. Mitigating the rates of conversion of natural habitat in specific cases.
3. Creating a more amenable and permeable "matrix" between habitat remnants compared to land uses with fewer trees.

Traditional agricultural practices often prioritize a singular ecosystem service—food production— through less complex large-scale monoculture, aiming for better economies (Varah et al. 2013). This approach diminishes species diversity, subsequently leading to lower functional diversity and a reduction in overall ecosystem functions. In contrast, agroforestry represents a strategic departure from this paradigm. When practicing agroforestry, a thoughtful decision is made to selectively and sustainably manage multipurpose trees. This intentional incorporation of diverse plant species in agroforestry systems transforms them into vibrant hubs that attract pollinators, benefiting both cultivated crops and the broader ecosystem. The specific tree species further act as natural hosts for predators of agricultural pests, fostering a natural balance and reducing the need for external interventions.

Agroforestry emerges as a deliberate sanctuary for wildlife, addressing the threats posed by rapid deforestation and fragmented habitats. The connectivity provided by human-dominated agroforestry systems plays a pivotal role in wildlife conservation. For the regional biota to thrive, it becomes imperative for agroforestry systems to function as potential habitats. This significance was underscored in a study conducted in Assam's agroforestry systems, revealing a rich diversity of 364 animal species, with home gardens exhibiting the highest diversity (311 spp.) (Yashmita-Ulman et al. 2021). Beyond supporting fauna, agroforestry systems also exhibit valuable contributions to soil health. In comparison to conventional crop and livestock practices, agroforestry systems foster significantly greater populations of soil organisms, including fungi, bacteria, and insects. The microbiota play supreme role in various biogeochemical processes, such as nutrient cycling, mineralization, and the degradation of chemicals. We can conclude that agroforestry is a holistic system nurturing biodiversity and ecological well-being from the ground up, while also promoting a resilient and thriving environment for generations to come.

7.3.1.4 Carbon Sequestration

Carbon sequestration involves extracting carbon from the atmosphere and securely storing it in reservoirs (Nair et al. 2010; Panwar et al. 2022). This method, emphasizes the transfer of carbon, especially carbon dioxide, and its long-term storage in resilient pools (UNFCCC 2007) such as soils, oceans, biomass etc. Traditionally, afforestation and reforestation of depleted natural forests have been recognized as effective strategies for carbon sequestration. However, agroforestry provides distinct benefits in this regard (Murthy et al. 2013) due to its dual-purpose approach. It optimizes land use, provides economic benefits through diverse products, enhances biodiversity, and improves resilience to climate change. An agroforestry system will encompass different types of trees grown in agroecosystems (Pandey 2002) where carbon sequestration primarily involves absorbing atmospheric carbon dioxide during photosynthesis and moving the captured carbon into vegetation, detritus, and soil pools for the purpose of long-term storage (Nair et al. 2010). The Intergovernmental Panel on Climate Change (IPCC) in 2019 had also recognized agroforestry as a leading land-use system with substantial potential for carbon sequestration. To the present day, it consistently stands out as a favourable method for carbon sequestration, not only augmenting carbon storage but also facilitating increased production (Kürsten 2000).

Globally, soils are recognized as the predominant terrestrial carbon reservoir, containing 1.5–3 times more carbon than vegetation (Dixon 1995). This understanding highlights the pivotal function of soil in the existing carbon cycle. The planned incorporation of woody perennials into agricultural landscapes emerges as a promising avenue for storing of carbon in plant biomass and soil. On average, agroforestry systems have the potential to sequester carbon at rates ranging from 0.2 to 3.1 tons per hectare annually (Watson et al. 2000). It has been projected that the agroforestry could reduce carbon emissions by 1.1–2.2 Pg C over the next five decades through biomass production and enhancement of the soil organic pool (Solomon et al. 2007; Sharma et al. 2021). Moreover, additional studies provide further insights into the carbon sequestration capabilities of agroforestry (Nair et al. 2009b; Nath et al. 2021; Panwar et al. 2022) indicating substantial increase in carbon stock over time. These findings not only underscore the significant carbon sequestration potential of agroforestry systems but also highlight their versatility in offering a gamut of valuable products and amenities, making them integral components of sustainable land management practices.

7.3.2 Role of Agroforestry in Livelihood Resilience

Agroforestry assumes a crucial role in enhancing the resilience of agricultural systems, mitigating vulnerability, and fortifying households against any climate-related risks, thereby ensuring livelihood security. This time-tested agricultural practice has been embraced by millions of farmers globally for millennia. According to a World Bank report in 2004, an estimated 1.2 billion rural individuals actively engage in agroforestry on their farms and within their communities, relying on its diversified products. In these settings, agroforestry assumes significance by meeting varied needs through the interconnected gains of its components—trees, crops, and livestock. Furthermore, it addresses the 6Fs: food, fruit, fuel, fodder, fertilizer, and fiber (Chavan et al. 2015), effectively maximizing the utility of limited land resources. A study conducted in Kilimanjaro, Tanzania, aimed to evaluate the potential of agroforestry in mitigating the impacts of climate change on smallholder farmers. The results unveiled a spectrum of benefits derived from agroforestry practices such as food (59.2%), fodder (58.2%), selling livestock (71%), fruits (54.4%), timber (27.2%) and fuelwood (45.7%) revealed to increase livelihood resilience during climatic severities (Charles et al. 2013). This holistic approach not only sustains agricultural productivity but also establishes a symbiotic balance between environmental conservation and community well-being. In regard to this Table 7.2 examines few more global studies that investigate the contribution of agroforestry to livelihood resilience.

7.3.2.1 Economic Significance

Agroforestry stands as a pioneering land-use approach, offering a transformative solution for income generation within local communities (Achiso and Masebo 2019). Scientifically defined as a dynamic and ecological natural resources management system, agroforestry integrates trees, pasture and/or livestock into farms and agricultural landscapes. The complex integration of different components becomes a catalyst for diversifying and sustaining production, unlocking substantial socio-economic and environmental advantages at all levels. It allows farmers to produce diverse range of products for personal use and sale without requiring additional land,

Table 7.2 Contribution of agroforestry systems to livelihood resilience

Types of agroforestry systems	Location	Contributions to livelihood resilience	Reference
Horticulture based agroforestry system	Bangladesh	High benefit cost ratio; high Land Equivalent Ratio; enhanced food and nutritional security	Das et al. (2020)
Traditional agroforestry systems	Southwestern Ethiopia	Household income generation	Jemal et al. (2018)
Acacia based agroforestry system	Sudan	Increased net revenue	Fadl and El sheikh (2010)
Agrihorticulture (Aonla based agroforestry system)	India	Livelihood security, raised living standards, improved financial status, food security	Dwivedi et al. (2007)

showcasing efficiency and sustainability. Trees in agroforestry systems offers alternative sources of income and food, such as fruits, fodder, or fuel (Charles et al. 2013; Garrity et al. 2010; Thorlakson and Neufeldt 2012). The systems often involve livestock alongside other components, which amplifies the economic impact by supporting sustainable livestock management. By seamlessly incorporating various other options such apiculture, sericulture, lac cultivation, gum and resin extraction, as well as the cultivation of medicinal and aromatic plants (Handa et al. 2016), agroforestry becomes a year-round source of diversified outputs. This multiplicity of products serves as a robust income stream for farmers, allowing them to navigate the challenges of market fluctuations and climatic uncertainties (Rahman et al. 2017). The economic impact is substantial, with agroforestry contributing to household incomes, as demonstrated by a study on coffee agroforestry systems in Sumatra, where produce constituted over 54% of income in contrast to traditional agricultural components contributing 12.5% (Suyanto et al. 2007).

Agroforestry's economic significance is further underscored by its substantial contribution to fulfilling crucial resource demands. It fulfils almost 72% of the fuelwood demand, supplying two-thirds of small timber, and providing 70–80% of wood for plywood, 60–80% of raw material for paper pulp, and 9–11% of green fodder requirement. This comprehensive support extends to meeting the vital needs of farming communities for food, fodder, fruit, fiber, fuel, and medicine (Dhyani and Handa 2013; Kumar 2016). This underscores the idea that agroforestry not only guarantees financial stability but also surpasses other traditional agricultural practices in economic returns. It stands out as a system that addresses citizens' demands while prioritizing sustainability. Clearly, it proves to be a dynamic and economically viable strategy, offering comprehensive benefits for a thriving system.

7.3.2.2 Food and Nutritional Security

Food and nutrition security (FNS) is a multifaceted concept, encompassing five crucial pillars: availability, access, utilization, stability, and sovereignty (Jemal et al. 2018). The global occurrence of undernourishment indicates that approximately 9.2% of the world populace suffered from hunger in 2022. Shockingly, 29.6% of the population around the world, equating to 2.4 billion people, experienced varying degrees of food insecurity, with 900 million facing severe food insecurity (11.3% of the world's population). Organisations like FAO, IFAD, UNICEF, WFP and WHO (2023) projected that nearly 600 million people may be undernourished by 2030. Addressing the challenges of food and nutritional security demands a comprehensive approach, involving strategies such as increasing crop yield, enhancing soil fertility, and cultivating a diverse range of plants providing edible fruits, vegetables, and nuts (Jamnadass et al. 2013). An effective solution to this complex issue is agroforestry, a system that encompasses more than two components and yields various food products, including grains, fruits, vegetables, spices, and livestock feed.

Agroforestry systems such as agri-silviculture, silvi-pastoral, agri-silvi-pastoral, agri-horticulture and others like aqua forestry and apiculture with tree species play a pivotal role in guaranteeing food security at the national level. Research indicates several substantial advantages of agroforestry systems. For instance, millet cultivation under *Faidherbia albida* in sub-Saharan Africa demonstrated significant increases in grain and protein yields (Mokgolodi et al. 2011). Also, the incorporation of nutrient rich fruit and other food crops into agroforestry systems directly contributes to improving the health and nutrition of global populations. Fruits, nuts, and leafy vegetables grown in agroforestry systems serve as major sources of micronutrients for farmers (Mbow et al. 2014).

In the pursuit of food security, as outlined in SDG2, the focus extends beyond mere crop yield to encompass resilience in the face of climate change and unforeseen shocks (Selvan et al. 2023). Agroforestry emerges as a crucial component, playing a vital role in bolstering crop resilience against adverse climate impacts such as drought and elevated temperatures by improving infiltration rate of water and its storage, while simultaneously mitigating temperature extremes (Charles et al. 2013; Garrity et al. 2010). It proves to be a forward-looking approach that not only addresses the immediate goal of augmenting food production but also aligns with the broader objective of ensuring sustained food security in defiance of climate change.

7.3.2.3 Social and Cultural Impacts

Agroforestry emerges not just as an agricultural practice but as a powerful driver for global social and cultural advancement. The active involvement of local communities in planning, executing, and managing agroforestry systems is pivotal, fostering empowerment, strengthening community bonds, and nurturing collaborative spirit. An expression of this collaboration is seen in the communal exchange of locally grown produce, a practice that cultivates familial bonds in rural villages (Nair et al. 2017).

Beyond its agricultural utility, agroforestry assumes the role of a guardian of indigenous wisdom, diligently preserving native plant species (Nair 2014). Within agroforestry systems, trees and plants acquire cultural significance, contributing to the fabric of community life. Some species hold historical or spiritual value, becoming custodians of cultural practices and traditions, thereby fostering a sustainable connection between people and their environment. The integration of traditional and indigenous practices into agroforestry safeguards cultural heritage, ensuring the seamless transmission of invaluable knowledge across generations. The Chania system in Greece stands as a testament to this, its rich cultural history reinforcing the identity of communities (Papanastasis et al. 2009). Similarly, the Sikkim Himalayan traditional farming systems exemplify indigenous management of homegarden agroforestry systems at small scale level, showcasing a harmonious interplay of socioecological, sociocultural, and socio-economic benefits (Sharma and Sharma 2017).

Agroforestry, extending beyond practicality, unveils its recreational dimension and eco-tourism (Pardini 2009). The traditional bamboo-pine system and rice-fish culture of the Apatani tribe in Arunachal Pradesh have gained UNESCO recognition for exceptional productivity and unique ecological preservation (Tangjang and Nair 2015, 2016). Furthermore, agroforestry emerges as an agent of livelihood diversification, breaking the chains of dependence on a single economic activity. This not only elevates the economic well-being of communities but also serves as a linchpin in poverty alleviation, contributing to improved nutrition and health (Gautam et al. 2003).

The collaborative ethos inherent in agroforestry systems not only bolsters social cohesion but equips communities with the adaptability to navigate changing climates. This collaborative effort transcends agricultural endeavours, birthing a supportive network that addresses an array of community needs. The shared responsibilities and benefits woven into agroforestry practices cultivate a profound sense of shared destiny among community members. In summation, agroforestry's impact on social and cultural upliftment transcends its agricultural roots. By empowering communities, preserving indigenous wisdom, upholding cultural legacies, fortifying economic resilience, and nurturing social cohesion, agroforestry emerges as a cornerstone in the journey towards comprehensive and sustainable development.

7.4 Challenges in Adoption of Agroforestry

The integration of agroforestry into mainstream agricultural practices is a crucial step toward achieving sustainable and resilient food systems. Despite the proven environmental and economic benefits associated with agroforestry, the widespread adoption of this practice faces various barriers. These barriers, spanning economic, social, and policy dimensions, hinder the seamless incorporation of agroforestry into agricultural landscapes. In exploring these challenges, it becomes evident that overcoming these barriers is essential for realizing the full potential of agroforestry in promoting sustainable agriculture, enhancing biodiversity, and mitigating climate change. Table 7.3 discusses the different barriers and challenges encountered in the adoption of agroforestry across various geographical areas.

7.5 Wayforward

As our climate shifts, it's vital to focus on making agroforestry resilient to ensure sustainable land use and livelihoods. To progress, we need to prioritize ongoing research and innovation, understanding the nuances of agroforestry across different environments. This includes the development of robust tree-crop combinations and agroecological practices aimed at maximizing productivity while mitigating environmental impact. Essential to building climate resilience is the empowerment of farmers, local communities, and agricultural practitioners. This necessitates

Table 7.3 Exploring challenges in the adoption of agroforestry practices

Authors	Location	Challenges in adoption
Mwase et al. (2015)	Southern Africa	High initial costs (herbicide, labor) Low extension capacity and insufficient training materials Limited seed availability compared to demand Limited knowledge of tree management Lack of availability of land for agroforestry practices Lack of cash incentives for agroforestry adoption Lack of comprehensive policies on agroforestry
Kiyani et al. (2017)	Rwanda	Absence of technical skills Inadequate supply of qualified seeds Lack of capital and manpower Inaccessible market
Sagastuy and Krause (2019)	Brazil	Uncertainty in the effectiveness of the system, Reduced yield of agricultural crops Absence of models and awareness in the region
Sollen-Norrlin et al. (2020)	Europe	High costs of implementation Inadequate Financial Incentives Limited Products and Marketing Lack of advisory service support Lack of education, field demonstrations
Jahan et al. (2022)	Bangladesh	No support of government Absence of technical knowledge & skills No training facility Lack of motivation & information Absence of capital Laborious Return is not satisfactory

comprehensive capacity-building initiatives, ranging from training programs and workshops to knowledge-sharing platforms, ensuring stakeholders possess the requisite skills for implementing sustainable land management practices. Aligning national and regional agricultural policies with climate resilience objectives can provide farmers with the necessary resources and incentives to seamlessly incorporate trees into their farming systems. Fundamental to the success of agroforestry initiatives is community engagement through participatory approaches. Integrating local communities into decision-making processes ensures that agroforestry practices resonate with their specific needs, cultural values, and traditional knowledge. Establishing robust monitoring and evaluation systems is equally critical for assessing the impact of agroforestry interventions on climate resilience, ecosystem health, and livelihoods. This data-driven approach informs adaptive management strategies, enhancing the efficacy of future initiatives. Financial support and incentives constitute integral components of the forward trajectory. Offering farmers subsidies, grants, or other financial mechanisms recognizes the enduring benefits of climate-resilient land use. Moreover, the integration of climate-smart technologies into agroforestry practices enhances adaptive capacity. Precision agriculture, agrometeorological forecasting, and smart irrigation systems contribute to the efficiency and effectiveness of agroforestry systems. Finally, the imperative of fostering international knowledge exchange and collaboration cannot be overstated. Sharing successful agroforestry models, experiences, and lessons learned on a global scale establishes a reservoir of best practices for climate-resilient agroforestry. This holistic approach, amalgamating research, community engagement, policy advocacy, and technological innovation, holds the potential to usher in a more resilient and sustainable future. In this envisioned future, agroforestry stands as a cornerstone for climate adaptation, ecosystem health, and thriving livelihoods.

7.6 Conclusion

In conclusion, the establishment of climate resilient agroforestry systems surfaces as a promising approach for attaining sustainable land utilization and securing economic sustenance. These models not only bolster the adaptability of ecosystems to climate shifts but also play an instrumental role in sustainable land management practices. By advocating for biodiversity, enhancing soil quality, and mitigating climate-induced risks and shocks, agroforestry manifests itself as an ingenious practice that promotes environmental, economic and social sustainability. The affirmative results emphasize the importance of further research, model refinement, community engagement, and policy support to scale up the adoption of climate-resilient agroforestry systems, thereby aiming to foster a synergistic equilibrium between environmental conservation and socio-economic well-being. Looking ahead, recognizing and investing in such a comprehensive approach is imperative for constructing a resilient and sustainable future for both the land and its inhabitants. It is a proactive step towards ensuring a balanced coexistence that nurtures environmental health, economic prosperity, and community well-being in a mutually reinforcing manner.

References

Achiso Z, Masebo N (2019) Parkland agroforestry practices on biodiversity conservation-a review. SSR Inst Int J Life Sci 5:2412–2420

Ahmad T, Gupta SK, Singh SK, Meraj G, Kumar P, Kanga S (2023) Unveiling nature's resilience: exploring vegetation dynamics during the COVID-19 era in Jharkhand, India, with the Google Earth Engine. Climate 11(9):187

Akamani K, Holzmueller EJ (2017) Socioeconomic and policy considerations in the adoption of agroforestry systems: an ecosystem-based adaptive governance approach. In: Dagar J, Tewari V (eds) Agroforestry: anecdotal to modern science, Springer, Singapore, pp 833–855. https://doi.org/10.1007/978-981-10-7650-3_35

Altieri MA, Nicholls CI, Henao A, Lana MA (2015) Agroecology and the design of climate change-resilient farming systems. Agron Sustain Dev 35(3):869–890

Arora NK (2019) Impact of climate change on agriculture production and its sustainable solutions. Environ Sustain 2(2):95–96

Ashraf M, Sanusi R, Zulkifli R, Tohiran KA, Moslim R, Ashton-Butt A, Azhar B (2019) Alley-cropping system increases vegetation heterogeneity and moderates extreme microclimates in oil palm plantations. Agric For Meteorol 276:107632

Atangana A, Khasa D, Chang S, Degrande A (2014) Agroforestry for soil conservation. In: Atangana A, Khasa D, Chang S, Degrande A (eds) Tropical agroforestry. Springer, Dordrecht, pp 203–216. https://doi.org/10.1007/978-94-007-7723-1_9

Awazi NP (2022) Agroforestry for climate change adaptation, resilience enhancement and vulnerability attenuation in smallholder farming systems in Cameroon. J Atmospheric Sci Res 5:25–33

Bekele SE (2018) Parkland agroforestry of Ethiopia; key to production, productivity, biodiversity conservation and climate change mitigation: a review. Open J For 8(04):472–488

Bhardwaj DR, Tahiry H, Sharma P, Pala NA, Kumar D, Kumar A (2021) Influence of aspect and elevational gradient on vegetation pattern, tree characteristics and ecosystem carbon density in Northwestern Himalayas. Land 10(11):1109

Bhardwaj DR, Kumar A, Pala NA, Sharma P, Kumar D, Kumar A, Zahoor S (2022) Carbon density and C-sequestration of tree plantation ecosystems in the mid-hills of the NW-Himalayas: implications for climate change mitigation. Land Degrad Dev 33(12):2115–2126

Bhardwaj DR, Sharma P, Kumar D, Panwar P, Kumar A, Pala NA, Rajput BS, Kumar R, Negi V, Rajput P, Salve A, Chisanga K, Tahiry H (2023a) Carbon stock inventory and biomass production in different land use systems of Northwestern Himalaya. In: Kumar A, Jong WD, Kumar M, Pandey R (eds) Climate change in the Himalayas. Academic Press Massachusetts pp 217–233

Bhardwaj DR, Salve A, Kumar J, Kumar A, Sharma P, Kumar D (2023b) Biomass production and carbon storage potential of agroforestry land use systems in high hills of north-western Himalaya: an approach towards natural based climatic solution. Biomass Convers Biorefinery 1–14. https://doi.org/10.1007/s13399-023-03952-0

Charles RL, Munishi PKT, Nzunda EF (2013) Agroforestry as adaptation strategy under climate change in Mwanga District, Kilimanjaro, Tanzania. Int J Environ Protect 3(11):29–38

Chavan SB, Keerthika A, Dhyani SK, Handa AK, Newaj R, Rajarajan K (2015) National Agroforestry Policy in India: a low hanging fruit. Curr Sci 108:1826–1834

Das AK, Rahman MA, Saha SR, Sarmin NS, Hoque MA, Bhuiyan F (2020) Transforming Malta orchard into agroforestry system with different crops for improving productivity, profitability and land uses. Ann Bangladesh Agric 24(1):113–125

Dey S, Prasad S, Tiwari P, Sharma P (2017) Effect of urea, KCl, Zinc placement & spray on growth of cowpea. J Pharmacogn Phytochem 6(6S):971–973

Dhyani SK, Handa AK (2013) Area under agroforestry in India: An assessment for present status and future perspective. Indian J Agrofor 15(1):1–11

Dhyani S, Murthy IK, Kadaverugu R, Dasgupta R, Kumar M, Adesh Gadpayle K (2021) Agroforestry to achieve global climate adaptation and mitigation targets: are South Asian countries sufficiently prepared? Forests 12(3):303

Dixon RK (1995) Agroforestry systems: sources of sinks of greenhouse gases? Agrofor Syst 31: 99–116

Dunn RM, Hawkins JM, Blackwell MS, Zhang Y, Collins AL (2022) Impacts of different vegetation in riparian buffer strips on runoff and sediment loss. Hydrol Process *36*(11):e14733

Dwivedi RP, Tewari RK, Kareemulla K, Chaturvedi OP, Rai P (2007) Agri-horticultural system for household livelihood-a case study. Management 2:2

Fadl KEM, El sheikh SE (2010) Effect of *Acacia Senegal* on growth and yield of groundnut, sesame and roselle in an agroforestry system in North Kordofan state. Sudan Agrofor Syst 78:243–252. https://doi.org/10.1007/s10457-009-9243-9

FAO (2014) Building a common vision for sustainable food and agriculture. Principles and approaches. http://www.fao.org/3/a-i3940e.pdf

FAO, IFAD, UNICEF, WFP, WHO (2023) In brief to the state of food security and nutrition in the world 2023. Urbanization, agrifood systems transformation and healthy diets across the rural–urban continuum. FAO, Rome

Frumkin H, Hess J, Luber G, Malilay J, McGeehin M (2008) Climate change: the public health response. Am J Public Health 98(3):435–445

Garima, Bhardwaj DR, Thakur CL, Kaushal R, Sharma P, Kumar D, Kumari Y (2021) Bamboo-based agroforestry system effects on soil fertility: Ginger performance in the bamboo subcanopy in the Himalayas (India). Agron J 113(3):2832–2845

Garrity DP, Akinnifesi FK, Ajayi OC, Weldesemayat SG, Mowo JG, Kalinganire A, Larwanou M, Bayala J (2010) Evergreen agriculture: a robust approach to sustainable food security in Africa. Food Secur 2:197–214

Gautam MK, Roberts EH, Singh BK (2003) Community based leasehold approach and agroforestry technology for restoring degraded hill forests and improving rural livelihoods in Nepal. In International Conference on Rural Livelihoods, Forests and Biodiversity, Bonn, pp 19–23

Gomiero T, Pimentel D, Paoletti MG (2011) Is there a need for a more sustainable agriculture? Crit Rev Plant Sci 30(1–2):6–23

Goncalves N, Andrade D, Batista A, Cullen L, Souza A, Gomes H, Uezu A (2021) Potential economic impact of carbon sequestration in coffee agroforestry systems. Agrofor Syst 95:419–430

Graves AR, Burgess PJ, Palma JHN, Herzog F, Moreno G, Bertomeu M, Dupraz C, Liagre F, Keesman K, van der Werf W, De Nooy AK (2007) Development and application of bio-economic modelling to compare silvoarable, arable, and forestry systems in three European countries. Ecol Eng 29(4):434–449

Gupta R, Somanathan E, Dey S (2017) Global warming and local air pollution have reduced wheat yields in India. Clim Chang 140(3–4):593–604

Gupta SR, Dagar JC, Sileshi GW, Chaturvedi RK (2023) Agroforestry for climate change resilience in degraded landscapes. In: Dagar JC, Gupta SR, Sileshi GW (eds) Agroforestry for Sustainable Intensification of Agriculture in Asia and Africa. Springer, Singapore, pp 121–174. https://doi.org/10.1007/978-981-19-4602-8_5

Haile MG, Wossen T, Tesfaye K, von Braun J (2017) Impact of climate change, weather extremes, and price risk on global food supply. Econ Disaster Clim Chang 1:55–75

Handa AK, Toky OP, Dhyani SK, Chavan SB, Toky ID (2016) Innovative agroforestry for livelihood security in India. World Agric 7:7–16

IPCC. IPCC Special Report on Climate Change, Desertifcation, Land Degradation, Sustainable Land Management, Food Security, and Greenhouse Gas Fluxes in Terrestrial Ecosystems; Intergovernmental Panel on Climate Change: Geneva, Switzerland, 2019

Jahan H, Rahman MW, Islam MS, Rezwan-Al-Ramim A, Tuhin MMUJ, Hossain ME (2022) Adoption of agroforestry practices in Bangladesh as a climate change mitigation option: Investment, drivers, and SWOT analysis perspectives. Environ Chall 7:100509

Jamnadass R, Place F, Torquebiau E, Malézieux E, Liyama M, Sileshi G, Kehlenbeck K, Masters E, McMullin S, Dawson I (2013) Agroforestry, food and nutritional security. ICRAF Working Paper No. 170. Nairobi, World Agroforestry Centre, pp 1–15

Jemal O, Callo-Concha D, Van Noordwijk M (2018) Local agroforestry practices for food and nutrition security of smallholder farm households in southwestern Ethiopia. Sustain For 10(8): 2722

Jinger D, Kumar R, Kakade V, Dinesh D, Singh G, Pande VC, Bhatnagar PR, Rao BK, Vishwakarma AK, Kumar D, Singhal V (2022) Agroforestry for controlling soil erosion and enhancing system productivity in ravine lands of Western India under climate change scenario. Environ Monit Assess 194(4):267

Joesidawati MI, Suntoyo S, Wahyudi W, Sambodho K (2017) Sea level rise on Tuban coast in East Java and its consistenty with MAGICC/SCENGEN prediction. Appl Mech Mater 862:83–89

Jose S (2009) Agroforestry for ecosystem services and environmental benefits: an overview. Agrofor Syst 76:1–10

Kanga S, Singh PK, Ohri A (2020) Evaluating the environmental sustainability of smart cities in India: The design and application of the Indian smart city environmental sustainability index. Sustain For 13(1):327

Kiyani P, Andoh J, Lee Y, Lee DK (2017) Benefits and challenges of agroforestry adoption: a case of Musebeya sector, Nyamagabe District in southern province of Rwanda. For Sci Technol 13(4):174–180

Kumar V (2016) Multifunctional agroforestry systems in tropics region. Nat Environ Pollut Technol 15(2):365

Kumar R, Gautam HR (2014) Climate change and its impact on agricultural productivity in India. J Clim Weather Forecasting 2(1):1–3

Kumar P, Singh SK, Gupta SK, Kanga S, Meraj G (2023) Uncovering the hydro-meteorological drivers responsible for forest fires utilizing geospatial techniques. Theor Appl Climatol. https://doi.org/10.1007/s00704-023-04497. Springer Nature

Kürsten E (2000) Fuelwood production in agroforestry systems for sustainable land use and CO2-mitigation. Ecol Eng 16:69–72

Lal R (2013) Food security in a changing climate. Ecohydrol Hydrobiol 13(1):8–21

Lobell DB, Gourdji SM (2012) The influence of climate change on global crop productivity. Plant Physiol 160(4):1686–1697

Mall RK, Gupta A, Sonkar G (2017) Effect of climate change on agricultural crops. In: Dubey SK, Pandey A, Sangwan RS (eds) Current developments in biotechnology and bioengineering. Elsevier, Netherlands, pp 23–46

Masebo N, Menamo M (2016) A review paper on: the role of agroforestry for rehabilitation of degraded soil. J Biol Agric Healthcare 6(5):128–136

Mbow C, Van Noordwijk M, Luedeling E, Neufeldt H, Minang PA, Kowero G (2014) Agroforestry solutions to address food security and climate change challenges in Africa. Curr Opin Environ Sustain 6:61–67

McNeely JA, Schroth G (2006) Agroforestry and biodiversity conservation–traditional practices, present dynamics, and lessons for the future. Biodivers Conserv 15:549–554

Meena NK, Gautam R, Tiwari P, Sharma P (2017) Nutrient losses in soil due to erosion. J Pharmacogn Phytochem 6(6S):1009–1011

Mendelsohn R (2009) The impact of climate change on agriculture in developing countries. J Nat Res Policy Res 1(1):5–19

Mokgolodi NC, Setshogo MP, Shi LL, Liu YJ, Ma C (2011) Achieving food and nutritional security through agroforestry: a case of Faidherbia albida in sub-Saharan Africa. For Stud China 13:123–131

Montagnini F, Metzel R (2017) The contribution of agroforestry to sustainable development goal 2: end hunger, achieve food security and improved nutrition, and promote sustainable agriculture. In: Integrating landscapes: Agroforestry for biodiversity conservation and food sovereignty, pp 11–45

Murthy IK, Gupta M, Tomar S, Munsi M, Tiwari R, Hegde GT, Ravindranath NH (2013) Carbon sequestration potential of agroforestry systems in India. J Earth Sci Climate Change 4(1):1–7

Mwase W, Sefasi A, Njoloma J, Nyoka BI, Manduwa D, Nyaika J (2015) Factors affecting adoption of agroforestry and evergreen agriculture in Southern Africa. Environ Nat Resour Res 5(2):148

Nair PR (2007) The coming of age of agroforestry. J Sci Food Agric 87(9):1613–1619

Nair PR (2014) Grand challenges in agroecology and land use systems. Front Environ Sci 2:1

Nair PKR, Mohan Kumar B, Nair VD (2009a) Agroforestry as a strategy for carbon sequestration. J Plant Nutr Soil Sci 172(1):10–23

Nair PR, Nair VD, Kumar BM, Haile SG (2009b) Soil carbon sequestration in tropical agroforestry systems: a feasibility appraisal. Environ Sci Pol 12(8):1099–1111

Nair PR, Nair VD, Kumar BM, Showalter JM (2010) Carbon sequestration in agroforestry systems. Adv Agron 108:237–307

Nair PR, Viswanath S, Lubina PA (2017) Cinderella agroforestry systems. Agrofor Syst 91: 901–917

Narain P, Singh RK, Sindhwal NS, Joshie P (1997) Agroforestry for soil and water conservation in the western Himalayan Valley Region of India 1. Runoff, soil and nutrient losses. Agrofor Syst 39:175–189

Nath AJ, Sileshi GW, Laskar SY, Pathak K, Reang D, Nath A, Das AK (2021) Quantifying carbon stocks and sequestration potential in agroforestry systems under divergent management scenarios relevant to India's Nationally Determined Contribution. J Clean Prod 281:124831

Navale MR, Bhardwaj DR, Bishist R, Thakur CL, Sharma S, Sharma P, Probo M (2022) Seasonal variations in the nutritive value of fifteen multipurpose fodder tree species: a case study of north-western Himalayan mid-hills. PLoS One 17(10):e0276689

Panda S, Bhardwaj DR, Sharma P, Handa AK, Kumar D (2021) Impact of climatic patterns on phenophase and growth of multi-purpose trees of north-western mid-Himalayan ecosystem. Trees For People 6:100143

Panda S, Bhardwaj DR, Thakur CL, Sharma P, Kumar D (2022) Growth response of seven multipurpose tree species to climatic factors: a case study from northwestern Himalayas. India J For Sci 68(3):83–95

Pandey DN (2002) Carbon sequestration in agroforestry systems. Clim Pol 2(4):367–377

Panwar P, Mahalingappa DG, Kaushal R, Bhardwaj DR, Chakravarty S, Shukla G, Thakur NS, Chavan SB, Pal S, Nayak BG, Srinivasaiah HT (2022) Biomass production and carbon sequestration potential of different agroforestry systems in India: a critical review. Forests 13(8):1274

Papanastasis VP, Mantzanas K, Dini-Papanastasi O, Ispikoudis I (2009) Traditional agroforestry systems and their evolution in Greece. In: Rigueiro-Rodróguez A, McAdam J, Mosquera-Losada MR (eds) Agroforestry in Europe: current status and future prospects, Springer, Dordrecht, pp 89–109. https://doi.org/10.1007/978-1-4020-8272-6_5

Pardini A (2009) Agroforestry systems in Italy: traditions towards modern management. In: Rigueiro-Rodróguez A, McAdam J, Mosquera-Losada MR (eds) Agroforestry in Europe: current status and future prospects. Springer, Dordrecht, pp 255–267. https://doi.org/10.1007/978-1-4020-8272-6_12

Plieninger T, Muñoz-Rojas J, Buck LE, Scherr SJ (2020) Agroforestry for sustainable landscape management. Sustain Sci 15:1255–1266

Rahman SA, Jacobsen JB, Healey JR, Roshetko JM, Sunderland T (2017) Finding alternatives to swidden agriculture: does agroforestry improve livelihood options and reduce pressure on existing forest? Agrofor Syst 91:185–199

Rigueiro-Rodríguez A, Fernández-Núñez E, González-Hernández P, McAdam JH, Mosquera-Losada MR (2009) Agroforestry systems in Europe: productive, ecological and social perspectives. In: Rigueiro-Rodróguez A, McAdam J, Mosquera-Losada MR (eds) Agroforestry in Europe: current status and future prospects. Springer, Dordrecht, pp 43–65. https://doi.org/10.1007/978-1-4020-8272-6_3

Sagastuy M, Krause T (2019) Agroforestry as a biodiversity conservation tool in the atlantic forest? Motivations and limitations for small-scale farmers to implement agroforestry systems in north-eastern Brazil. Sustain For 11(24):6932

Santiago-Freijanes JJ, Mosquera-Losada MR, Rois-Díaz M, Ferreiro-Domínguez N, Pantera A, Aldrey JA, Rigueiro-Rodríguez A (2021) Global and European policies to foster agricultural sustainability: agroforestry. Agrofor Syst 95(5):775–790

Schoeneberger MM, Bentrup G (2017) Agroforestry: enhancing resiliency in US agricultural landscapes under changing conditions. Gen. Tech. Report WO-96. US Department of Agriculture, Forest Service, Washington, DC, 96

Selvan T, Panmei L, Murasing KK, Guleria V, Ramesh KR, Bhardwaj DR, Thakur CL, Kumar D, Sharma P, Digvijaysinh Umedsinh R, Kayalvizhi D, Deshmukh HK (2023) Circular economy in agriculture: unleashing the potential of integrated organic farming for food security and sustainable development. Front Sustain Food Syst 7:1170380. https://doi.org/10.3389/fsufs.2023.1170380

Sharma G, Sharma E (2017) Agroforestry systems as adaptation measures for sustainable livelihoods and socio-economic development in the Sikkim Himalaya. In: Dagar J, Tewari V (eds) Agroforestry: anecdotal to modern science. Springer, Singapore, pp 217–243. https://doi.org/10.1007/978-981-10-7650-3_8

Sharma P, Singh MK, Tiwari P (2017a) Agroforestry: a land degradation control and mitigation approach. Bull Environ Pharmacol Life Sci 6(5):312–317

Sharma P, Singh MK, Tiwari P, Verma K (2017b) Agroforestry systems: opportunities and challenges in India. J Pharmacogn Phytochem 6(6S):953–957

Sharma P, Bhardwaj DR, Singh MK, Verma K, Vishwakarma SP, Kumar D (2021) Carbon trading in agroforestry: potential prospects and constraints. In: Yadav L, Yadav SK (eds) Climate change: adaptation, loss and damage. Neel Kamal Prakashan, Delhi, pp 185–197

Sharma P, Singh M, Verma K, Prasad SK (2022) Soil weed seedbank under different cropping systems of middle Indo-Gangetic Plains. Plant Soil Environ 68(11):542–551

Sharma P, Bhardwaj DR, Singh MK, Nigam R, Pala NA, Kumar A, Thakur P (2023) Geospatial technology in agroforestry: status, prospects, and constraints. Environ Sci Pollut Res 30(55): 116459–116487. https://doi.org/10.1007/s11356-022-20305-y

Singh BK, Delgado-Baquerizo M, Egidi E, Guirado E, Leach JE, Liu H, Trivedi P (2023) Climate change impacts on plant pathogens, food security and paths forward. Nat Rev Microbiol 21(10): 640–656. https://doi.org/10.1038/s41579-023-00900-7

Skendžić S, Zovko M, Živković IP, Lešić V, Lemić D (2021) The impact of climate change on agricultural insect pests. Insects 12(5):440

Sollen-Norrlin M, Ghaley BB, Rintoul NLJ (2020) Agroforestry benefits and challenges for adoption in Europe and beyond. Sustain For 12(17):7001

Solomon S, Qin D, Manning M, Chen Z, Marquis M, Averyt KB, Tignor MMHL, Miller HL (2007) The physical science basis. Contribution of working group I to the fourth assessment report of the intergovernmental panel on climate change, 996, p 2007

Suyanto S, Khususiyah N, Leimona B (2007) Poverty and environmental services: case study in Way Besai watershed, Lampung Province, Indonesia. Ecol Soc 12(2)

Tangjang S, Nair PR (2015) Rice+ fish farming in homesteads: sustainable natural-resource management for subsistence in Arunachal Pradesh, India. J Environ Sci Eng A 4:545–557

Tangjang S, Nair PR (2016) Integrated bamboo+ pine homegardens: a unique agroforestry system in Ziro Valley of Arunachal Pradesh, India. Int J Environ Agric Res 2(2):25–34

Tesfaye K, Zaidi PH, Gbegbelegbe S, Boeber C, Rahut DB, Getaneh F, Seetharam K, Erenstein O, Stirling C (2017) Climate change impacts and potential benefits of heat-tolerant maize in South Asia. Theor Appl Climatol 130:959–970

Thorlakson T, Neufeldt H (2012) Reducing subsistence farmers' vulnerability to climate change: evaluating the potential contributions of agroforestry in western Kenya. Agric Food Secur 1:1–13

Tomar P, Singh SK, Kanga S, Pattanaik A (2023) Water bodies mapping and monitoring using high-resolution satellite images. Sustain Agri Food Environ Res 11:1–18. https://doi.org/10.7770/safer-V11N1-art2673

Tripathi G, Shakya A, Upadhyay RK, Singh SK, Kanga S, Pandey SK (2023) Landslide susceptibility mapping of Tehri reservoir region using geospatial approach. In: Sharma S, Kuniyal JC, Chand P, Singh P (eds) Climate change adaptation, risk management and sustainable practices in the Himalaya. Springer, Cham, pp 135–156. https://doi.org/10.1007/978-3-031-24659-3_7

Tsedeke RE, Dawud SM, Tafere SM (2021) Assessment of carbon stock potential of parkland agroforestry practice: the case of Minjar Shenkora; North Shewa, Ethiopia. Environ Syst Res 10: 1–11

Udawatta RP, Rankoth LM, Jose S (2021) Agroforestry for biodiversity conservation. In: Udawatta RP, Jose S (eds) Agroforestry and ecosystem services. Springer, Cham, pp 245–274. https://doi.org/10.1007/978-3-030-80060-4_10

UNFCCC (2007) Report of the conference of parties on its thirteenth session, Bali, Indonesia. In "United Nations Framework Convention on Climate Change" Geneva, Switzerland, UN

United Nations (2022) World population prospects: the 2022 revision. United Nations Population Division, New York

Vacek Z, Řeháček D, Cukor J, Vacek S, Khel T, Sharma RP, Kučera J, Král J, Papaj V (2018) Windbreak efficiency in agricultural landscape of the Central Europe: multiple approaches to wind erosion control. Environ Manag 62:942–954

Varah A, Jones H, Smith J, Potts SG (2013) Enhanced biodiversity and pollination in UK agroforestry systems. J Sci Food Agric 93(9):2073–2075

Verma K, Sharma P, Kumar D, Vishwakarma SP, Meena NK (2021) Strategies sustainable management of agroforestry in climate change mitigation and adaptation. Int J Curr Microbiol App Sci 10(01):2439–2449

Verma K, Prasad SK, Singh MK, Sharma P (2023a) Assessment of agronomic zinc biofortification of Alley Cropped Pearl Millet. Bangladesh J Bot 52(1):203–209

Verma K, Prasad SK, Singh MK, Sharma P (2023b) Response of alley-cropped pearl millet (Pennisetum glaucum) to nitrogen and zinc schedules under semi-arid regions. Indian J Agron 68(1):105–109

Verma K, Sharma P, Bhardwaj DR, Kumar R, Kumar NM, Singh AK (2023c) Land and environmental management through agriculture, forestry and other land use (AFOLU) system. In: Raj A, Jhariya MK, Banerjee A, Nema S, Bargali K (eds) Land and environmental management through forestry. https://doi.org/10.1002/9781119910527.ch10

Verma T, Bhardwaj DR, Sharma U, Sharma P, Kumar D, Kumar A (2023d) Agroforestry systems in the mid-hills of the north-western Himalaya: a sustainable pathway to improved soil health and climate resilience. J Environ Manag 348:119264

Watson RT, Noble IR, Bolin B, Ravindranath NH, Verardo JD, Dokken DJ (2000) Land use, land-use change and forestry, IPCC special report. Cambridge University Press, Cambridge. 388 pp. http://www.grida.no/climate/ipcc/landuse/index.htm

Wiryono W, Kristiansen P, de Bruyn LL, Saprinurdin S, Nurliana S (2023) Ecosystem services provided by agroforestry home gardens in Bengkulu, Indonesia: smallholder utilization, biodiversity conservation, and carbon storage. Biodivers J Biol Divers 24(5):2657–2665

World Meteorological Organization (1992) International meteorological vocabulary, 2nd edn. WMO, Geneva, Switzerland

Yashmita-Ulman S, M., Kumar, A. et al (2021) Conservation of wildlife diversity in agroforestry systems in eastern Himalayan biodiversity hotspot. Proc Zool Soc 74:171–188. https://doi.org/10.1007/s12595-021-00361-x

Young A (1989) Agroforestry for soil conservation. CAB international, Wallingford. Retrieved November 21, 2023, from http://www.worldagroforestry.org/Units/Library/Books/PDFs/03_Agroforestry_for_soil_conservation.pdf

Zasada I, Weltin M, Zoll F, Benninger SL (2020) Home gardening practice in Pune (India), the role of communities, urban environment and the contribution to urban sustainability. Urban Ecosyst 23(2):403–417

Zhu X, Liu W, Chen J, Bruijnzeel LA, Mao Z, Yang X, Cardinael R, Meng FR, Sidle RC, Seitz S, Nair VD (2020) Reductions in water, soil and nutrient losses and pesticide pollution in agroforestry practices: a review of evidence and processes. Plant Soil 453:45–86

Zomer RJ, Trabucco A, Bossio DA, Verchot LV (2008) Climate change mitigation: a spatial analysis of global land suitability for clean development mechanism afforestation and reforestation. Agric Ecosyst Environ 126(1–2):67–80

Zomer RJ, Trabucco A, Coe R, Place F, Van Noordwijk M, Xu JC (2014) Trees on farms: an update and reanalysis of agroforestry's global extent and socio-ecological characteristics. Working Paper 179. (ed WACISARPD WP14064. PDF). Bogor, Indonesia, pp 54

Zomer RJ, Neufeldt H, Xu J, Ahrends A, Bossio D, Trabucco A, Van Noordwijk M, Wang M (2016) Global tree cover and biomass carbon on agricultural land: the contribution of agroforestry to global and national carbon budgets. Sci Rep 6(1):29987

Chapter 8
Climate Change Impacts on Crop Productivity and Food Security: An Overview

Aamir Raza, Muhammad Safdar, Muhammad Adnan Shahid, Ghulam Shabir, Aka Khil, Shahbaz Hussain, Mudassar Khan, Saeed Ur Rehman Aziz, Sidra Azam, Jaffar Sattar, Nalain E. Muhammad, Rana Muhammad Waqas, and Hafiz Muhammad Bilawal Akram

Abstract This chapter presents an in-depth examination of the effects of the changing climate on crop yield and food security. It starts by highlighting the importance of recognizing these implications in the context of shifting climate patterns and agriculture's vital role in global food systems. The chapter goes into the causes of climate change, such as human activity and emissions of greenhouse gases, as well as the observed changes in the patterns of precipitation and temperature. The chapter's primary focus is on the influence of the changing climate on

A. Raza · S. U. R. Aziz · S. Azam · N. E. Muhammad
Department of Irrigation and Drainage, University of Agriculture, Faisalabad, Pakistan

M. Safdar (✉) · M. Adnan Shahid
Department of Irrigation and Drainage, University of Agriculture, Faisalabad, Pakistan

Agricultural Remote Sensing Lab (ARSL), University of Agriculture, Faisalabad, Pakistan

G. Shabir
Department of Plant Breeding and Genetics, University of Agriculture, Faisalabad, Pakistan

A. Khil · H. M. B. Akram
Department of Agronomy, University of Agriculture, Faisalabad, Pakistan

S. Hussain
Department of Structures and Environmental Engineering, University of Agriculture, Faisalabad, Pakistan

M. Khan
Department of Agronomy, University of Trás-os-Montes e Alto Douro, Vila Real, Portugal

J. Sattar
Department of Irrigation and Drainage, University of Agriculture, Faisalabad, Pakistan

Department of Agricultural Engineering, Khwaja Fareed University of Engineering and Information Technology, Rahim Yar Khan, Pakistan

R. M. Waqas
Institute of Agricultural Extension, Education and Rural Development, University of Agriculture, Faisalabad, Pakistan

S. Kanga et al. (eds.), *Transforming Agricultural Management for a Sustainable Future*, World Sustainability Series, https://doi.org/10.1007/978-3-031-63430-7_8

crop productivity. It investigates how rising temperatures affect the process of photosynthesis, respiration, & phenological variations in crops. Variations in patterns of precipitation, including floods and droughts, as well as the prevalence of severe weather conditions like heatwaves and hurricanes, are often explored in terms of their impact on crop yields. The chapter also discusses the larger consequences for worldwide food security. It exposes food system vulnerabilities, particularly those affecting small-scale farmers and supply linkages. The nutrition, dietary alterations, and socioeconomic effects of decreased crop quality and interrupted supply are investigated. In addition, the chapter discusses adaptation and mitigation measures ranging from genetic resiliency to legislative initiatives. To address the problems posed by climate change on crop yield and food security, the abstract continues by emphasizing the necessity of coordinated efforts among diverse stakeholders, including researchers, policymakers, and communities. Continuous research, new solutions, and worldwide collaboration have all been acknowledged as critical components of addressing these complex concerns.

Keywords Climate change · Crop yield · Food security · Shifting climate patterns · Agriculture's role · Human activity · Adaptation and mitigation

8.1 Introduction

"Climate change" is the enduring alterations observed in the average temperature and weather patterns of a certain location over a prolonged period of time span (Ahmad et al. 2023). The alterations in question have been brought about by human activity, specifically the combustion of fossil fuels, the increase in greenhouse gases in Earth's atmosphere leads to longer heat retention, thereby increasing global temperatures (Anderson et al. 2016). The impact of climate change is already significant on crop output and food security. The acceleration of plant growth due to elevated temperatures has led to various challenges, including premature crop maturation and heightened vulnerability to pests and diseases (Balafoutis et al. 2017) Alterations in precipitation patterns are exerting an influence on agricultural productivity, whereby certain areas are witnessing a rise in arid conditions while others are encountering excessive inundation (Boretti and Rosa 2019).

The problems caused by climate change are complicated and have many different parts. They include:

- Increasing occurrence and intensity of heat waves, floods, and droughts
- Rising sea levels, which can flood coastal areas and force people to move.
- Precipitation patterns changing, which can hurt agricultural production.
- Loss of biodiversity, which can make food systems less reliable.
- Rising demand for food as the world's population keeps growing.

Crop production is how much food can be grown on a certain amount of land. Food security means that there is always safe, healthy food for everyone to eat. Food security and crop production are both important for the health and well-being of people (Singh et al. 2023). When crops don't grow well, food prices can go up, making it hard for people to buy food. This can cause people to go hungry, become malnourished, or even die.

This part will give an overview of how climate change affects the number of crops grown and how safe food is. It will talk about what causes climate change, how rain and temperature trends have changed, and how these changes have affected crop growth. The chapter will also discuss the threats that climate change poses to agricultural security, as well as the significance of mitigating its effects. The focus of this chapter will be on global trends, but it will also illustrate how climate change affects crop growth and food security in various regions of the globe (Sharma et al. 2023). At the conclusion of the chapter, we will discuss how everyone must collaborate to address climate change-related problems.

8.2 Climate Change and Its Drivers

8.2.1 What Is Climate Change?

The phenomena known as "climate change" is defined by long-lasting changes in the weather and temperature within a particular geographical region. Human activities are responsible for the alterations in the environment. The subject of this investigation concerns the utilization of fossil fuels, this causes greenhouse gases to be released into the atmosphere of Earth. These are responsible for the retaining of thermal energy, resulting in an elevation of the Earth's overall temperature. The worldwide occurrence of climate change is currently exerting discernible effects on individuals throughout many regions of the planet (Kumar et al. 2023). The rise in sea levels, the frequency of severe weather conditions, the shifts in agricultural output, and the reduction in biodiversity are all examples of the negative consequences (Debnath et al. 2023).

Climate change is substantiated by a wealth of scientific evidence. Based on the results of the Intergovernmental Panel on Climate Change (IPCC), a prominent worldwide organization responsible for evaluating climate change, it has been determined that human activities are very probable to have contributed to over 50% of the documented rise in global average surface temperature between the years 1951 and 2010 (Davin and de Noblet-Ducoudré 2010). Figure 8.1 shows the global impact of climate change, which is expected to worsen in the coming years, emphasizing the need for policies to reduce greenhouse gas emissions and mitigate its effects (Tripathi et al. 2023).

8.2.2 The Greenhouse Effect

The phenomenon known as the greenhouse effect pertains to the ability of certain gases present in the Earth's atmosphere to trap and retain solar radiation. Greenhouse

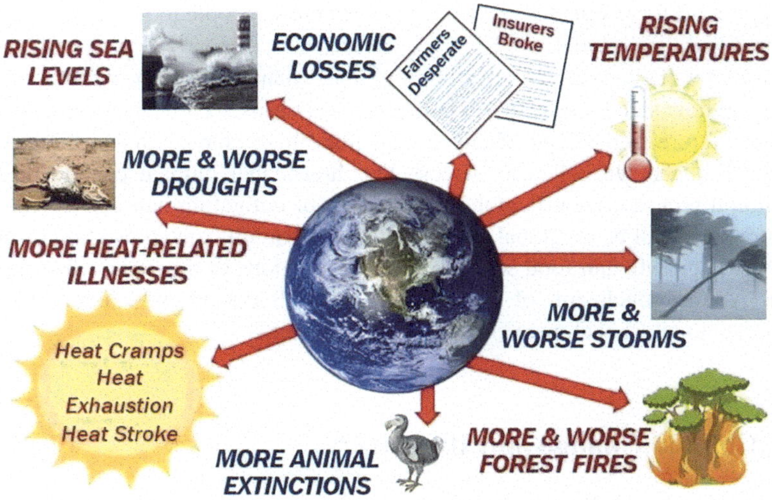

Fig. 8.1 The effects of the changing climate

gases facilitate the transmission of sunlight through the Earth's atmosphere, while concurrently capturing a portion of the thermal radiation emitted by the planet's surface. Global warming is caused by the retention of heat by greenhouse gases such as methane, nitrous oxide, carbon dioxide, and ozone on Earth's surface and atmosphere (Tomar et al. 2023). The occurrence of these gases may be traced back to natural sources; however, human activities have made a significant contribution to the growth in the quantities of these gases in the atmosphere of the Earth. Fossil fuel combustion, human activities like deforestation, agriculture, and industry, are the primary contributors to greenhouse gas emissions, with human activities such as deforestation, agriculture, and industry also playing a significant role (Kanga et al. 2022).

The greenhouse effect is one of several elements that affect life on Earth. It is essential to maintaining Earth's temperature for life. However, Earth's temperature rise is accelerating due to growing greenhouse gas concentrations. Sea levels are rising, extreme weather is getting worse, and agricultural output is changing due to climate change. Scientists agree that the greenhouse effect is real and caused by humans (Calzadilla et al. 2013). The IPCC, a leading global climate change authority, believes human activities contributed to over 50% of the documented rise in average global surface temperature between 1951 and 2010 (Ceccarelli et al. 2010). This conclusion was reached because the IPCC found that human activities caused over 50% of the global temperature rise.

Climate change is already having global repercussions, and the future of climate change is expected to worsen, necessitating urgent action to reduce greenhouse gas emissions and slow down its effects. Strategies such as reducing deforestation, increasing energy efficiency, and utilizing renewable energy sources are crucial for mitigating these negative impacts. Climate change adaptation strategies include growing drought-resistant crops, building seawalls in coastal areas, and improving early warning systems for extreme weather. Allocating funds to R&D for breakthrough technology can reduce greenhouse gas emissions and climate change.

8.2.3 Human Activities as Drivers of Climate Change

Climate change, an incontrovertible modification of Earth's climatic patterns, represents a significant worldwide challenge in the present day. While natural processes certainly play a crucial role in shaping the climate, it is predominantly the consequence of human activities that the climate has experienced rapid and substantial transformations in recent years. The alterations observed in the Earth's climate system can be attributed to significant factors such as changes in land usage and greenhouse gas emissions, which are mostly caused by human activity (Costello et al. 2009). This section delves into the scientific intricacies of the correlation between human activities and change in climate, along with the resulting consequences that follow. Including the effects of global variability in climate and its interplay with the agricultural system is depicted inside the Coupled Human-Natural System (CHNS) diagram. The CHNS diagram in Fig. 8.2 depicts the progression of global climate change, showcasing the augmented combustion of fossil fuels inside the human system as a causal factor leading to heightened atmospheric CO_2 levels.

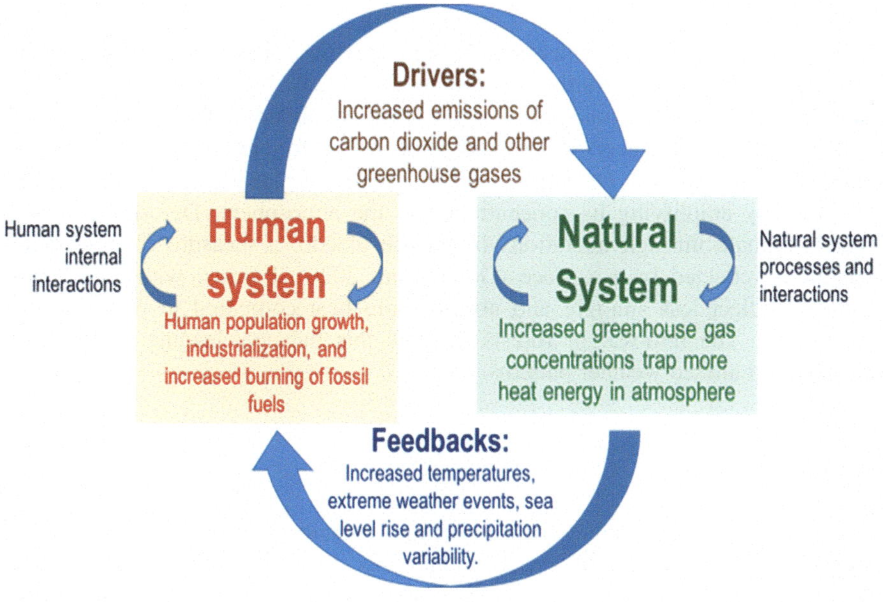

Fig. 8.2 Coupled Human-Natural System diagram illustrating the progression of global climate change

In the natural system, an increase in heat energy results in a greater amount of energy being retained. The feedback mechanism that impacts the human system entails an elevation in temperature, in addition to other consequential effects of climate change.

8.2.3.1 Drivers of Climate Change

Greenhouse Gas Emissions

Human activities, such as human activities, are primarily influenced by greenhouse gases (GHGs), which are substances that trap heat energy in the Earth's atmosphere, thereby creating the greenhouse effect (Daryanto et al. 2016). The natural occurrence under consideration plays a vital role in sustaining a favorable climate; nevertheless, the exponential rise in greenhouse gas emissions attributable to human actions has amplified this phenomenon, leading to a global warming trend. The combustion of fossil fuels, including coal, oil, and natural gas, for the purpose of generating energy and facilitating transportation, is the predominant origin of human-induced carbon dioxide (CO_2) emissions. As these fossil fuels combust, carbon stored for millions of years in the form of hydrocarbons is rapidly released into the atmosphere in the form of CO_2. This excess CO_2 amplifies the greenhouse effect, leading to an elevation in global temperatures, a phenomenon commonly referred to as the "enhanced greenhouse effect."

Deforestation and Land Use Changes

Beyond the realm of direct emissions, human activities related to land use change, particularly deforestation, further exacerbate the climate crisis (Feng et al. 2012). Forests, acting as carbon sinks, play a pivotal role in mitigating the accumulation of atmospheric CO_2. However, widespread deforestation for agriculture, urbanization, and industrial expansion has substantially reduced the planet's capacity to absorb CO_2, thereby amplifying its concentration in the atmosphere. Deforestation also impacts climate through alterations in surface albedo, a measure of how much sunlight is reflected by a surface (Gobu et al. 2020). Forests, with their dense canopies, reflect less sunlight and absorb more heat compared to open land or urban areas. The removal of forests disrupts this balance, leading to localized warming and altered weather patterns.

Industrial Production

Agriculture, livestock production, cement and chemicals production, and landfill disposal contribute to greenhouse gas emissions. Livestock production produces methane, a potent greenhouse gas. Industrial operations also contribute to the release of greenhouse gases into the atmosphere. The disposal of rubbish also accelerates methane emissions from landfills.

8.2.4 The Impacts of Climate Change on Temperature and Precipitation

The Earth's temperature is increasing, and precipitation patterns are becoming more erratic due to climate change (Hatfield and Prueger 2015). Changes in crop yields, livestock production, and water availability are having a significant effect on the agricultural sector. Rising temperatures are causing plants to grow faster, which can lead to problems such as crop maturity before harvest and increased susceptibility to pests and diseases. They are also causing changes in the phenology of plants, such as when they flower and fruit. This can disrupt pollination and seed production, which can further reduce crop yields. Changes in precipitation patterns make it harder to predict when and where it will rain. This makes it challenging for producers to plan crops and irrigation systems. In some regions, droughts are becoming more frequent and severe, whereas in others, inundation is increasing in frequency. The availability of water for livestock and other agricultural purposes is also being impacted by these changes.

The effects of climate change on crop yields differ by type of crop and region. Wheat and maize, for example, are more susceptible to heat and drought than other crops (Huber and Gulledge 2011). In many regions of the globe, higher temperatures and more erratic precipitation patterns are anticipated to reduce crop yields. Climate change has a significant effect on livestock production as well. Rising temperatures make it more difficult for livestock to maintain a calm body temperature, which can result in heat stress and decreased productivity. Changes in precipitation patterns make it more challenging to provide sufficient water for livestock. Additionally, the availability of water, which is critical to agricultural production, is negatively impacted by climate change. The amount of water that is available for use in irrigation and other processes is reduced when temperatures rise because more water is lost to the atmosphere through evaporation. Because of shifting patterns of precipitation, it is becoming increasingly difficult to forecast when and where water will be available. Climate change's impact on agricultural practices poses a significant food supply risk, necessitating simultaneous reduction of greenhouse gas emissions and adaptation to its effects.

8.3 Impact on Crop Productivity

8.3.1 Climate Change Affect Crop Growth and Development

Climate change affects agricultural growth in several ways. Rising temperatures accelerate plant growth, which can cause crop maturity before harvest and increased pest and disease risk (Ju et al. 2013). They are also changing plant phenology, including flowering and fruiting. This can hinder pollination and seed production, lowering agricultural productivity. Predicting rain is getting harder due to precipitation patterns changing. Farmers are finding it harder to organize crop output and irrigation systems. The frequency and intensity of droughts and flooding are increasing in some places. These changes are reducing water availability for cattle and other agricultural uses.

Climate change affects crop growth and maturation differently in several crops and locales. Heat and drought damage wheat and maize more than other crops. Rising temperatures and more unpredictable precipitation will lower crop production worldwide. Climate change affects agricultural growth by rising temperatures, precipitation patterns, and CO_2 levels (Karl et al. 2009). Climate change's effects on plant growth and maturation threaten food security.

8.3.2 Impact of Temperature on Crop Yields

Rising temperatures are having a significant impact on crop yields (Liu et al. 2021). In general, higher temperatures are expected to reduce crop yields, as plants are unable to grow and develop as efficiently. This is especially true for crops that are sensitive to heat, such as wheat and maize (Loboguerrero et al. 2019). As temperatures rise, plants are less able to photosynthesize, which is the process by which they convert sunlight into energy. This can lead to reduced crop yields. As temperatures rise, plants also respire more rapidly, which is the process by which they release energy. This can also lead to reduced crop yields. As temperatures rise, plants lose more water through transpiration. This can lead to water stress, which can damage plant tissues and reduce crop yields. Pests and diseases are more likely to thrive in warmer conditions. This can lead to increased crop losses.

Climate change is predicted to cause a 10% decrease in maize yields by the end of the century, as temperature impacts crop production differently based on crop and area (Lynas et al. 2021). But according to a Chinese study, climate change would cause rice harvests to rise by 10% by the end of the century (Mac Kinnon et al. 2018). Heat stress and drought are exacerbated by temperature, causing damage to plant tissues and decreased photosynthesis, while drought restricts water availability for irrigation. There are a number of things that can be done to mitigate the impact of temperature on crop yields, including, developing heat-tolerant crop varieties, improving irrigation systems, changing planting dates, using shade nets.

8.3.3 Impact of Precipitation Changes on Crop Yields

Precipitation, the amount of water that falls from the atmosphere, is crucial for crop cultivation as it provides hydration for plant growth. Changes in precipitation patterns can significantly impact agricultural productivity. Droughts can cause physiological strain in plants, reducing crop output. Floods can also cause physical damage and damage crops, reducing yields. The impact of precipitation changes on agricultural productivity varies depending on crop type and geographical location. For instance, crops grown in dry climates are more susceptible to drought, while those in flood-prone areas are more susceptible. Therefore, the impact of precipitation on agricultural productivity is multifaceted and varies depending on the specific crop type and location.

Climate change is worsening the effects of precipitation changes on crop yields, leading to increased evaporation and less precipitation, making droughts more severe. To mitigate this, improving irrigation systems, using drought-tolerant crop varieties, and managing pests and diseases are essential steps. It is crucial to take action to ensure sustainable agriculture and mitigate the negative effects of climate change.

The timing of precipitation is also important for crop production. For example, crops that need water during the flowering stage are more likely to be affected by drought if they do not receive enough rain during that time. The quality of precipitation can also affect crop yields. For example, crops that are sensitive to salt can be damaged by salty water. Climate change is expected to cause changes in precipitation patterns. Changes in weather patterns, including increased droughts, floods, and precipitation timings, are expected to significantly affect crop yields.

8.3.4 Impact of Extreme Weather Events on Crop Yields

The phenomenon of climate change is contributing to a rise in both the frequency and intensity of extreme weather occurrences, including but not limited to droughts, floods, and heat waves (Mbow et al. 2014). These events can have a substantial impact on agricultural yields because they can cause crop damage and make it difficult for farmers to cultivate crops (Kumar 2016). Droughts are extended periods of arid weather that can result in crop failure. In addition, they can cause water shortages, making it difficult for producers to irrigate their crops. Periods of heavy precipitation or snowmelt can produce flooding. In addition to damaging crops and infrastructure, flooding can impede farmers' access to their fields. Heat waves are extended periods of high temperatures that can burden crops and increase their susceptibility to pests and diseases. Additionally, heat surges can reduce crop yields. There is also a consistent rise in the yield of the main staple crops in tons per hectare. Since both overall food production as well as the yield of important staple crops tons per hectare have increased, the explanation for this may come from either scientific or management breakthroughs as shown in Fig. 8.3.

Extreme weather events affected the crop productivity with respect to crop and region. Arid-region crops are more sensitive to drought than humid-region crops. Flooded agricultural produce is also more subject to damage. Climate change exacerbates the negative effects of extreme weather on agricultural productivity. The rise in temperatures caused by climate change has increased evaporation and decreased precipitation. This is increasing drought frequency and severity. Various measures can alleviate harsh weather's negative effects on grain yields. These

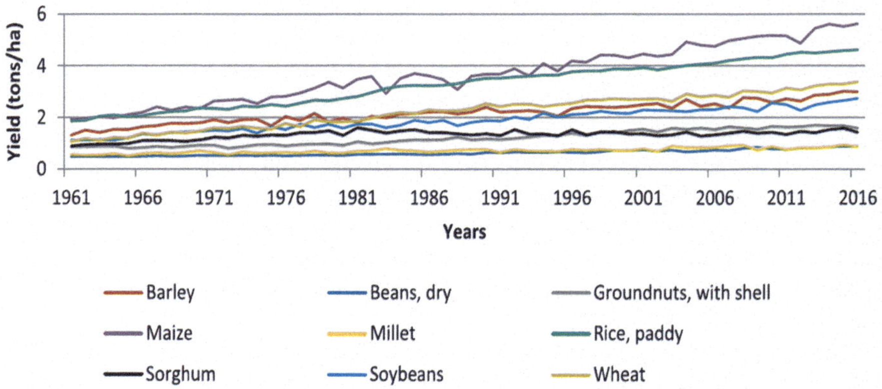

Fig. 8.3 Food production from major staple crops (Ton/Ha) 1961–2016. (Source: FAO Statistical Database 2016 crop statistics)

include improving irrigation, diversifying crop selection, adopting drought-resistant crops, and setting up early warning systems.

Additionally, extreme weather occurrences are becoming more severe. The potential repercussions for agriculture are likely to intensify, even without a change in extreme weather frequency. The duration of extreme weather events is also rising. This suggests that agricultural crops are exposed to extended harsh weather, making them more vulnerable. Extreme weather events are projected to grow with climate change. These events will significantly impair agricultural productivity.

8.4 Effects on Food Security

8.4.1 How Does Climate Change Affect Food Security?

Food security is seriously threatened by climate change (McMichael and Lindgren 2011). Already, agricultural yields, livestock production, and fisheries are being impacted. As the planet continues to warm, these effects are projected to deteriorate in the future. Crops are maturing earlier due to rising temperatures, which can cause issues such as crop maturity before harvest and increased susceptibility to pests and diseases. They are also altering the phenology of plants, such as the timing of flowering and fruiting. This can disrupt pollination and seed production, reducing crop yields further. Changes in precipitation patterns make it harder to predict when and where it will rain. This makes it challenging for producers to plan crops and irrigation systems. In some regions, droughts are becoming more frequent and severe, whereas in others, inundation is increasing in frequency. The availability of water for livestock and other agricultural purposes is also being impacted by these changes.

Additionally, increased CO_2 levels in the atmosphere impact crop growth and development. CO_2 is a plant nutrient, and increased CO_2 levels can stimulate photosynthesis and plant growth. However, the advantages of increased CO_2 are frequently outweighed by other factors, such as heat exhaustion and drought. Climate change is contributing to an escalation in the occurrence and intensity of extreme weather phenomena, including but not limited to droughts, floods, and heat waves. The occurrence of these events can exert a significant influence on agricultural production due to their potential to induce crop damage and impede farmers' cultivation efforts. Climate change is a significant factor in the reduction of biodiversity. This matter is of concern due to the indispensable role of biodiversity in the production of subsistence. Pollinators, such as bees, are of utmost importance for the process of crop pollination, while fish have a critical position within the food chain.

Climate change is posing a significant threat to global food availability, necessitating immediate measures to mitigate greenhouse gas emissions and adopt strategies to cope with its effects. The data presented represents the annual production of significant staple crops from 1961 to 2016, emphasizing the need for prompt implementation of these measures. Failure to act will result in significant consequences. Figure 8.4 illustrates the rise in the overall production of main staple crops over the years to demonstrate the consistent increase in food production over time. This may be due to the expansion of the growing area for these important staple crops, the outcome of higher yields per acre, or a combination of the two.

8.4.2 *The Vulnerability of Global Food Systems*

The extent to which global food systems are vulnerable to climate change is its vulnerability. Global agriculture is already suffering from climate change. Rising temperatures, changing precipitation patterns, and more intense weather are making

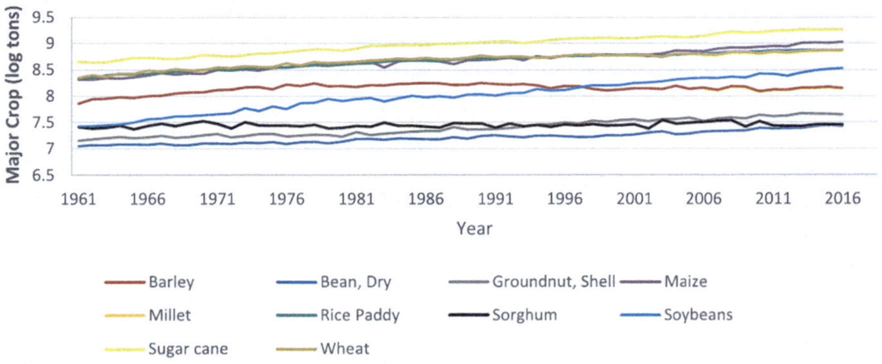

Fig. 8.4 Major staple crop production 1961–2016 (ton). (Source: FAO Statistical Database 2016 crop statistics)

food production harder. The world population will grow 9.4–10.2 billion by 2050 (Nelson et al. 2010). More people will need food, straining world food production. Individuals living in poverty are particularly susceptible to the adverse impacts of climate change and the issue of food insecurity. They are more likely to live in climate-affected areas and have fewer resources to mitigate it. Conflict can make food systems more vulnerable. Conflict can impair agriculture, transport, and distribution. Additionally, government policies might affect agriculture system vulnerability. Sustainable agriculture and climate change adaptation measures can strengthen food systems.

It is anticipated that climate change would result in a global reduction in agricultural yields. The increasing temperatures and altering precipitation patterns pose challenges to crop cultivation. Food prices may rise due to lower agricultural yields. Because food supply will fall, and demand will soar. Climate change will likely worsen food insecurity. Due to their heightened vulnerability to climate change and limited access to resources for mitigation and adaptation, those belonging to the lowest socioeconomic strata are disproportionately affected.

8.4.3 The Socioeconomic Implications of Climate

Climate change has extensive and intricate social implications. Climate change is expected to lead to decreased agricultural productivity in many regions across the globe. The cause of this phenomenon can be ascribed to the rise in temperatures and alterations in precipitation patterns, which pose challenges to the cultivation of crops. It is anticipated that there will be an increase in food prices due to a fall in agricultural productivity. This phenomenon can be attributed to the decrease in food supply coupled with the persistent increase in demand. The exacerbation of food insecurity is expected to occur due to the impacts of climate change. The heightened vulnerability of the world's most impoverished individuals to the impacts of climate change can be attributed to their limited capacity to cope with the resulting repercussions. It is anticipated that a substantial number of individuals will experience displacement as a result of the consequences arising from climate change. The challenges faced by human populations in specific places are attributed to the escalating sea levels, intensified weather phenomena, and other consequences associated with climate change. Climate change is predicted to increase heat stress, respiratory diseases, and waterborne illnesses, posing a significant threat to the global economy. This phenomenon can be attributed to the potential disruption of economic activity, inflationary effects on the prices of essential commodities such as food, and the subsequent negative impact on employment rates.

The socioeconomic ramifications of climate change pose a significant worldwide concern. Taking proactive measures to mitigate climate change and effectively adapt to its ramifications is of utmost importance. The socioeconomic impacts of climate change will differ based on geographical location and the specific impacts of the climate change. The impacts of rising sea levels are anticipated to be more pronounced in coastal regions, whilst the consequences of drought are expected to be more severe in arid areas.

8.5 Adaptation and Mitigation Strategies

8.5.1 How Can We Adapt to Climate Change?

The phenomenon of climate change is currently exerting a discernible impact on the Earth, with projections indicating that its effects will intensify in the future. In order to mitigate the deleterious effects of climate change, it is necessary that proactive actions be undertaken to facilitate adaptation. As the observable and widespread impacts of climate change continue to escalate, the imperative to effectively respond and adjust to these changes has emerged as an important imperative. Proactive measures are crucial to ensure the resilience and prosperity of communities, industries, and ecosystems due to climate change's changing patterns, intensifying catastrophic weather events, and ecosystem dynamics. A holistic strategy covering multiple sectors, scales, and disciplines is necessary for effective adaptation. The creation of a more resilient global environment, capable of withstanding the adversities presented by climate change, can be achieved through the integration of cutting-edge technologies, nature-oriented approaches, and community-led strategies. The crucial factor is not only in the role of individuals to navigate these changes, but also in our collective capacity to do so, thereby guaranteeing a sustainable and resilient future for subsequent generations. It is imperative to understand that the concept of adaptation lacks a universally applicable solution. The efficacy of adaptation strategies may vary based on the specific circumstances of individual regions or groups. In their study, (Wolfe et al. 2018) conducted an estimation of the potential consequences of climate change on food security, as well as the techniques employed for adaptation. They assess the likelihood of yield decline and increased food insecurity, while also identifying opportunities for adaptation at various levels. Table 8.1 presents adaptation strategies for agriculture to address environmental and climatic challenges at various levels. On-farm adaptations focus on improving fertilizer and water management, with advantages like better crop varieties and higher water productivity. Field-level adaptations include selecting appropriate crops, advanced irrigation technologies, and preventing flooding and erosion. Irrigation policies emphasize increasing rootzone moisture retention, reducing evaporation, and enhancing farm-level practices. At the river basin and national levels, strategies involve policy development, managing drought and flooding risks, and maintaining aquifer recharge. The study highlights the need for more research on measuring the implications of climate change on food security.

Table 8.1 Climate change adaptation in agriculture and food production

Adaptation	Major adaptation strategies	Advantages
On farm adaptation	Fertilizer management Water management on farm	(1) Better crop varieties; (2) Higher water productivity; (3) Agroforestry, (4) High-value crop diversification;
At irrigation system/ field level adaptation	Selection of crops and crop growth schedule. Farm irrigation technologies preventing of flooding and erosion Commercial farming Water distribution System efficiency Using calendars and cropping patterns Combined use of groundwater and surface water Measures of irrigation policy	Three strategies for optimizing agricultural water usage include (1) Increasing rootzone moisture retention capacity, (2) Reducing evaporation, and (3) Improving farm-level irrigation practices and adaptability.
At river basin & national levels adaptation	Policy for the agriculture sector Managing a drought risk Managing flooding events Maintaining of aquifer recharge	

Source: Adapted from Turral et al. (2008)

8.5.2 *Crop Breeding and Genetic Resilience*

Crop breeding and genetic resilience are essential climate change adaptation strategies (Pachauri et al. 2014). The effects of climate change on temperature, precipitation, and other environmental factors are making it difficult to cultivate crops. Crop breeding can contribute to the development of new, more resistant crop varieties. There are several methods for breeding climate-resistant crops. Selecting traits that are already present in the produce is one method. For instance, certain plants have innate resistance to drought or heat duress. Introducing new DNA into a crop is another method of breeding for climate resistance. This can be carried out via genetic engineering or conventional breeding techniques.

Through genetic engineering, scientists are able to insert genes from other organisms into a crop. This can be used to introduce genes that confer resistance to drought, thermal stress, and other challenges related to climate change. Traditional breeding techniques involve crossing two distinct crop varieties to produce a new variety with the desired characteristics. This can be a lengthy and difficult procedure, but it is the most natural method to introduce new genes into a crop.

Crop propagation is a crucial adaptation strategy to climate change. Crop breeding is a complex and difficult procedure, but it is a crucial tool for coping with climate change. By creating new crop varieties that are more tolerant of a changing environment, we can help ensure a future food supply that is sustainable.

8.5.3 *Precision Agriculture and Technology Adoption*

Precision agriculture is a farming management concept that optimizes the production and efficacy of agricultural practices through the use of information technology (Schimmelpfennig 2016). Using sensors, GPS, and other technologies to gather information about the soil, vegetation, and weather. This information is then used to make decisions regarding fertilizer application, irrigation, and sowing, among other things. Precision agriculture can assist producers in increasing crop yields, decreasing input costs, and conserving natural resources. It can also aid in adaptation to climate change by improving the efficiency and sustainability of agricultural practices.

Precision agriculture utilizes several technologies. Sensors are utilized to gather information about the soil, vegetation, and weather. This information can be gathered on the ground, in the air, or in space. GPS is used to monitor the location of crops and equipment. This information can be used to make decisions regarding fertilization and irrigation, for example. Computers are used for data storage and analysis. This information can be used to make crop management decisions. Software is used to automate and control equipment.

The discipline of precision agriculture is expanding rapidly. There are a number of obstacles that must be addressed, including the price of technology and the dearth of farmer training. Precision agriculture can contribute to a 20% increase in crop production. This is because it enables producers to more precisely target their inputs, such as fertilizer and irrigation. Precision agriculture can reduce input costs by as much as 10% (Wolfe et al. 2018). It enables farmers to utilize less fertilizer, water, and other inputs. By decreasing pollution and greenhouse gas emissions, precision agriculture can help improve environmental sustainability. Precision agriculture can contribute to a secure food supply in the future by enhancing the efficiency and sustainability of agricultural practices.

8.5.4 *Water Management and Irrigation Innovations*

Water management and irrigation innovations can be used to adapt to and mitigate climate change by making water use more efficient, reducing water pollution, and conserving water resources. For example, drip irrigation can help to conserve water by delivering water directly to the roots of plants, where it is most needed. Micro-irrigation can also help to improve water efficiency by using small emitters to deliver water to the soil. Soil moisture sensors can be used to measure the amount of water in the soil, which can help farmers to determine when and how much to irrigate. Weather forecasting can be used to predict rainfall and other weather events, which can help farmers to plan irrigation schedules and water conservation measures. Water harvesting can help to increase water availability during dry periods by collecting rainwater or other water for agricultural use. Water recycling can help to reduce water pollution and conserve water resources by treating and reusing wastewater for agricultural use. Water pricing can be used to encourage water conservation by charging higher prices for water, which can make people more likely to use water more efficiently.

These technologies can help provide a sustainable food supply in the future. Water management and irrigation innovations can help farmers adapt to climate change by improving water efficiency. This could reduce the effects of droughts and water scarcity. Water management and irrigation advances can reduce water pollution and GHG, mitigating climate change. This could slow climate change and its effects. It is important to recognize that water management and irrigation improvements do not solve climate change. However, these organizations can help us adapt to the mitigation of climate change.

8.5.5 Agroforestry and Sustainable Land Use Practices

Agroforestry and sustainable land use practices can effectively reduce and adapt to climate change by promoting sustainable land use practices. This can be achieved by the implementation of several strategies, such as the cultivation of trees and shrubs, the utilization of cover crops, and the adoption of conservation agricultural techniques. These measures have the potential to enhance soil quality, promote water conservation, and mitigate erosion. Additionally, they possess the capacity to effectively trap and sequester carbon within the soil, thereby contributing to the amelioration of climate change. Agroforestry has the potential to enhance soil quality through the provision of shade and shelter for crops, thereby mitigating erosion and promoting water penetration. The implementation of cover cropping has been shown to have a positive impact on soil quality through the dual mechanisms of weed suppression and organic matter provision. The implementation of conservation agriculture practices, such as reduced tillage and cover cropping, can contribute to the preservation of water resources.

The use of agroforestry techniques and sustainable land use practices holds the capacity to lessen the negative consequences of climate change and foster the development of a more sustainable trajectory for the future. Agroforestry and sustainable land use techniques ought not to be thought of as a magic bullet for combating climate change. Nevertheless, these entities possess the potential to assume a significant function in facilitating our ability to mitigate and adapt to the effects of climate change. It is essential to customize these processes to the unique circumstances prevailing in each respective locale. The absence of a universally applicable solution is evident. It is imperative to allocate resources towards research and development in order to enhance these processes. This will facilitate the enhancement of their effectiveness and efficiency. It is imperative to disseminate knowledge and provide education to farmers and other individuals involved in land utilization regarding these techniques. This measure will contribute to the broad adoption and widespread implementation of the initiatives. By implementing these measures, we may enhance the efficacy of agroforestry and sustainable land use practices as a means of adapting to and mitigating the impacts of climate change.

8.5.6 Policy Interventions and International Cooperation

Policy interventions and international cooperation are essential strategies for addressing and mitigating the impacts of climate change. These approaches involve collaborative efforts to formulate and execute policies, agreements, and partnerships that facilitate effective climate action. These endeavors can contribute to the mitigation of emissions, enhance adaptability to climate change, and foster the advancement of sustainable development. It is imperative for national governments to formulate comprehensive policies aimed at regulating emissions, allocating resources towards the development and implementation of renewable energy sources, and providing substantial support for research and development initiatives. International agreements have the capacity to establish specific objectives for the reduction of emissions and offer financial aid to underdeveloped nations. Public-private partnerships facilitate collaboration between governmental entities, commercial enterprises, and other relevant groups, enabling the pooling of resources and knowledge. Education and raising awareness have the potential to foster popular endorsement for climate action. The technology transfer process can facilitate the adoption of clean technologies in underdeveloped nations, while financial support significantly aids in their ability to adapt to and mitigate the negative impacts of climate change.

Implementing these strategies can mitigate the impacts of climate change and promote a more sustainable approach. It is imperative to customize policy interventions and foster international collaboration in accordance with the distinct requirements of certain countries or regions. The absence of a universally applicable solution is evident. It is imperative to cultivate public consciousness and garner widespread endorsement for initiatives aimed at addressing climate change. This will aid in guaranteeing the presence of the necessary political determination to undertake necessary measures. By implementing these measures, we can enhance the efficacy of policy interventions and international collaboration is an effort to lessen and adapt to the effects of climate change.

8.6 Future Projections

8.6.1 What Are the Projected Impacts of Climate Change?

The phenomenon of climate change is presently exerting a substantial influence on our world, with projections indicating that the severity of its repercussions will escalate in the coming years. The mechanism of glacial & ice sheet melting is responsible for the rise in sea levels; Fig. 8.5 below illustrates this shift in the global sea level. The phenomenon is resulting in the occurrence of coastal floods, erosion, as well as the intrusion of saltwater into freshwater aquifers.

The World Resources Institute predicts that increasing water use and rising temperatures will worsen water stress in agricultural areas by 2025. This is due to the growing demand for freshwater in agriculture and the melting of snowpack and glaciers, which are crucial water sources for agriculture. Climate change can disrupt rainfall patterns, leading to droughts and soil erosion. Warmer temperatures can also increase pest and disease pressure, potentially reducing agricultural productivity. Water stress occurs when water demand exceeds available supply, resulting in reduced yields, crop failures, and food shortages. The Institute stresses the need for urgent action on water management and climate change to ensure the sustainability of agriculture in the face of these challenges as shown in Fig. 8.6.

The frequency and intensity of extreme weather events, such as heat waves, droughts, floods, and storms, are increasing due to climate change. These incidents

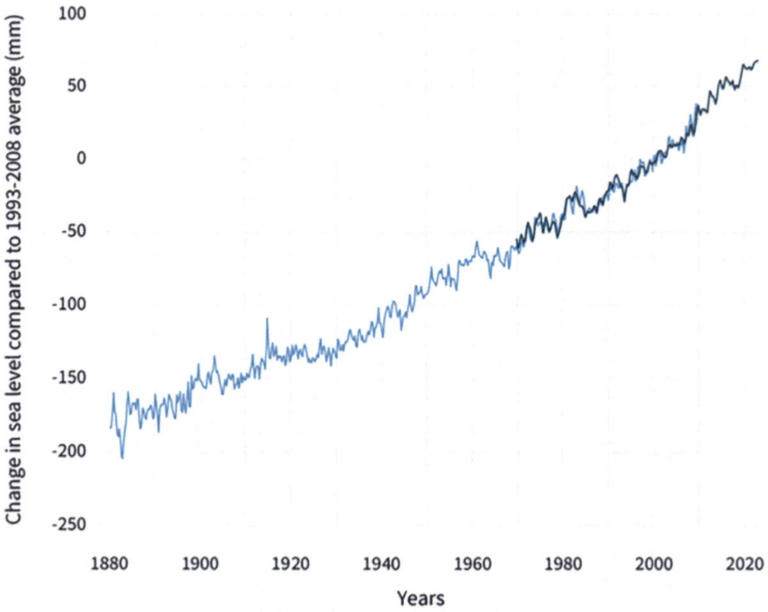

Fig. 8.5 Change in Global Sea level. (Source: NOAA Climate.gov)

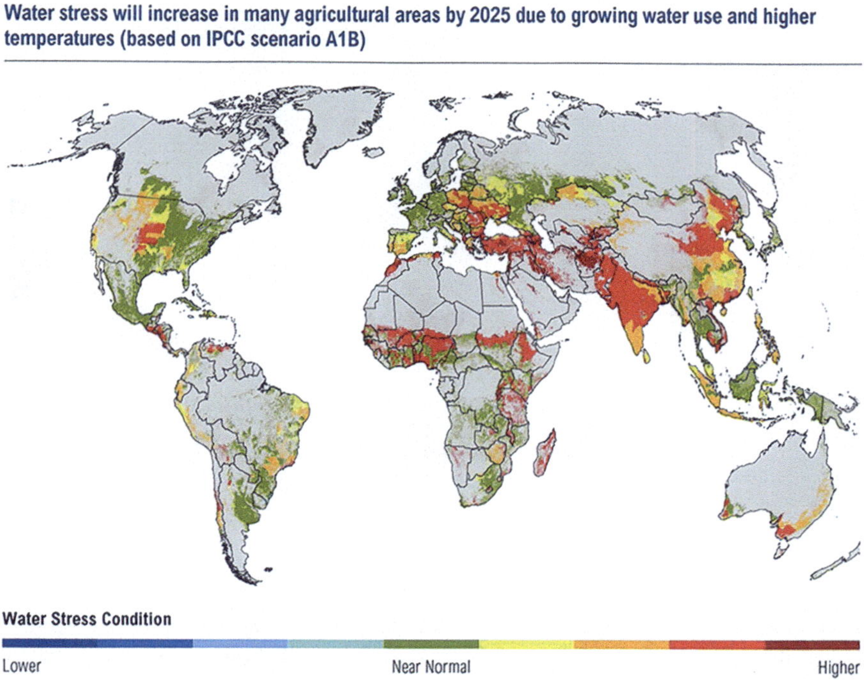

Fig. 8.6 Rising temperatures and increasing water demand are predicted to worsen water stress in various agricultural areas by 2025. (Source: World Resources Institute n.d.)

possess the capacity to lead to substantial devastation and notable human casualties. In a recent study undertaken by experts at University College London, it was found that the occurrence of climate change had the capacity to initiate sudden and extensive alterations. The agricultural sector in certain locations has seen difficulties in food production as a consequence of the influence of climate change. This phenomenon possesses the capacity to lead to a scarcity of food and subsequent increase in costs. Figure 8.7 depicts a cartographic representation demonstrating the projected declines in agricultural production by the year 2080, resulting from the impacts of climate change.

There is a connection between climate change and biodiversity loss, as the changing environmental conditions pose challenges for the adaptation of plants and animals. The potential consequences of this phenomenon may have a cascade impact on ecosystems, resulting in a reduced capacity to withstand and recover from additional stressors. Climate change is exacerbating the susceptibility of human populations to various health concerns, including but not limited to heat stress, respiratory ailments, and infections transmitted by vectors. Furthermore, it is exacerbating the challenge of obtaining clean water. The phenomenon of human displacement is being observed as a consequence of climate change, wherein individuals are compelled to relocate from their residences to regions that are

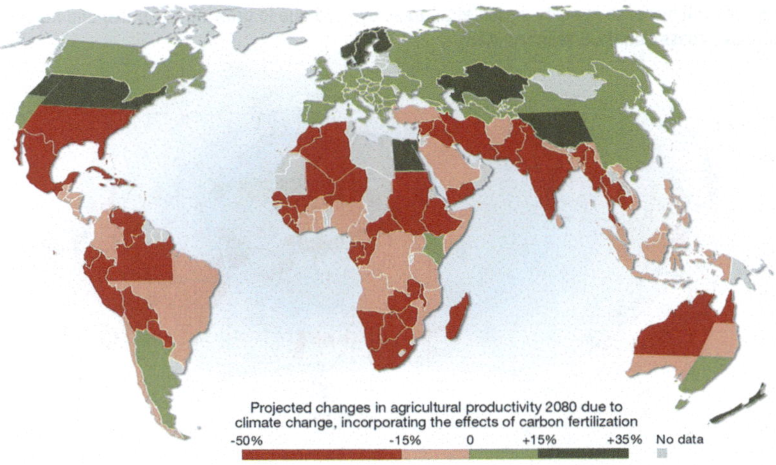

Fig. 8.7 2080 food production losses anticipated owing to climate change. (Source: UNEP)

comparatively less susceptible to the negative consequences of global warming. The aforementioned represent a subset of the anticipated repercussions of climate change. The specific implications of climate change will show variation contingent upon the geographical region and the particular climate change scenario. Climate change threatens a significant peril to both the Earth and its inhabitants. It is essential to put policies into place that will both lessen the effects of climate change and help us cope with its aftermath. It has been projected that crop yields in northern regions, like Canada or northern Europe, may grow by up to 100%, while crop yields in southern countries are predicted to decrease by up to 50% as shown in Fig. 8.8.

8.6.2 The Importance of Ongoing Research and Monitoring

The ongoing evaluation and observation of climate change are imperative in order to comprehend its fundamental origins and ramifications, as well as to formulate efficacious approaches for mitigating and adjusting to its effects. Research plays a pivotal role in enhancing our comprehension of the intricate mechanisms that underpin climate change, specifically in elucidating the methodologies via which greenhouse gases effectively retain heat within the Earth's atmosphere. Furthermore, it has the capability to facilitate the anticipation of the possible effects that climate change may have on various geographical areas and socioeconomic domains. Monitoring plays a crucial function in enabling the evaluation of advancements in climate change and the identification of developing trends and regions of heightened importance. Furthermore, it has the potential to contribute to the evaluation of the efficacy of mitigation and adaptation strategies.

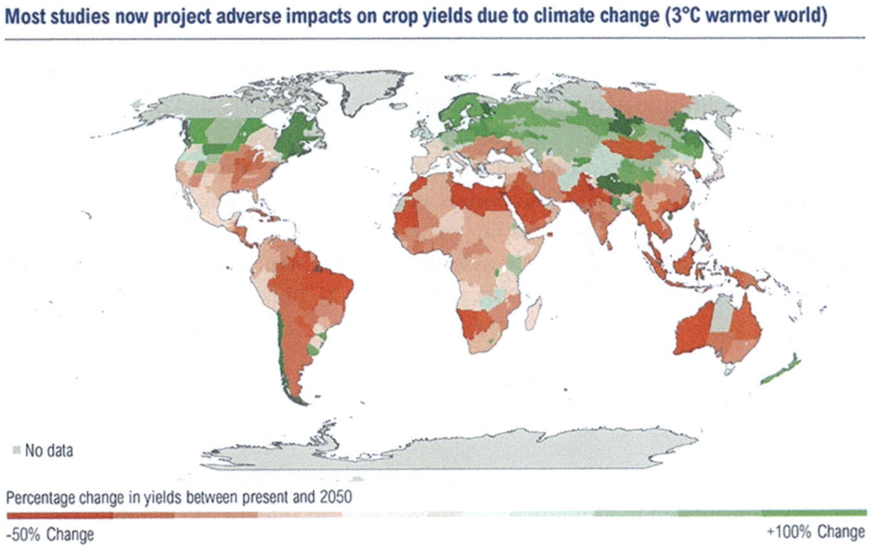

Fig. 8.8 Percentage variations in crop production projected by 2050 because of climate change effects. (Source: World Resources Institute n.d.)

Participating in research endeavors enhances one's comprehension of the underlying elements and consequences linked to climate change, concurrently fostering the formulation of enhanced approaches to mitigate and adapt to its effects. The practice of monitoring has a crucial role in identifying developing patterns and places with significant levels of activity, as well as in promptly disseminating warnings about the effects of climate change. Research and monitoring play a vital role in the advancement of more efficient strategies for reducing and reacting to diverse issues. The proposed remedies encompass a range of strategies, including the construction of seawall infrastructure and the decrease of greenhouse gas emissions, in order to safeguard coastal towns against the negative effects of rising sea levels. The significance of research and monitoring cannot be overstated in facilitating informed decision-making on strategies aimed at minimizing the impacts of climate change. This encompasses the allocation of resources towards investments in renewable energy and the advancement of drought-resistant agricultural development.

The significance of continuous study and monitoring is easily evident. By persisting in the allocation of resources towards these endeavors, we might enhance our comprehension and administration of the myriad issues associated with climate change. The challenges associated with continuous study and monitoring encompass various dimensions, including financial implications, complexity, and time commitment. Climate change is a multifaceted and persistent issue that necessitates a significant duration for the examination and assessment of research and monitoring initiatives.

The concept of political will refers to the determination and commitment exhibited by political actors, such as government authorities and policymakers, to actively pursue and implement actions. Numerous locations worldwide exhibit a deficiency in political determination when it comes to properly tackling the issue of climate change.

Notwithstanding the existence of these problems, it is imperative to persist in allocating resources towards continuous study and surveillance endeavors. By actively engaging in this undertaking, we can enhance our comprehension and capacity to tackle the complex issues associated with climate change, while fostering the development of a more sustainable future for both humanity and the Earth.

8.7 Conclusions

Climate change poses a significant threat to the planet and its inhabitants, with its effects already evident worldwide and expected to worsen in the near future. Human activity, particularly the burning of fossil fuels, is largely blamed for the phenomenon. The consequences include rising sea levels, intensifying extreme weather events, changes in agricultural productivity, biodiversity loss, and increased risks to human health. Implementing policies to minimize climate change and address its effects is crucial.

Climate change can be addressed through various strategies, including renewable energy sources, improving energy efficiency, and addressing deforestation. Funding for research and development is crucial for developing innovative technologies to reduce emissions and support climate adaptation efforts. Public awareness of climate change is essential, and political resolve is needed to implement policies. Climate-resilient infrastructure, such as seawalls and levees, can mitigate climate change's impacts. Agricultural practices that promote crop growth and livestock resilience to drought conditions can also help. Early warning technologies can improve efficiency, and aiding those relocated due to climate change is crucial. The urgency of addressing climate change and implementing necessary measures is paramount, as the current situation requires prompt action. The current situation requires immediate resolution and action.

Governments must set ambitious goals to reduce greenhouse gas emissions while allocating resources towards renewable energy and sustainable technology. Businesses should also reduce their emissions and adopt sustainable practices. Individuals can also contribute by reducing their reliance on cars, conserving energy, and reducing meat consumption. Everyone has a responsibility to contribute to climate change. Collaborative efforts can create a more sustainable future for humanity and the Earth. Fostering collaboration among sectors and transcending geographical boundaries is crucial for effective resolutions. Trust and collaboration among diverse stakeholders are also essential. By working together, a future characterized by sustainability can be built for all individuals.

References

Ahmad T, Gupta SK, Singh SK, Meraj G, Kumar P, Kanga S (2023) Unveiling nature's resilience: exploring vegetation dynamics during the COVID-19 era in Jharkhand, India, with the Google Earth Engine. Climate 11(9):187

Anderson TR, Hawkins E, Jones PD (2016) CO_2, the greenhouse effect and global warming: from the pioneering work of Arrhenius and Callendar to today's earth system models. Endeavour 40(3):178–187

Balafoutis A, Beck B, Fountas S, Vangeyte J, Van der Wal T, Soto I et al (2017) Precision agriculture technologies positively contributing to GHG emissions mitigation, farm productivity and economics. Sustain For 9(8):1339

Boretti A, Rosa L (2019) Reassessing the projections of the world water development report. NPJ Clean Water 2(1):15

Calzadilla A, Rehdanz K, Betts R, Falloon P, Wiltshire A, Tol RS (2013) Climate change impacts on global agriculture. Clim Chang 120:357–374

Ceccarelli S, Grando S, Maatougui M, Michael M, Slash M, Haghparast R et al (2010) Plant breeding and climate changes. J Agric Sci 148(6):627–637

Costello A, Abbas M, Allen A, Ball S, Bell S, Bellamy R et al (2009) Managing the health effects of climate change: lancet and University College London Institute for Global Health Commission. Lancet 373(9676):1693–1733

Daryanto S, Wang L, Jacinthe PA (2016) Global synthesis of drought effects on maize and wheat production. PLoS One 11(5):e0156362

Davin EL, de Noblet-Ducoudré N (2010) Climatic impact of global-scale deforestation: radiative versus nonradiative processes. J Clim 23(1):97–112

Debnath J, Sahariah D, Lahon D, Nath N, Chand K, Meraj G, Kumar P, Singh SK, Kanga S, Farooq M (2023) Assessing the impacts of current and future changes of the planforms of river Brahmaputra on its land use-land cover. Geosci Front 14(4):101557. Elsevier

FAO (Food and Agricultural Organization of the United Nations) Statistical Database, 2016. Crops. [Data file]. Retrieved from. http://www.fao.org/faostat/en/#data/QC

Feng S, Oppenheimer M, Schlenker W (2012) Climate change, crop yields, and internal migration in the United States, No. w17734. National Bureau of Economic Research, Cambridge, MA

Gobu R, Shiv A, Kumar CA, Basavaraj PS, Harish D, Adhikari S et al (2020) Accelerated crop breeding towards development of climate resilient varieties. Climate change and Indian Agriculture: Challenges and Adaptation Strategies. Not Available

Hatfield JL, Prueger JH (2015) Temperature extremes: effect on plant growth and development. Weather Climate Extremes 10:4–10

Huber DG, Gulledge J (2011) Extreme weather and climate change: understanding the link, managing the risk. Pew Center on Global Climate Change, Arlington

Ju H, van der Velde M, Lin E, Xiong W, Li Y (2013) The impacts of climate change on agricultural production systems in China. Clim Chang 120:313–324

Kanga S, Singh PK, Ohri A, Singh RS (2022) Development of environmental decision support system for sustainable smart cities in India. Environ Prog Sustain Energy 41:e13817

Karl TR, Melillo JM, Peterson TC (2009) Global climate change impacts in the United States: a state of knowledge report from the US Global Change Research Program. Cambridge University Press, Cambridge

Kumar M (2016) Impact of climate change on crop yield and role of model for achieving food security. Environ Monit Assess 188:1–14

Kumar P, Singh SK, Gupta SK, Kanga S, Meraj G (2023) Uncovering the hydro-meteorological drivers responsible for forest fires utilizing geospatial techniques. Theoretical and Applied Climatology. https://doi.org/10.1007/s00704-023-04497. Springer Nature

Liu W, Shao XF, Wu CH, Qiao P (2021) A systematic literature review on applications of information and communication technologies and blockchain technologies for precision agriculture development. J Clean Prod 298:126763

Loboguerrero AM, Campbell BM, Cooper PJ, Hansen JW, Rosenstock T, Wollenberg E (2019) Food and earth systems: priorities for climate change adaptation and mitigation for agriculture and food systems. Sustain For 11(5):1372

Lynas M, Houlton BZ, Perry S (2021) Greater than 99% consensus on human caused climate change in the peer-reviewed scientific literature. Environ Res Lett 16(11):114005

Mac Kinnon MA, Brouwer J, Samuelsen S (2018) The role of natural gas and its infrastructure in mitigating greenhouse gas emissions, improving regional air quality, and renewable resource integration. Prog Energy Combust Sci 64:62–92

Mbow C, Smith P, Skole D, Duguma L, Bustamante M (2014) Achieving mitigation and adaptation to climate change through sustainable agroforestry practices in Africa. Curr Opin Environ Sustain 6:8–14

McMichael AJ, Lindgren E (2011) Climate change: present and future risks to health, and necessary responses. J Intern Med 270(5):401–413

Nelson GC, Rosegrant MW, Palazzo A, Gray I, Ingersoll C, Robertson R et al (2010) Food security, farming, and climate change to 2050: scenarios, results, policy options, vol 172. International Food Policy Research Institute

Pachauri RK, Allen MR, Barros VR, Broome J, Cramer W, Christ R et al (2014) Climate change 2014: synthesis report. Contribution of Working Groups I, II and III to the fifth assessment report of the intergovernmental panel on climate change. Ipcc, p 151

Schimmelpfennig D (2016) Farm profits and adoption of precision agriculture (No. 1477-2016-121190)

Sharma M, Upadhyay RK, Tripathi G, Kishore N, Shakya A, Meraj G, Kanga S, Singh SK, Kumar P, Johnson BA, Thakur SN (2023) Assessing landslide susceptibility along India's National Highway 58: a comprehensive approach integrating remote sensing, GIS, and logistic regression analysis. Conservation 3(3):444–459

Singh V, Stevanović M, Jha CK, Beier F, Ghosh RK, Lotze-Campen H, Popp A (2023) Assessing policy options for sustainable water use in India's cereal production system. Environmental Research Letters 18(9):094073. https://doi.org/10.1088/1748-9326/acf9b6

Tomar P, Singh SK, Kanga S, Pattanaik A (2023) Water bodies mapping and monitoring using high-resolution satellite images. Sustain Agri Food Environ Res 11:100–116. https://doi.org/10.7770/safer-V11N1-art2673

Tripathi G, Shakya A, Upadhyay RK, Singh SK, Kanga S, Pandey SK (2023) Landslide susceptibility mapping of Tehri reservoir region using geospatial approach. In: Climate change adaptation, risk management and sustainable practices in the Himalaya. Springer, Cham, pp 135–156

Turral H, Burke J, Faures J-M (2008) Climate change, water and food security, FAO water reports 36. FAO, Rome

Wolfe DW, DeGaetano AT, Peck GM, Carey M, Ziska LH, Lea-Cox J et al (2018) Unique challenges and opportunities for northeastern US crop production in a changing climate. Clim Chang 146:231–245

World Resources Institute (n.d.) http://www.wri.org/blog/2013/12/global-foodchallenge-explained-18-graphics. Accessed 13 May 2018

Chapter 9
Nanotechnology- A Novel Tool to Enhance Agricultural Productivity and Quality of Life of Farmers

Alka Kataria, Archana Kumari, Ramgopal Dudwal, Surendra Singh, Vishnu D. Rajput, and Mukesh Kumar Jat

Abstract Nanotechnology has been proved as a substitute to the existing chemical fertiliser which has a lot of negative impact on the quality of crop as well as on the environment. Use of nanoparticles in adequate quantity can be used as biosensors, agrochemicals like nano pesticides, nano fertilizers, soil enhancers and stress tolerant contributing to increasing the crop productivity and ensuring food security. In this review paper updated uses of nanotechnology in all facets of agriculture are discussed and attention is also given to new guidelines proposed by the Government of India for safe use of nanotechnology in agriculture sector. Irrespective of so many advantages, nanoparticles are not being used widely in Indian agricultural sector. In order to increase farmers' acceptance of nanotechnology in agriculture, it is crucial to educate the farmers about this technology through trainings or workshops.

Keywords Agricultural sector · Crop productivity · Food security · Nanoparticles · Nanotechnology

---

A. Kataria (✉)
Department of Environmental Science, IIS (Deemed to be University), Jaipur, Rajasthan, India
e-mail: alka.kataria@iisuniv.ac.in

A. Kumari
Department of Home Science, IIS (Deemed to be University), Jaipur, Rajasthan, India

R. Dudwal · M. K. Jat
School of Agriculture, Suresh Gyan Vihar University, Jagatpura, Jaipur, Rajasthan, India

S. Singh
Department of Agronomy, College of Agriculture, Kotputli, Rajasthan, India

V. D. Rajput
Academy of Biology and Biotechnology, Southern Federal University, Rostov-on-Don, Russia

S. Kanga et al. (eds.), *Transforming Agricultural Management for a Sustainable Future*, World Sustainability Series, https://doi.org/10.1007/978-3-031-63430-7_9

9.1 Introduction

The use of pesticides and chemical fertilisers increased crop productivity as expected with the introduction of the Green Revolution in India in 1960s but there tremendous use, lack of understanding of how to apply the technology, market risks and ignoring the use of natural resources in a sustainable and efficient way, brought lots of negative consequences (Dudwal et al. 2016, 2023, 2024; Kataria et al. 2022) like increase in the level of NPK to almost double (8.2: 3.2: 1) than optimal ratio (4:2:1) (Chand and Pavithra 2015) contamination in the groundwater, eutrophication in surface water (Huang et al. 2017), removal of macro and micronutrients from the soil due to intensive farming (Adhikari and Ramana 2019) and disturbance in the soil structure (Guo et al. 2018). Not only this, the surplus of nitrogen in the applied fertilizers has been slowly affecting human health with increased prevalence of diseases like methemoglobinemia, gastric and bladder cancer (Iqbal 2019). These consequences are further aggravated by growing food demand due to the increase in population. A few studies claim that by 2050, food demand will increase by 70% and there will be 9.6 billion people on Earth (Bindraban et al. 2018). All these problems raised a growing concern among scientists and agriculturalists for developing new innovative types of slow-release fertilizers which can help in sustainable agriculture (Manikandan and Subramanian 2016). In this context, use of 'Nantechnology' has gained much popularity, which involves materials being practiced at the nanoscale size (1–100 nm) (Chaudhry et al. 2018). High surface area-to-volume ratio, small size, and distinctive optical characteristics have enabled widespread use of nanoparticles. Especially in agricultural sector like in crop protection, soil enhancement, stress tolerance, crop improvement and crop production (Joshi et al. 2019). India's government has also played a significant role in nanotechnology research by financing and creating scientific institutes to use nanotechnology for the benefit of humanity (Mishra et al. 2014). Few government agencies working for the advancement of nanotechnology in agricultural sector are Department of Biotechnology (DBT), Department of Science and Technology (DST), The Energy and Resources Institute (TERI), National Centre for Nano-Structured Materials (NCNSM), Nanoscience and Technology Initiative (NSTI).

9.1.1 Use of Nanoparticles in Agriculture

Nanotechnology has a lot of potential to change the existing fertiliser situation through the genetic development of plants with environmental safety and ecological sustainability (Mahanta et al. 2019). In the case of plants, nanotechnology boosts growth rate, dry weight, chlorophyll and photosynthesis rate etc. (Singh et al. 2017). As a result, these agriculturally useful nanoparticles produced, using nanotechnology may be used across the whole agriculture production system's value chain. The most frequently used nanoparticles in the agricultural industry are carbon nanoparticles (Single-Walled Carbon Nanotubes), metallic nanoparticles (Ag, Fe, ZnO, TiO_2), polymeric nanoparticles, magnetic nanoparticles (Iron Oxide nanoparticle) and dendritic nanoparticles (Bose 2021).

Although in India, the use of nanoparticles in the agricultural sector (in plant protection, nutrition and governance of farm practices) began in the early twenty-first century (Duhan et al. 2017), but still it is at the budding stage. The factors such as lack of awareness among farmers about the benefits of using nanoparticles in agriculture in terms of time, cost and lack of easy availability of nanoparticles in the market for agricultural products may be the few obstacles for wide use of nanoparticles in the agriculture sector. Lots of research, awareness campaigns and training of farmers are required so that they can fully utilise nanoparticles in the agricultural sector for the improvement of crop, to boost its production. It guarantees to strengthen crop quality and its productivity at every stage of agriculture that too in an environmental friendly manner.

Nanoparticles contain unique features that make them useful in a range of sectors including electronics, pharmacology, medicine, and agriculture. The application of nanotechnology in the agriculture sector can be done in precision farming, crop production, crop improvement and food industries (Fig. 9.1).

Precision Farming It is also called as prescription farming, soil specific farming, variable rate farming or site specific management (Wolf and Buttel 1996). It is defined as a management strategy that employs thorough, site-specific data to precisely control production inputs (Searcy 1997). This strategy is used to increase the output from the crops while reducing fertiliser, pesticide, and herbicide inputs by

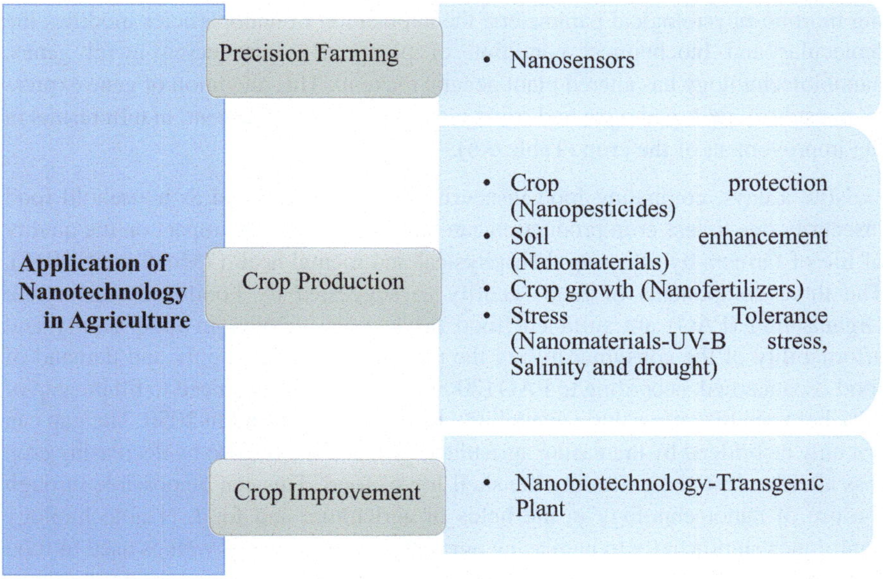

Fig. 9.1 Application of nanotechnology in agriculture. (Source: Hamad et al. 2020)

monitoring environmental factors and taking focused action (Subramanian and Tarafdar 2011). The detection of agro-contaminants, microbial load etc. as well as quick temperature, traceability, and humidity monitoring, are some of the sensor applications in agriculture. The list of nanoparticles that are used as nanosensors in precision farming to detect food contaminants are given in Table 9.1.

For real-time monitoring, autonomous sensors linked to a GPS system are being used in precision farming. It is one such solution to most of the challenges that farmers experience in cultivation since it monitors, detects, and alerts farmers, eventually increase agricultural output by providing exact data, allowing farmers to take better decisions (Subramanian and Tarafdar 2011). The key features of nanotechnology in precision farming are monitoring the condition of soil with nanosensors, slow release of agrochemicals, enhancement of shelf-life and quality of agricultural products, act as a growth regulator, and help in retention of soil water and minimise the agricultural waste that ultimately reduces environment pollution (Duhan et al. 2017).

Role of Nanoparticles in the Improvement of Crop Production Nanotechnology is one of the most important developments in contemporary agriculture since it allows the development of innovative agrochemical agents and delivery systems to improve soil health and increase crop production. The crop productivity is increased by developing, (i) nanoformulations of agrochemicals (nanopesticides/nanoherbicides and nanofertilizers) (Tables 9.2 and 9.3) (ii) green synthesized nanoparticles for agricultural use (stress tolerance- UV-B stress, Salinity and drought) (Table 9.4) (iii) nanoparticles for soil enhancement (Table 9.5).

Improvement of Quality of Crops Through Nanoparticles In addition to improving morpho-physiological parameters, the application of nanoparticles modifies the molecular and biochemical condition of plants. By introducing novel genes, nanobiotechnology has altered plant genetic makeup. This alteration of gene expression results in pest, pathogen and stress resistant crop development, in turn results in the improvement of the crop (Table 9.6).

Now a days, combating food insecurity is in great demand as household food insecurity due to less crop production can have a considerable impact on the quality of life of farmers by affecting their physical and mental health (Moafi et al. 2018). The three components of food security as suggested by Food and Agriculture Organisation (FAO) are sufficient food production, stability in food supply and affordability of the consumer to buy the product. As far as supply and demand of food is concerned, according to FAO (2006), there is an urgent need to fill the gap of 70% between the crop calorie availability in 2006 and demand in 2050. The gap can not only be bridged by increasing agriculture production; but also by decreasing crop loss after harvest and extending the shelf life of food. This can be possible through the use of nanotechnology in the fields of agriculture and food. Nanotechnology contributes significantly to increasing agriculture production as well as used in food processing, packaging, increasing shelf life of food and encapsulating nutrients (Sekhon 2014). The Canadian Institute of Food Security Research Fund and the

Table 9.1 Use of nanoparticles as nanosensors to detect food contaminants

Nanoparticles as nanosensors	Size (nm)	Food contaminant detected	Principle	Conclusion	References
Magnetic nanoparticles (MNP)	–	Organophosphate insecticides and nerve agents	A colour reaction is produced when MNPs with the enzymes acetylcholinesterase (AChE) and choline oxidase catalyse the oxidation of colorimetric substrates in the presence of H_2O_2	Reduced colour intensity is caused by decreased H_2O_2 production and inhibition of AChE enzymatic activity.	Liang et al. (2013)
Gold nanoparticles (AuNPs) coated carbon nanotube (CNT)	30–60	Triazophos as a contaminant in food samples and vegetables	The electrochemical response of triazophos was investigated with cyclic voltammetry	Triazophos' reductive capabilities were successfully enhanced by AuNPs electrode, which also eliminated pesticide interference.	Li et al. (2012)
Silver nanoparticles (AgNPs)	<50	Methyl-parathion pesticide	Scanning electron microscopy (SEM) and surface-enhanced Raman scattering (SERS) were used to characterise a constructed Ag nanosized monolayer film	Synthesized SERS-active substrate detected methyl-parathion, and detection limit was 10^{-7} M.	Zhang (2013)
Liposome (enzyme)-based nano-biosensor	300	Organophosphorus pesticides, dichlorvos and paraxon	Immobilization of pH-sensitive fluorescent pyranine within the liposomes for the optical transduction of enzyme activity	The signal of the liposome biosensors decreases, when paraoxon and dichlorvos concentrations decline to 10^{-10} M levels.	Vamvakaki and Chaniotakis (2007)
Gold nanoparticles (Au NPs)	13	Melamine	Melamine causes Au NPs to aggregate and change colour from wine-red to purple, providing a platform for quick melamine colorimetric detection.	The method uses naked eye inspection to detect melamine concentration in liquid form of milk and infant formulation (detection limits of 1.0 and 4.2 ppm), in 30 min, without the need of any sophisticated apparatus or extensive preprocessing.	Guo et al. (2010)

(continued)

Table 9.1 (continued)

Nanoparticles as nanosensors	Size (nm)	Food contaminant detected	Principle	Conclusion	References
Copper nanoparticles-modified gold electrode	–	Salicyclic acid (SA) level in oil seed rape when infected with pathogen *Sclerotinia sclerotiorum*	The anodic oxidation and electro-chemical behavior of SA was reported using the modified electrode	Using the modified electrode, the electrochemical behavior of salicyclic acid was substantially improved and detection of salicyclic acid levels in oilseed rape (infected with the fungal pathogen *Sclerotinia sclerotiorum*) was extremely comparable to that obtained by HPLC.	Wang et al. (2010)

Table 9.2 Use of nanoparticles for crop growth

Nanaparticles	Concentration	Crop plant	Effect	References
Zn Nano-fertilizer	10 mg/L	Pearl millet (*Pennisetum americanum* L.)	Grain yield was increased by 37.7%	Tarafdar et al. (2014)
ZnS and ZnO Nano particles	20–60 mg/L	Sunflower (*Helianthus Annuus* L.)	Enhanced the growth, yield and oil content of sunflower	Patel et al., (2019)
ZnO Nano particles	2 mg/L	Corn-SC704	Improved growth, leaf area and leaf dry weight	Taheri et al. (2015)
SiO_2 Nano-particles	100, 200, 300, and 400 mg/L	Common bean (*Phaseolus vulgaris)*	Improved the germination and growth of the test species	Alsaeedi et al. (2017)
TiO_2 nano particles	10, 100, 1000, 1200, 1500 and 1700 mg/L	Wheat (*Triticum aestivum* L.)	Accelerated the germination of wheat seeds and enhanced its vigor when added in proper concentration but if added in higher concentration TiO_2 nanoparticle has stimulatory effect	Mahmoodzadeh and Aghili (2014)
Ag Nano-particles	0, 25, 50, 75, 100, 125, 150 mg/L	Wheat variety NARC-2009	Improved the majority of wheat's growth and yield characteristics as well as NPK absorption and nutrient uptake effectiveness	Jhanzab et al. (2015)
Nanofertilizer super micro plus (NSMP)	100 and 150 ml/ 100 L	Wheat cultivar Ebba 99	Had a favourable impact on wheat yield and growth parameters	Al-Juthery et al. (2018)
Zeolite nanoparticle	–	Maize (NK-6240)	Nanozeourea treatment resulted in greater growth, yield, quality and nutrient absorption than conventional urea treatment	Manikandan and Subramanian (2016)

(continued)

Table 9.2 (continued)

Nanaparticles	Concentration	Crop plant	Effect	References
Nanoparticles-Fe_2O_3	Nanoparticles of 9 nm-at 2 mg/L, 20 mg/L, 50 mg/L and 18 nm-at 2 mg/L, 20 mg/L, 50 mg/L	Watermelon (*Citrullus lanatus*)	Proper concentration of nano-Fe_2O_3 improved seed growth and seedling germination, in addition to improving tolerance capacity to environmental challenges and physiological function of watermelon	Li et al. (2013)
Nanoparticles-$CoFe_2O_4$	0–1000 mg/L	Tomato plants (*Lycopersicon lycopersicum*)	Increased root and shoot length	López-Moreno et al. (2016)

institution Developmental Research Centre are working in collaboration with five countries including India for preservation of fruits by developing a nanotechnology smart hexanal-impregnated packaging with coating made of banana leaves and other agriculture waste. Improvement in production, storage, processing and quality of food remarkably contributes in increasing income of farmers. Ensuring food security along with increased farm income ultimately results in improvement in their work and life condition which means their quality of life (Arbuckle, & Kast, 2007).

9.1.2 New Guidelines Proposed by the Indian Government for Safe Use of Nanotechnology in Agriculture

Despite of numerous benefits of nanotechnology like lower costs and higher efficiency, agricultural nanoproducts have not been able to reach to the agricultural fields and market. The possible reasons could be lack of available data on correct methods of using nanoparticles and doses of nanofertilizers for the different plants, regulatory uncertainty, fast accumulation in the living tissues and public opinion for nanoproducts in the market (Adholeya and Singh 2020). A new Guidelines for Evaluation of Nano-Agri input Products (NAIP) and Nano Agriproducts (NAP) was proposed by India in 2020 with the aim to monitor the use and spread of NAIP and NAP, and to avoid the toxicity of nanoparticles in both humans and the environment. According to the guidelines, it is more difficult to follow the rules for evaluating nanoproducts in agriculture and food than to use the current techniques for assessing the safety of fertilisers, pesticides, and foods.

NAIPs and NAPs are submicroscopic products used in agricultural and associated sectors for a variety of purposes, including crop production, pest and disease

Table 9.3 Use of nanoparticles for crop protection

Nanoparticles	Concentration	Pest	Type of application	Effect	References
Cu nano-particles	0, 0.1, 0.25, 0.5, 0.75 and 1.0 mg/mL	*F. Solani*, *Neofusicoccum* sp., and *F. oxysporum*	Fungicide	Useful for controlling pathogenic fungi like *F. Solani*, *Neofusicoccum* sp., and *F. oxysporum* that affect crop and forest species.	Pariona et al. (2019)
S nano-particles (SNPs)	˜35, 55, 87 and 200 nm	*Fusarium solani* and *Venturia inaequalis*	Fungicide	SNP-based fungicides especially small sized particles (˜35 nm) protect the essential crops from a variety of fungal diseases.	Rao and Paria (2013)
Nanoparticle: TiO_2	–	Maize southern leaf spot, litchi downy blight and rice blast	Fungicide	Diseases like litchi downy blight, maize southern leaf spot and rice blast reduced in *L. chinensis* Sonn. (Litchi), *Oryza Sativa* L. (Rice) and *Zea Mays* L. (Maize) by 81.7%, 67.5% and 38.6%, respectively.	Lu et al. (2006)
Ag nanoparticles with leaf extract (aqueous) of *Aristolochia indica* L.	5, 10, 20, 30, 40 and 50 mL/L	*H. armigera*	Insecticide	With LC50 values of 365.72 and 309.98 mg/mL; 623.45 and 766.54 mg/mL, respectively, synthesised Ag NPs may have a 92.40% antifeedant and a 100% larvicidal effect against *H. armigera*. The least damaging effects on human HeLa cells were seen with the aqueous and synthetic Ag NPs extracts of *A. indica*.	Siva and Kumar (2015)
TiO_2 nanoparticles synthesized from the extract of *Moringa Oleifera*	75 mg/L	Red palm weevil (*Rhynchophorus Ferrugineus*)	Insecticide	The produced solution of natural extract with TiO_2 nanoparticles improved the mortality rate of the larvae stage of red palm weevil.	Al-Bartya and Hamzab (2015)

Table 9.4 Use of nanoparticles for stress tolerance

Nanoparticles	Concentration	Crop	Abiotic stress	Effect	References
SiO_2	0, 10, 50 and 100 mg/L	*Prunus mahaleb* (Mahaleb)	Drought	SNPs pre-treated seedlings maintained the root growth and had less effect on root dry mass accumulation (caused due to severe drought).	Ashkavand et al. (2018)
ZnO and Fe_3O_4 nanoparticles	30, 60 and 90 mg/L	*Moringa peregrine*	Salinity	Nanoparticles after foliar application, resulted in a significant decrease in the concentrations of Na^+ and Cl^- and an increase in N, P, K^+, Ca^{2+}, Mg^{2+}, Fe, Zn, total chlorophyll, carotenoids, crude protein, proline, and enzyme- and non-enzyme-dependent antioxidants.	Soliman et al. (2015)
Al_2O_3 nanoparticles	5, 50 and 500 mg/L	*Glycine max* L. cv. Enrei (Soyabean)	Flooding	The fresh weight of the plant and the length of the roots increased when seedlings were exposed to 50 ppm of Al_2O_3 nanoparticles, although cell death at the root tip reduced.	Mustafa et al. (2015)
Silicon nanoparticles	10 μm	Wheat (*Triticum aestivum*) seedlings	UV-B	SiNp may protect wheat seedlings from reactive oxygen species damage by activating an antioxidant defence mechanism triggered by NO.	Tripathi et al. (2017)

prevention, feeding, improving the efficiency of nutrients, growth management, agrochemical delivery and efficient packing and storage (Aggarwal 2019).

These guidelines would encourage the scientific community in India to use scientific approaches for assuring quality, evidence based-safety, and effectiveness of nanoproducts for its acceptance in the market and will significantly benefit the technologies like National Mission on Sustainable Agriculture and doubling farmers' income by 2022 (TERI 2020). These nano innovations will not only benefit the producers and suppliers but also farmers and consumers especially in improving

Table 9.5 Use of Nanoparticles for Soil enhancement

Nanoparticles	Pollutants	Effect	References
Nanoparticles-Fe_3O_4	2,4-dichlorophenoxyacetic acid (2,4-D)	Fe_3O_4 nanoparticles coupled with soil microbes resulted in a greater 2,4-D breakdown efficiency. It enhanced the microbial populations and enzyme activities in the soil.	Fang et al. (2012)
Zero-valent nano iron (ZVNI) particles	Chlorpyrifos	ZVNI (particle size 1–100 nm) destroyed 90% of chlorpyrifos in soil, however only 32% of chlorpyrifos was destroyed by mZVI (micro zero-valent iron).	Reddy et al. (2012)
Nanoparticles-Al_2O_3 and SiO_2	–	Increased nano additions of Al_2O_3 and SiO_2 resulted in a considerable increase in unconfined compressive strength.	Mir and Reddy (2021)
nZVI and Ni/Fe nanoparticles	Phenol	The BFN (*Bacillus fusiformis*) strain grew faster and improved biodegradation of phenol when exposed to both, nZVI and Ni/Fe nanoparticles.	Kuang et al. (2013)
Nano-hydroxyapatite (average particle size: 60 nm)	Heavy metals-Cadmium (Cd), Copper (Cu), Zinc (Zn) and Lead (Pb)	Cd, Cu and Zn concentrations in soil were dropped by 66.6–98.4% for one-year and 62.7–96.5% for three-year of treatment. In available Pb the concentration dropped by 61.3–74.2% in one-year and 69.2–74.4% in three-years of treatment.	Xing et al. (2016)
Iron nanoparticles (Fe-NPs)	Cadmium	The rate of photosynthesis, yield, and Fe concentrations in the crop plant were improved after the application of Fe-NPs in the polluted soil, however the Cd content in the tissues of wheat was decreased.	Adrees et al. (2020)

yields by optimising water and nutrient management, decreasing the nutrient losses in fertilization and the amount of sprayed chemicals.

The Energy and Resources Institute (TERI), with assistance from the Department of Biotechnology, Ministry of Science and Technology, Ministry of Agriculture and Farmers' Welfare, Food Safety and Standards Authority of India, and Ministry of Health and Family Welfare, has jointly prepared the guidelines after consulting with various stakeholders (Press Information Bureau (PIB), 2020). Thus, these guidelines are the result of concerted inter-ministerial efforts by numerous ministries and government entities.

Table 9.6 Use of nanoparticles for improvement of the crop

Nanoparticles	Concentration	Crop	Effect	References
TMA-OH (tetra methyl ammonium hydroxide) coated magnetic nanoparticles	10, 50, 100, 150, 200 and 250 microL/L	Maize	Small ferrofluid concentration (10–50 microl/L) increased chlorohyll a level up to 13% and nucleic acid level, up to 10%, in maize plantlets during their first days of life in turn increasing their productivity. On the other hand, higher concentration of ferrofluid decreased the chlorophyll a level and ratio of chlorophyll a/chlorophyll b by 35% in both cases.	Racuciu and Creanga (2007)
Cobalt nanoparticles	0.25, 2.5, and 5 mg/L	*Artemisia annua*	They observed 2.25-fold i.e. 113.35 mg/g dw, higher level of artemisinin production (2.25 fold i.e. 113.35 mg/g dw), than that of the control, after 24 h of treatment of 5 mg/L nano cobalt. On the other side, the expression levels of SQS and DBR2 genes were found to be decreased.	Ghasemi et al. (2015)
Silver nanoparticles and silver ions	5 mg/L AgNPs (20 nm) and 5 mg/L Ag^+	*Arabidopsis thaliana*	Out of the total upregulation of 286 genes and downregulation of 81 genes, three highly upregulated genes belonged to the thalianol biosynthetic pathway (involved in the plant defense system).	Kaveh et al. (2013)
Nano-anatase TiO_2	0.25% for 48 h at 10 °C	*Spinacia oleracea*	Exposure of nanoparticle increased RuBisCo activase activity and protein levels in spinach.	Gao et al. (2006)
Mn_3O_4 nanozymes	1 and 5 mg per plant for 7 days	Cucumber (*Cucumis sativus)*	At 1 mg per plant Mn_3O_4 NPs increased leaf photosynthetic pigment content, net photosynthesis, and biomass.	Lu et al. (2020)

9.2 Conclusion

The potential of nanotechnology in improving agricultural productivity and quality of life of farmers has already been evidenced in many researches. The review of various available literature related to use of nanotechnology in agriculture as bio-sensors, agrochemicals like nanopesticides, nanofertilizers, soil enhancers and stress tolerant shows that, when nanoparticles are used in appropriate quantity, it can improve the crop productivity and its quality. The increased production and enhanced quality of crops leads to economic use of time and effort on the part of farmers and hence contributes in the improvement of their quality of life. It is also concluded that the use of nanoparticles in many agricultural and food areas need more research into their effectiveness at field level. With new nanoproducts and processes, there are still a lot of possibilities to investigate, in the realm of agriculture. Irrespective of so many advantages, farmers and experts are hesitant to spend in this field, may be due to high starting production cost. Thus, arduous efforts from all stakeholders are needed to tackle these obstacles. Raising farmers' awareness of nanotechnology through trainings or workshops, conducting extensive research to determine the minimum concentration dose, and thoroughly comprehending the bio-accumulation capability of nanoparticles in field applications, are all important factors in influencing acceptance of nanotechnology in agriculture.

Acknowledgments The authors are grateful to IIS (Deemed to be University), Jaipur that provided internet services in order to download and analyse the online databases for writing this review article.

References

Adhikari T, Ramana S (2019) Nano fertilizer, its impact on crop growth and soil health. J Res PJTSAU 47(3):1–11

Adholeya A, Singh R (2020) Guidelines for evaluation of nano-based agri-input & food products in India. Science Reporter. December, pp 52–53. Retrieved from https://dbtindia.gov.in

Adrees M, Khan ZS, Ali S, Hafeez M, Khalid S, Ur Rehman MZ, Hussain A, Hussain K, Chatha SAS, Rizwan M (2020) Simultaneous mitigation of cadmium and drought stress in wheat by soil application of iron nanomaterials. Chemosphere 238:124681

Aggarwal M (2019, August 29) Government proposes new guidelines for safe use of nanotechnology in: agriculture. THE WIRE. Retrieved from https://thewire.in/agriculture/govt-proposes-new-guidelines-for-safe-use-of-nanotechnology-in-agriculture

Al-Bartya AM, Hamzab RZ (2015) Larvicidal, antioxidant activities and perturbation of Transminases activities of titanium dioxide nanomaterials synthesized using *Moringa oleifera* leaves extract against the red palm weevil (*Rhynchophorus ferrugineus*). Eur J Pharm Med Res 2(6):49–54

Al-Juthery HW, Habeeb KH, Altaee FJK, Al-Taey DK, Al-Tawaha ARM (2018) Effect of foliar application of different sources of nano-fertilizers on growth and yield of wheat. Biosci Res 15(4):3988–3997

Alsaeedi A, El-Ramady H, Alshaal TA, Almohsen M (2017) Enhancing seed germination and seedlings development of common bean (*Phaseolus vulgaris*) by SiO_2 nanomaterials. Egypt J Soil Sci 57(4):407–415

Arbuckle JG, Kast C (2007) Quality of life on the agricultural treadmill: individual and community determinants of farm family well-being. J Rural Soc Sci 27(1):84–113

Ashkavand P, Zarafshar M, Tabari M, Mirzaie J, Nikpour A, Bordbar SK, Struve D, Striker GG (2018) Application of SiO_2 nanomaterials as Pretreatment alleviates the impact of drought on the physiological performance of *Prunus Mahaleb* (Rosaceae). BolSoc Argent Bot 53(2): 207–219

Bindraban PS, Dimkpa CO, Angle S, Rabbinge R (2018) Unlocking the multiple public good services from balanced fertilizers. Food Secur 10(2):1–13

Bose P (2021, February 8) How Agricultural Nanotechnology Will Influence the Future of Farming Sustainability. Retrieved from https://www.azonano.com/article.aspx?ArticleID=5647

Chand R, Pavithra S (2015) Fertiliser use and imbalance in India, analysis of states. Econ Polit Wkly 50(44):98–104

Chaudhry N, Dwivedi S, Chaudhry V, Singh A, Saquib Q, Azam A, Mussarrat J (2018) Bio-inspired nanomaterials in agriculture and food: current status, forseen applications and challenges. Microb Pathog 123:196–200

Dudwal R, Jat BL, Babu SR, Kataria A (2016) Bio-efficacy of newer insecticides and biopesticides against Thrips, Thrips tabaci in garlic. Indian Journal of Ecology 43(1):141–144

Dudwal RG, Jakhar BL, Pathan ARK, Jan I, Kakralya BL, Dhaka SR, Kataria A, Yadav AK, Choudhary SK (2023) Dissipation kinetics, risk assessment, and waiting period of spiromesifen on chili fruits using gas chromatography–electron capture detector. Biomed Chromatogr 37: e5577. https://doi.org/10.1002/bmc.5577

Dudwal R, Jakhar BL, Pathan ARK, Kataria A, Dhaka SR, Jan I, Sayyed RZ, Khan A, Wong LS, Kumarasamy V, Gupta G, Subramaniyan V, Malik NA (2024) Impact of different decontamination methods on the reduction of spiromesifen residue in chilli fruits. Heliyon 10(9):e30065. https://doi.org/10.1016/j.heliyon.2024.e30065

Duhan JS, Kumar R, Kumar N, Kaur P, Nehra K, Duhan S (2017) Nanotechnology: the new perspective in precision agriculture. Biotechnol Rep 15:11–23

Fang G, Si Y, Tian C, Zhang G, Zhou D (2012) Degradation of 2,4-D in soils by Fe_3O_4 nanomaterials combined with stimulating indigenous microbes. Environ Sci Pollut Res Int 19(3):784–793

FAO (2006) World agriculture: towards 2030/2050—interim report. Rome. http://www.fao.org/fileadmin/user_upload/esag/docs/Interim_report_AT2050web.pdf

Gao F, Hong F, Liu C et al (2006) Mechanism of nano-anatase TiO_2 on promoting photosynthetic carbon reaction of spinach. Biol Trace Elem Res 111:239–253. https://doi.org/10.1385/BTER:111:1:239

Ghasemi B, Hosseini R, Nayeri FD (2015) Effects of cobalt nanomaterials on artemisinin production and gene expression in *Artemisia annua*. Turk J Bot 39(5):769–777

Guo L, Zhong J, Wu J, Fu F, Chen G, Zheng X, Lin S (2010) Visual detection of melamine in milk products by label-free gold nanomaterials. Talanta 82(5):1654–1658

Guo H, White JC, Wang Z, Baoshan X (2018) Nano enabled fertilizers to control the release and use efficiency of nutrients. CurrOpin Environ Sci Health 6:77–83

Hamad HT, Al-Sharify ZT, Al-Najjar SZ, Zainab AGA (2020) Review on nanotechnology and its applications on fluid flow in agriculture and water resources. Mater Sci Eng 870(1):012038

Huang J, Xu C, Ridoutt BG, Wang X, Ren P (2017) Nitrogen and phosphorus losses and eutrophication potential associated with fertilizer application to cropland in China. J Clean Prod 159:171–179

Iqbal MA (2019) Nano-fertilizers for sustainable crop production under changing climate: a global perspective. In: Hasanuzzaman M et al (eds) Sustainable crop production. Intechopen, London, pp 1–13. Retrieved from https://www.intechopen.com/chapters/68927

Jhanzab HM, Razzaq A, Jilani G, Rehman A, Hafeez A, Yasmeen F (2015) Silver nano-particles enhance the growth, yield and nutrient use efficiency of wheat. Int J Agron Agri Res 7(1):15–22

Joshi H, Choudhary P, Mundra S (2019) Future prospects of nanotechnology in agriculture. Int J Chem Stud 7(2):957–963

Kataria A, Sharma J, Jhamaria C (2022) A review on biofertilizers with special reference to liquid biofertilizers. Indian J Nat Sci 13(73):45525–45537

Kaveh R, Li YS, Ranjbar S, Tehrani R, Brueck CL, Van Aken B (2013) Changes in Arabidopsis thaliana gene expression in response to silver nanoparticles and silver ions. Environ Sci Technol 47:10637–10644. https://doi.org/10.1021/es402209w

Kuang Y, Zhou Y, Chen Z, Megharaj M, Naidu R (2013) Impact of Fe and Ni/Fe nanomaterials on biodegradation of phenol by the strain Bacillus fusiformis (BFN) at various pH values. Bioresour Technol 136:588–594

Li H, Xie C, Li S, Xu K (2012) Electropolymerized molecular imprinting on gold nanoparticle-carbon nanotube modified electrode for electrochemical detection of triazophos. Colloids Surf B: Biointerfaces 89:175–181

Li J, Chang PR, Huang J, Wang Y, Yuan H, Ren H (2013) Physiological effects of magnetic iron oxide nanomaterials towards watermelon. J Nanosci Nanotechnol 13(8):5561–5567

Liang M, Fan K, Pan Y, Jiang H, Wang F, Yang D, Lu D, Feng J, Zhao J, Yang L (2013) Fe_3O_4 magnetic nanoparticle peroxidase mimetic-based colorimetric assay for the rapid detection of organophosphorus pesticide and nerve agent. Anal Chem 85(1):308–312

López-Moreno ML, Avilés LL, Pérez NG, Irizarry BÁ, Perales O, Cedeno-Mattei Y, Román F (2016) Effect of cobalt ferrite ($CoFe_2O_4$) nanomaterials on the growth and development of *Lycopersicon lycopersicum* (tomato plants). Sci Total Environ 550:45–52

Lu JW, Li FB, Guo T, Lin LW, Hou MF, Liu TX (2006) TiO_2 photocatalytic antifungal technique for crops diseases control. J Environ Sci 18(2):397–401

Lu L, Huang M, Huang Y, Corvini PF-X, Ji R, Zhao L (2020) Mn_3O_4 nanozymes boost endogenous antioxidant metabolites in cucumber (*Cucumis sativus*) plant and enhance resistance to salinity stress. Environ Sci Nano 7:1692–1703. https://doi.org/10.1039/D0EN00214C

Mahanta N, Dambale A, Rajkhowa M (2019) Nutrient use efficiency through nano fertilizers. Int J Chem Stud 7(3):2839–2842

Mahmoodzadeh H, Aghili R (2014) Effect on germination and early growth characteristics in wheat plants (*Triticum aestivum* L.) seeds exposed to TiO_2 nanomaterials. J Chem Health Risks 4(1): 29–36

Manikandan A, Subramanian S (2016) Evaluation of zeolite based nitrogen nano-fertilizers on maize growth, yield and quality on Inceptisols and Alfisols. Int J Plant Soil Sci 9(4):1–9

Mir BA, Reddy SH (2021) Enhancement in shear strength characteristics of soft soil by using nanomaterials. In: Reddy KR et al (eds) Sustainable environment and infrastructure, lecture notes in civil engineering 90. Springer, Cham, pp 421–435

Mishra S, Singh A, Keswani C, Singh HB (2014) Nanotechnology: exploring potential application in agriculture and its opportunities and constraints. Biotech Today 4(1):9–14

Moafi F, Kazemi F, Siboni FS, Alimoradi Z (2018) The relationship between food security and quality of life among pregnant women. BMC Pregnancy Childbirth 18(1):1–9

Mustafa G, Sakata K, Komatsu S (2015) Proteomic analysis of flooded soybean root exposed to aluminum oxide nanomaterials. J Proteome 128:280–297

Pariona N, Mtz-Enriquez AI, Sánchez-Rangel D, Carrión G, Paraguay-Delgado F, Rosas-Saito G (2019) Green-synthesized copper nanomaterials as a potential antifungal against plant pathogens. RSC Adv 9(33):18835–18843

Press Information Bureau (2020, July 7) Guidelines for evaluation of nano-based agri-input and food products in India [Press release]. Ministry of Science and Technology. Retrieved from https://pib.gov.in/PressReleasePage.aspx?PRID=1637011

Racuciu M, Creanga D (2007) TMA-OH coated magnetic nanoparticles internalized in vegetal tissue. Rom J Phys 52(3/4):395

Rao KJ, Paria S (2013) Use of sulfur nanomaterials as a green pesticide on *Fusarium solani* and *Venturia inaequalis* phytopathogens. RSC Adv 3(26):10471–10478

Reddy AVB, Madhavi V, Reddy KG, Madhavi G (2012) Remediation of Chlorpyrifos-contaminated soils by laboratory-synthesized zero-valent nano iron particles: effect of pH and aluminium salts. J Chem 2013:1–7

Searcy SW (1997) Precision farming: a new approach to crop management. Texas Agricultural Extension Service. USA. The Texas A&M University System, College Station

Sekhon BS (2014) Nanotechnology in agri-food production: an overview. Nanotechnol Sci Appl 7: 31–53

Singh MD, Gautam C, Patidar OP, Meena HM, Prakasha G, Vishwajith (2017) Nano-fertilizers is a new way to increase nutrients use efficiency in crop production. Int J Agri Sci 9(7):3831–3833

Siva C, Kumar MS (2015) Pesticidal activities of ecofriendly synthesized silver particles using Aristolochia indica extract against Helicoverpa armigera Hubner (Lepidoptera: Noctuidae). Int J Adv Sci Tech Res 2:197–226

Soliman AS, El-feky SA, Darwish E (2015) Alleviation of salt stress on Moringa peregrine using foliar application of nanofertilizers. J Horticult For 7(2):36–47

Subramanian KS, Tarafdar JC (2011) Prospects of nanotechnology in Indian farming. Indian J Agric Sci 81(10):887–893

Taheri M, Qarache HA, Qarache AA, Yoosefi M (2015) The effects of zinc-oxide nanomaterials on growth parameters of corn (SC704). STEM Fellowship J 1(2):17–20

Tarafdar JC, Raliya R, Mahawar H, Rathore I (2014) Development of zinc nanofertilizer to enhance crop production in Pearl Millet (Pennisetum americanum). Agric Res 3(3):257–262

TERI Web Desk (2020, August 6) Guidelines for evaluation of nano-based agri-input and food products in India. Teri-The Energy and Resources Institute. Retrieved from https://www.teriin.org/policy-brief/guidelines-evaluation-nano-based-agri-input-and-food

Tripathi DK, Singh S, Singh VP, Prasad SM, Dubey NK, Chauhan DK (2017) Silicon nanomaterials more effectively alleviated UV-B stress than silicon in wheat (*Triticum aestivum*) seedlings. Plant Physiol Biochem 110:70–81

Vamvakaki V, Chaniotakis NA (2007) Pesticide detection with a liposome-based nano-biosensor. Biosens Bioelectron 22(12):2848–2853

Wang Z, Wei F, Liu SY, Xu Q, Huang JY, Dong XY, Yu JH, Yang Q, Zhao YD, Chen H (2010) Electrocatalytic oxidation of phytohormone salicylic acid at copper nanomaterials-modified gold electrode and its detection in oilseed rape infected with fungal pathogen Sclerotinia sclerotiorum. Talanta 80(3):1277–1281

Wolf SA, Buttel FH (1996) The political economy of precision farming. Am J Agric Econ 78(5): 1269–1274

Xing JF, Cang L, Li QGE, Zhou DM (2016) Long term stability of immobilizing remediation of a heavy metal contaminated soil with nano-hydroxyapatite. Agro Environ Sci 35:1271–1277

Zhang L (2013) Self-assembly Ag nanoparticle monolayer film as SERS substrate for pesticide detection. Appl Surf Sci 270:292–294

Chapter 10
Innovative Soil Management Technologies for a Changing Climate

Shazia M. Jamal, Rajesh Kumar Yadav, and Jaya Dayal

Abstract The chapter "Innovative Soil Management Technologies for a Changing Climate" discusses the impact of climate change on soil health and presents advanced technologies and practices for sustainable soil management. Key areas covered include carbon sequestration, soil remediation, and the use of microbial amendments to enhance soil fertility and reduce greenhouse gas emissions. The chapter also explores the potential of bioremediation and phytoremediation for soil restoration, alongside innovative approaches like nanoremediation and electrokinetic remediation to combat soil contamination. Technological advancements such as remote sensing, AI-driven irrigation platforms, and the development of artificial soils for urban agriculture are examined for their roles in improving soil management. Additionally, strategies like controlled traffic farming and the use of autonomous vehicles to prevent soil compaction are highlighted. The integration of biosensors and AI for real-time irrigation management is discussed as a method to optimize water usage and crop health. The chapter concludes that sustainable soil management is crucial for mitigating the adverse effects of climate change on agricultural productivity and environmental health. Implementing these innovative practices can significantly contribute to food security, soil conservation, and overall ecosystem resilience.

Keywords Climate change · Soil management · Carbon sequestration · Remediation · Microbial · Bioremediation · AI irrigation

10.1 Introduction of Climate Change

The term "climate change" describes alterations to the average atmospheric condition, encompassing factors like temperature, humidity, precipitation, cloud cover, and wind patterns, as well as variations in the frequency or intensity of these

S. M. Jamal (✉) · R. K. Yadav · J. Dayal
S.S. Jain Subodh PG College, Jaipur, Rajasthan, India

S. Kanga et al. (eds.), *Transforming Agricultural Management for a Sustainable Future*, World Sustainability Series, https://doi.org/10.1007/978-3-031-63430-7_10

conditions. These changes are attributed to a combination of natural reasons like the orbit of earth's revolution, volcanic activity, crustal movements, sea water temperature, ice cap distribution, westerly waves and atmospheric waves as well as anthropogenic factors such as a rise in greenhouse gas concentrations (Singh et al. 2023). Climate change is the result of an average increase in global temperature, also known as "global warming". It is anticipated that this global trend would significantly alter the global environment in the future (Ahmad et al. 2023).

Throughout Earth's history, the climate has fluctuated in cycles that last for incredibly long times. This is a natural process. In the recent geological history of Earth, the ice age cycles, which consist of cold glacial periods followed by shorter warm periods, are the largest global-scale climate change (Choudhury et al. 2023). These natural cycles have been recurring roughly every 100,000 years for the past several. They are primarily driven by modest changes in Earth's orbit, which affect the way the Sun's energy is distributed on Earth with latitude and season (Debnath et al. 2023). These orbital alterations have been very small during the past few centuries, and they are not large enough to affect the entire Earth orbit of the Earth's revolution or to account for the observed amount of temperature rise since the Industrial Revolution (Climate Change Evidence and Causes 2020).

These days, "climate change" is the term most often used to describe the remarkably fast changes in climate that we have experienced over the last 50 years or so. Global scientists from all over the world constitute the Intergovernmental Panel on Climate Change (IPCC), which was founded by the World Meteorological Organization and the United Nations Environment Programme (UNEP) with the mission of evaluating the scientific evidence related to climate change and global warming. They have compared climate model simulations of the consequences of greenhouse gas (GHG) emissions to historical climate changes, as well as natural factors such as solar and volcanic activity. They concluded that these changes are not the product of prolonged natural climatic cycles (Kumar et al. 2023). Rather, human activities and global warming by the increase of greenhouse gases are the primary causes of them (Fig. 10.1).

Causes of Climate Change Since the beginnings of the Industrial Revolution over 200 years ago, human activities have introduced massive amounts of greenhouse gases (GHG) to the Earth's atmosphere. These GHGs work like a greenhouse to retain the sun's energy and heat rather than allowing it to bounce back into space. The earth's temperature rises beyond the range of normal variability when the concentration of greenhouse gases is too high, trapping too much heat. There are numerous GHG, and each one has a unique half-life in the atmosphere as well as a varied capacity to trap heat (referred to as its "global warming potential"). The GHG most responsible for global warming to date is carbon dioxide (CO_2). The incomplete combustion of fossil fuels, such as coal, oil, and gas, which are used in the manufacturing of electricity, transportation, and industrial activities, releases the majority of CO_2. More than 80% of the CO_2 released into the atmosphere is attributed to these three activities combined (Singh et al. 2023).

Black carbon, methane, nitrous oxide, and other fluorinated gases are significant additional greenhouse gases. Compared to CO_2, these gases release fewer emissions,

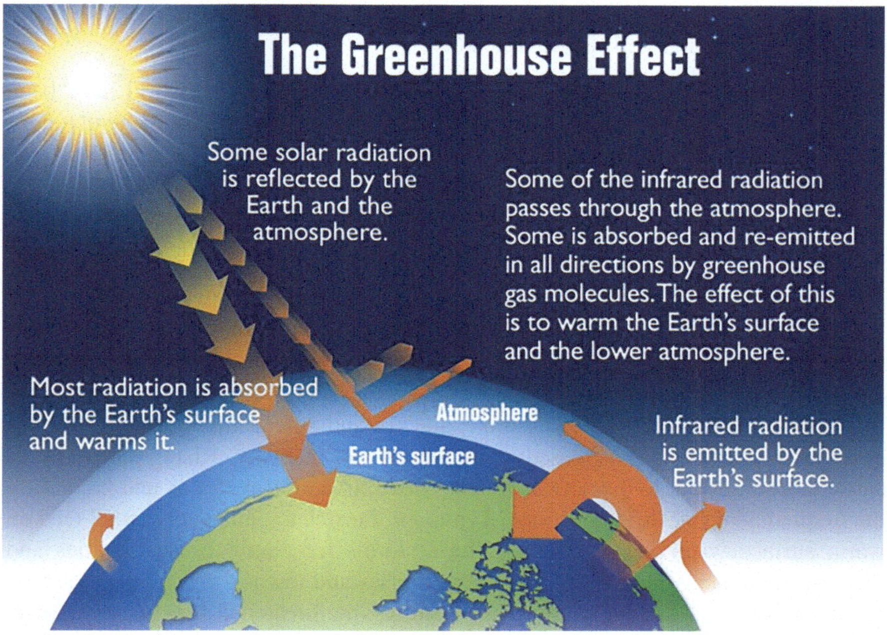

Fig. 10.1 Greenhouse effect. (Source: EPA (U.S. Environmental Protection Agency))

but they retain more heat in the atmosphere (Rathore et al. 2023). Global Warming Potential (GWP) is a unit of measurement for heat-trapping capacity. All other GHG warming potentials are compared to CO_2, since it is the most prevalent and plentiful greenhouse gas with a GWP of 1. For example, the greenhouse gas pressure of fluorinated gases is hundreds of times higher than that of CO_2, implying that these gases contribute far more to climate change per unit of energy (Climate Change 101: climate science basics, 2016).

The primary source of emissions from manufacturing and industry is the combustion of fossil fuels to produce energy for the creation of various products, including electronics, iron, plastics, steel and textiles. Building and mining both release gases into the atmosphere along with other industrial processes. The industrial sector is a significant global contributor to greenhouse gas emissions. Deforestation for agriculture, grazing, or other purposes results in emissions because they absorb CO_2 from the atmosphere, trees aid in regulating the temperature. When trees are cut down, the greenhouse effect increases because the beneficial effect is lost and the carbon stored in the trees is released into the atmosphere.

Methane is produced in huge quantities by the manufacture and use of fertilizers and manure for growing crops, by the digestion of food by cows and sheep, and by the energy required to run farming machinery, which is typically done using fossil fuels. All of these play a significant role in climate change. Approximately 25% of carbon dioxide emissions connected to energy is produced worldwide by transportation. Additionally, patterns suggest that during the next few years, energy use for transportation will rise significantly.

Potential Impacts of Climate Change A lot of things on our world have been impacted by climate change. According to the IPCC's Sixth Assessment report, which was released in 2021, from 1850–1900, human emissions of gases that trap heat have already caused the climate to increase by almost 2° Fahrenheit (1.1 °C). Within the nextfew decades, it is predicted that the average global temperature would rise to 1.5 °C, or roughly 3° Fahrenheit. All areas of the planet will be impacted by these developments.

The global surface temperature rises in tandem with greenhouse gas concentrations. The warmest decade on record is the one that ended in 2020. Elevated temperatures contribute to heat-related health issues and complicate outdoor labour. In warm weather, wildfires ignite more quickly and spread faster. In many areas, destructive storms have increased in frequency and intensity. Rising temperatures cause more moisture to evaporate, causing flooding and heavy rains and increasing the frequency of catastrophic storms. Typhoons, hurricanes, and cyclones all feed on warm surface waters of the ocean. These storms frequently cause fatalities and significant financial damage by destroying houses and towns.

The survival of marine and terrestrial animals is threatened by climate change. The higher the temperature, the greater these risk. The globe is losing species at a rate 1000 times faster than at any other point in recorded human history, which has been accelerated by climate change. There is a one million species that could go extinct in the next few decades. Among the numerous risks associated with climate change are invasive pests and illnesses, severe weather patterns, and forest fires. While certain species may move and thrive, others cannot.

The greatest threat to human health that we are currently confronting is climate change. Climate change is already having a negative influence on people's health through increased hunger and poor nutrition in areas where people cannot grow or get enough food, as well as air pollution, disease, extreme weather events, forced displacement, and mental health strains. The number of diseases is rising due to changing weather patterns, and health care systems are finding it harder to keep up with the increase in deaths caused by extreme weather events.

Impact of Climate Change on Soil An essential part is being played by changes in rainfall patterns and rising average temperatures brought on by climate change. An increase in global temperature speeds up the release of carbon from soils, increasing the atmospheric concentration of carbon dioxide. Of course, the altered patterns of rainfall would also exacerbate erosion in susceptible soils, many of which already have low levels of organic matter. Thus, in addition to increasing the pressure currently placed on soil quality, climate change will also raise the risk of desertification and land degradation, which is already having an impact in many places.

The paedogenic qualities of soils are impacted by climate change. The primary variables that could alter soil-forming conditions as a direct result of climate change are the availability of organic matter from biomass, the temperature regime of the soil, and soil hydrology—the latter due to variations in rainfall zones and potential evapotranspiration. The mineral content, organic mattercontent, and structural stability of many soils are projected to act as quite strong buffers against other climate change effects, such as temperature and precipitation. However, a drop in crop cover or vegetation brought on by insufficient rainfall may result in a decrease in porosity and deterioration of the soil's structure, as well as increased runoff and erosion on sloping sites and the ensuing more extensive and quick sedimentation.

The impact of climate change on soil moisture levels will manifest in various ways, including direct interactions with the weather (e.g., precipitation, temperature variations on evaporation), changes in vegetation due to the climate, plant growth rates, rates at which plants drain water from the soil, and increased CO_2 levels on plant transpiration. By reducing available moisture, changing circulation patterns, and raising air temperatures, changes in soil water fluxes may also have a knock-on effect on the climate and perhaps accelerate drought levels (Pareek 2017).

The chemical composition of the soil, the position of an ion with respect to the root surface, and the distance a nutrient must travel through the soil before reaching the root surface all affect how readily available nutrients are for plants. Temperature and moisture regimes in the root zone are significantly impacted by variations in precipitation and air temperature. Given the established knowledge that temperature and soil moisture are the main factors influencing nutrient availability, root growth, and development, as well as the fact that carbon transfer to roots controls nutrient acquisition, it is logical to assume that the results of these processes will be influenced by the changing climate.

Sustainable Soil Management Soil, air and water are vital for social life on earth. The consequence of soil is only now being appropriately documented. As the most noteworthy medium for cultivation, soil is the chief enable for feeding and nourishing everyone on earth. Soil has multitudinous economic, environmental and social significance—as a habitat for billions of living beings, a source of biofuel energy, medicines and clothing, by providing ecosystem services such as filtering water, and by naturally sequestering carbon to help build resilience against events caused byclimate changes. As per the report of Food and Agriculture Organization of the United Nations (FAO Soils Portal (2020)), "Soil management is sustainable if the associate, provisioning, regulating and cultural services provided by soil are sustained or improved without significantly impairing either the soil utilities that empower those amenities or biodiversity.

Sustainable development may be explained as "to meet the needs of the present without compromising the ability of the future generations to meet their own needs from the land" World Commission on Environment and Development–WCED report (Brundtland, G.H. 1987). The above-mentioned depiction has been reformed for Sustainable soil management by Smith and Powlson (2007) as "soil management that meets the needs of the present without compromising the ability of the forthcoming generations to come across their own necessities from that soil". Later, 'the Intergovernmental Technical Panel on Soils' (ITPS) states soil health as "the ability of the soil to sustain the productivity, diversity, and environmental services of

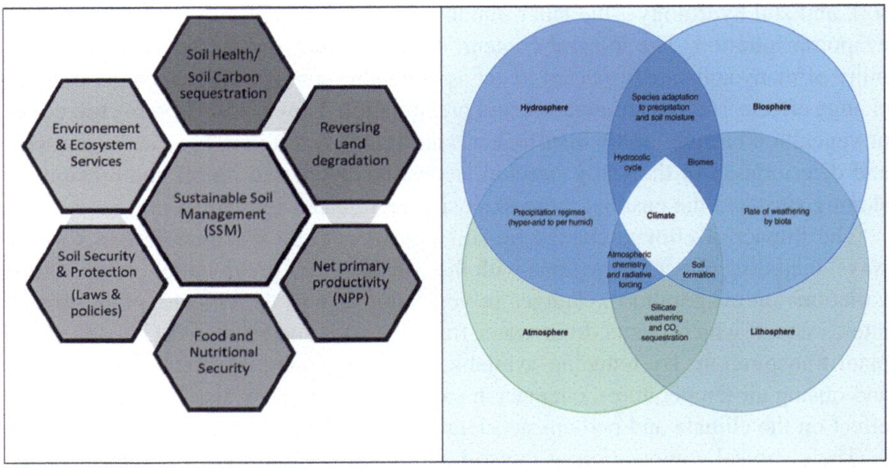

Fig. 10.2 Sustainable Soil Management in close nexus with Soil functions and Climate

terrestrial ecosystems". In the ITPS Soil Letters (it is stated that "In managed systems, soil health may be maintained, stimulated or recovered through the implementation of sustainable soil management practices" (UN 2015; FAO/ITPS 2020).

Similar to this, Lal (2009) put up a few rules for sustainable management techniques that have a low carbon and water footprint while also reversing soil deterioration and restoration processes. Intimidating the soil biological population through management techniques might also pose a threat to soil sustainability by decreasing the soils' future adaptability. Soil sustainability is exposed by many management practices such as conversion of naturalvegetation to arable cropping using inappropriate management practices, over-cultivation/fertilization/use of chemicals, crop residue burning, low addition of organic inputs (manures and crop residue), which lead to the decline in soil organic matter. (Smith and Powlson 2007; Powlson et al. 2011). Therefore, SSM is one stop solution, essential for attaining food and nutritional security and ecosystem services (Fig. 10.2).

Role of Sustainable Soil Management in Climate Change The ultimate goal of the sustainable soil management is to add in sustainable land use and good soil quality, to ensure efficient use of raw materials, mitigate climate change, and reduce nutrient losses in agronomy. Sustainable soil administration reduces polygonal benefits to meeting the United Nations Sustainable Development Goals (Hou et al. 2020; Keesstra et al. 2016).

Its inferences for climate change alleviation have been illustration specific interest in recent years (FAO 2017; Poulton et al. 2018). However, soil signifies the leading carbon pool in the Earth's native system, nearly five times that of the atmospheric carbon pool (Jansson and Hofmockel 2020; Le Quere et al. 2018); on the other hand, N_2O, CH_4 and CO_2 discharges from soils represent momentous fluxes that affect greenhouse gas balance in the atmosphere (Lal 2019; Tian et al. 2019). Soil microorganisms play a precarious part in the cycling of soil organic carbon, but we still lack adequate scientific information to build computable predictive models in the milieu of soil use and management (Jansson and Hofmockel 2020). There is a crucial necessity to cognize how soil use and management affect climate change, due to both carbon sequestration and greenhouse gas emissions (Nath et al. 2023).

Climate change also has a pivotal effect on soil carbon balance (Davidson and Janssens 2006). On a worldwide measure, rising temperatures have resulted in the higher heterotrophic respiration of SOC and thus CO_2 emissions; however, this is partly due to increased plant growth and autotrophic fixation (Jansson and Hofmockel 2020). Due to the multifarious response loops involved in these processes, the impact of climate change on the net flux of greenhouse gas releases still remains unknown (Bradford et al. 2016; Davidson and Janssens 2006). To virtuous recognize these interactive dynamics, it is imperious to have additional field monitoring data and an in-depth analysis of statistics from unlike sources.

Advance Technologies for Sustainable Soil Management in a Changing Climate The climate disaster is presently intimidating the efficiency of agrifood systems, affecting the most prone people and exacerbating inequalities. Soil deprivation, loss of water quality, and water insufficiency are on the increase, limiting access to lush and vigorous soils and water assets, and contributing to food and nutrition uncertainty (Tomar et al. 2021). Integrated climate-smart land, soil, and water management technologies and practices are considered essential for improving efficiency and enhancing soil productivity. Revolutions in recarbonizing soils, refining soil health, and augmenting water use efficacy. Strategies for promoting integrated control of natural resources and handling water scarcity, with the aim of strengthening the resilience of agrifood systems in the face of the climate crisis. Judicious fertilizer use, reduced pesticide use, and improvements in soil health could lead to average crop yield increases of 80%.

Ways to drive climate action and growth across key areas such as:

- Navigating and winning the net-zero transition
- Adapting to climate risks and ensuring an inclusive transition to net-zero
- Accelerating innovation in climate technologies
- Closing the net-zero financing gap
- Decarbonizing industry and accelerating the energy transition
- Decarbonizing the mobility and built environment sectors
- Hyperscaling green businesses
- Scaling carbon management
- Solving the nature nexus: Food, agriculture, land use, and water.
- The materials transition: Sustainable, circular, and scalable.

Working towards Innovative Climate Solutions Increasing public awareness, monitoring soil moisture, adopting sustainable practices using adapted plant and accounting for green water. Innovative methodologies of reinstating despoiled soils, which can bring synergistic benefits for food security, climate, biodiversity and water availability. Real-time water and soil information at scale through remote sensing, particularly focusing on root zone soil moisture, fast and cost-effective soil analysis, including handheld devices for on-site, hyper-localized information, and The practice of geospatial implements for detecting viable locations for rainwater harvesting.

Carbon Requisitioning or Sequestration Carbon sequestration increases the capability of soils to engross and sustain carbon from plant biomass. Soil-based carbon sequestration also helps reestablish despoiled soils, making the crops more resilient in contradiction of both droughts and heavy rainfall. Soil management startups encourage and incentivize soil-based carbon sequestration using various platforms to measure, monitor, verify, and reward farmers for adopting practices that increase soil carbon.

Soil Sequestration Through Microbes Natural consortia of microbes that farmers use as a molecular cover crop. Microorganisms fix nitrogen, appropriate soil carbon, and increase soil health by improving nutrient availability and water retention. Microbes mixed in the soil post-harvest to gain these agronomic benefits (Singh et al. 2020). The microbes later sequester atmospheric carbon during the growing season and stock it in the soil as organic matter, dropping greenhouse gas (GHG) radiations and enhancing crop yield. This is a low-cost, environmentally friendly, and scalable way to expand crop yields and soil quality, reduces greenhouse gas emissions and fertilizer inputs.

Soil Remediation and Restoration Soil contamination occurs because of industrial processes, improper disposal of hazardous wastes, and agricultural practices. Bioremediation, phytoremediation etc. are innovative solutions, to break down contaminants in the soil and restore soil structure and microbial diversity. Nano remediation using nanoparticles and electro kinetic remediation via electric currents to neutralize soil sources.

Contamination of soil occurs due to industrial processes, hazardous wastes, and agricultural practices. Innovative solutions, like bioremediation and phytoremediation, to break down contaminants in the soil and restore soil structure and microbial diversity. Nanoremediation through nanoparticles/nanomaterial and electrokinetic remediation by electric currents to decontaminate soil sources. In Soil remediation and restoration help reduce these risks and improve public health whilst ensuring better yield.

Soil Compaction Control Compacted soil reduces root growth, water infiltration, crop quality and yield, as well as more erosion and runoff. Soil compaction control innovations improve the water-retention capacity of soil by escaping/reducing soil compaction. Controlled traffic farming (CTF) to limit heavy machinery traffic to specific lanes reducing soil compaction. Invention of autonomous or semi-autonomous vehicles that perform precision farming tasks without compacting the soil. New techniques, such as soil aeration, allow air, water, and nutrients to penetrate the soil more easily and reduce compaction.

Direct Plant Sensing Irrigation management platform using direct plant sensing technology softwares are in trends. Using biosensors, measures of real-time water consumption and irrigation needs of crops can possible. The platform also integrates the plant sensor, soil, satellite, and real-time weather data and applies AI to drive farm and field-level irrigation decisions. Treetoscope solution provides irrigation visions and recovers soil stress level management. By continuously monitoring, provides valuable data for optimizing crop health and productivity.

Artificial Soil Formation Agriculture is made possible in urban areas where natural soils are insufficient or contaminated by artificial soil development.. Farmers use biogeochemical interfaces that startups create by mixing mineral materials, organic matter, and microorganisms to support vegetation growth and soil functions. Instead of using as much soil for small-scale farming, startups are adopting soil substitutes like coir and peat, which give the earth more time to heal. Soilless farming techniques, such as hydroponics and aquaponics, utilize water or other substrates instead of soil to grow plants, often in controlled environments like vertical farms. In addition to separating polluted soil from plants, this artificial soil or soil substitutes lower plant exposure to contaminants.

Carbon-Based Artificial Soil Carbon-based artificial soil for agriculture. It consists of plant residue biochar, charcoal, and microbes. This artificial soil converts organic fertilizers into inorganic nutrients, creating good soil conditions and fixing carbon dioxide. Unlike conventional alkaline biochar, the pH of the soil is almost neutral, which makes it better suited for long-term agricultural practices. With TOWING's soil, farmers may increase crop output, improve nutrient absorption, and improve soil quality at a reasonable cost.

It is concluded that, in terms of scale and scope, climate change presents significant interconnected difficulties for the future that were not foreseen a century ago. Predictably, the most significant alteration in soils that would arise from these alterations would be a steady improvement in soil fertility and physical characteristics in humid and subhumid environment. To reduce the impact of climate change on our soils, the environment, and the economy, careful soil management will be required. To enhance soil health, sustainable soil management methods must be applied. Enhancing carbon sequestration in soil and lowering net emissions of N_2O, CO_2, and CH_4 are two ways that soil and soil research can help mitigate climate change. The implementation of intervention measures aimed at enhancing soil organic matter (SOC) can have significant benefits in guaranteeing food security and soil health, since it can also boost crop production, improve soil biodiversity, and immobilize some soil contaminants. Therefore, sustainable soil management and use techniques could benefit humanity in many ways by reducing the effects of climate change.

References

Ahmad T, Gupta SK, Singh SK, Meraj G, Kumar P, Kanga S (2023) Unveiling Nature's resilience: exploring vegetation dynamics during the COVID-19 era in Jharkhand, India, with the Google earth engine. Climate 11(9):187

Bradford MA, Wieder WR, Bonan GB, Fierer N, Raymond PA, Crowther TW (2016) Managing uncertainty in soil carbon feedbacks to climate change. Nat Clim Chang 6:751–758

Choudhury U, Singh SK, Kumar A, Meraj G, Kumar P, Kanga S (2023) Assessing land use/land cover changes and urban heat Island intensification: a case study of Kamrup Metropolitan District, Northeast India (2000–2032). Earth 4(3):503–521

Climate Change 101: climate science basics (2016) Public Health Institute/Center for Climate Change and Health

Climate Change Evidence & Causes: Update 2020 (2020) An overview from the Royal Society and the US National Academy of Sciences

Davidson EA, Janssens IA (2006) Temperature sensitivity of soil carbon decomposition and feedbacks to climate change. Nature 440:165–173

Debnath J, Sahariah D, Lahon D, Nath N, Chand K, Meraj G, Kumar P, Singh SK, Kanga S, Farooq M (2023) Assessing the impacts of current and future changes of the planforms of river Brahmaputra on its land use-land cover. Geosci Front 14(4):101557. Elsevier

FAO (2017) Agriculture: driving action across the 2030 agenda for sustainable development, Rome

FAO (2020) Food and agriculture organization of the United Nations report: protocol for the assessment of sustainable soil management

FAO/ITPS (2020) Towards a definition of soil health, intergovernmental technical panel on soils in soil letters. September 20

Hou D, Bolan NS, Tsang DCW, Kirkham MB, O'Connor D (2020) Sustainable soil use and management: an interdisciplinary and systematic approach. Sci Total Environ 729:138961

Jansson JK, Hofmockel KS (2020) Soil microbiomes and climate change. Nat Rev Microbiol 18: 35–46

Keesstra SD, Bouma J, Wallinga J, Tittonell P, Smith P, Cerdà A, Montanarella L, Quinton JN, Pachepsky Y, van der Putten WH, Bardgett RD, Moolenaar S, Mol G, Jansen B, Fresco LO (2016) The significance of soils and soil science towards realization of the United Nations sustainable development goals. Soil 2:111–128

Kumar P, Singh SK, Gupta SK, Kanga S, Meraj G (2023) Uncovering the hydro-meteorological drivers responsible for forest fires utilizing geospatial techniques. Theor Appl Climatol. https://doi.org/10.1007/s00704-023-04497. Springer Nature

Lal R (2009) Laws of sustainable soil management. In: Lichtfouse E, Navarrete M, Debaeke P, Souchère V, Alberola C (eds) Sustainable Agriculture. Springer Science+Business Media B.V. EDP Sciences 2009, Dordrecht, Holland, pp 1–5

Lal R (2019) Accelerated soil erosion as a source of atmospheric CO_2. Soil Tillage Res 188:35–40

Le Quere C, Andrew RM, Friedlingstein P, Sitch S, Hauck J, Pongratz J, Pickers PA, Korsbakken JI, Peters GP, Canadell JG, Arneth A, Arora VK, Barbero L, Bastos A, Bopp L, Chevallier F, Chini LP, Ciais P, Doney SC, Zheng B (2018) Global carbon budget 2018. Earth Syst Sci Data 10:2141–2194

Nath N, Sahariah D, Meraj G, Debnath J, Kumar P, Lahon D, Chand K, Farooq M, Chandan P, Singh SK, Kanga S (2023) Land use and land cover change monitoring and prediction of a UNESCO world heritage site: Kaziranga eco-sensitive zone using cellular automata-Markov model. Land 12(1):151

Pareek N (2017) Climate change impact on soils: adaptation and mitigation. MOJ Eco Environ Sci 2(3):136–139. https://doi.org/10.15406/mojes.2017.02.00026

Poulton P, Johnston J, Macdonald A, White R, Powlson D (2018) Major limitations to achieving "4 per 1000" increases in soil organic carbon stock in temperate regions: Evidence from longterm experiments at Rothamsted research, United Kingdom. Glob Chang Biol 24:2563–2584

Powlson DS et al (2011) Soil carbon sequestration to mitigate climate change: a critical re-examination to identify the true and the false. Eur J Soil Sci 62(1):42–55

Rathore SS, Mathur SC, Mathur S, Sharma S, Singh SK (2023) Late cretaceous–early paleocene geodiversity of Fatehgarh formation of petroliferous Barmer Basin, Western Rajasthan, India: a potential Geopark Site. Geoheritage 15:54

Singh RP, Kumar S, Rana PS (2020) UN SDGs and context of holy-heritage cities in India: a study of Ayodhya and Varanasi. In: Geoecology of landscape dynamics. Springer, Singapore, pp 187–206

Singh S, Singh SK, Kumar R, Shrama A, Kanga S (2022) Finding alternative to river sand in building construction. Evergreen 9(4):973–992

Singh SK, Kotecha MJ, Kanga S, Pankhaniya SK, Agrawal S, Meraj G (2023) Urban Heat Island (UHI) resilience plan in varying climatic conditions using geospatial approach: a case study of Rajkot City. In: Climate Change, Agriculture and Society, vol 1, pp 307–338

Smith P, Powlson DS (2007) Sustainability of soil management practices – a global perspective in book: soil biological fertility, pp 241–254

Tian H, Yang J, Xu R, Lu C, Canadell JG, Davidson EA, Jackson RB, Arneth A, Chang J, Ciais P, Gerber S, Ito A, Joos F, Lienert S, Messina P, Olin S, Pan S, Peng C, Saikawa E, Zhang B (2019) Global soil nitrous oxide emissions since the preindustrial era estimated by an ensemble of terrestrial biosphere models: magnitude, attribution, and uncertainty. Glob Chang Biol 25: 640–659

Tomar P, Singh SK, Kanga S, Pattanaik A (2021) Water bodies mapping and monitoring using high-resolution satellite images. Sustain Agric Food Environ Res 11(2):100–116

UN (2015) Transforming our world: the 2030 agenda for sustainable development. Resolution Adopted by the General Assembly on 25 September 2015, 42809, pp 1–13

WCED Report: Brundtland GH (1987) Our common future: report of the world commission on environment and development, Geneva, UN

Chapter 11
Millets as a Source of Nutrients and Nutraceuticals for a Sustainable Future

B. Shivani, Surjeet Singh Dhaka, Chidanand Patil, and Khetan Shevkani

Abstract Millets hold an important place as a source of food and feed in drought prone semi-arid and arid regions of the world. They are promising in achieving food/nutrition security and a sustainable future. Millets not only are rich in slowly digestible and resistant starch, non-starch polysaccharides, dietary proteins and essential nutrients but also exert antioxidative, antiinflammatory, antihyperlipidaemic, anticancer, neuroprotective and low-glycaemic index characteristics owing to the presence of appreciable quantities of an array of nutraceuticals (phenolics, saponins, phytates/phytic acids, carotenoids, tocols and bioactive peptides). Moreover, millet cultivation presents many agro-economic advantages, for example, short crop duration, ability to grow in marginal soils under rain-fed conditions, resistance to high temperatures, water stress and diseases/pests, lesser requirements of agricultural inputs and lower labour-intensive farm operations compared to conventionally grown food crops. This chapter provides an overview of the nutritional/nutraceutical properties and agro-economic advantages of millets in order to highlight their importance in achieving a sustainable future.

Keywords Agro-economic advantages · Health benefits · Millets · Nutrition · Sustainability

11.1 Introduction

Millets include small seeded drought-resistant cereal crops. Millet grains exhibit highly varied physicochemical properties though they are generally characterized by a small size and round shape. They are one of the oldest domesticated crops and evidence of their consumption is documented as far back as the Indus valley civilization (Kheya et al. 2023). Based on popularity and cultivation quantities, millets are broadly categorized as major and minor millets. The major millets include

B. Shivani · S. S. Dhaka · C. Patil · K. Shevkani (✉)
Department of Applied Agriculture, Central University of Punjab, Bathinda, Punjab, India
e-mail: khetan.shevkani@cup.edu.in

S. Kanga et al. (eds.), *Transforming Agricultural Management for a Sustainable Future*, World Sustainability Series, https://doi.org/10.1007/978-3-031-63430-7_11

sorghum (*Sorghum bicolor*), pearl millet (*Pennisetum glaucum*) and finger millet (*Eleusine coracana*), while foxtail millet (*Setaria italica*), proso millet (*Panicum miliaceum*), little millet (*Panicum sumatrense*), kodo millet (*Paspalum scrobiculatum*) and barnyard millet (*Echinochloa esculenta*) are examples of minor millets.

Millets contain 55–75% carbohydrates, 7–14% proteins, 7–20% dietary fibre and 2–5% lipids. They also provide essential vitamins and minerals (vitamin A, B vitamins, Fe, Ca, Mg, K, Zn, Cu and Se. Furthermore, they contain lower amounts of available carbohydrates than rice/wheat and naturally free from gluten and cholesterol. In addition to nutrients, millets also contain appreciable quantities of several health-benefitting constituents including phenolic compounds, saponins, carotenoids, tocols, phytosterols as well as bioactive peptides/proteins, slowly digestible and resistant starch (Bello et al. 2022; Gowda et al. 2022; Majid and Poornima Priyadarshini 2020; Chandra et al. 2016; Annor et al. 2017; Kumar et al. 2016). Many studies have also shown positive effects of millet-based diets consumption on reduced risks of chronic/lifestyle-related diseases such as diabetes, hypertension, obesity, dyslipidaemia/coronary artery diseases, etc. (Amarakoon et al. 2021; Xu et al. 2021; Kumari et al. 2020).

There has been a considerable decline in the cultivation of millets globally during the last few decades due to the increased popularity of rice and wheat (Meena et al. 2021). However, millet crops have several agro-economic advantages over conventional food crops and are promising in achieving food/nutrition security for a sustainable future. They require much lesser resources than many other food crops. The shorter growing season as well as botanical features of plants (C4 photosynthetic pathway, large leaf area index despite small leaf area, straight leaf type, high radiation use efficiency, thick cell walls, dense root system, etc.) make millet crops capable of tolerating biotic stresses and attaining growth with much lesser requirements of water, agrochemicals, time and energy (Li and Brutnell 2011; Sage and Zhu 2011). Various studies have reported the ability of millet crops to tolerate water stress, drought conditions, salinity, climate change, pests and diseases (Krishnamurthy et al. 2014; Kumar et al. 2018; Saxena et al. 2018; Vetriventhan et al. 2012; Zhang et al. 2021). Therefore, the present article provides an overview of the nutritional/nutraceutical properties and agro-economic advantages of millets in order to highlight their importance in achieving a sustainable future.

11.2 Production

Millets serve as an important food source in developing countries, particularly in the arid/semi-arid areas of Africa and Asia. Worldwide, millet production spans over 93 countries (Meena et al. 2021). Asia and Africa are the largest producers with India (40%), Niger (11%), China (9%), Nigeria (7%), Mali (6%), Sudan (5%), Ethiopia (4%), Burkina Faso (3%), Senegal (3%) and Chad (2%) being the top ten millet producing countries in the world (https://ipad.fas.usda.gov/cropexplorer/cropview/

commodityView.aspx?cropid=0459100). Sorghum is the most cultivated millet and the fifth most important cereal crop after maize, wheat, rice, and barley. Sorghum predominantly serves as a staple food crop in arid regions of Africa, China and India (Ajeigbe et al. 2018; Mrema et al. 2017). In 2022, global sorghum production has reached over 62 million tonnes (https://www.fao.org/faostat/en/#data/QCL). Nigeria, Sudan, the USA, Mexico and Ethiopia with production quantities of 6.80, 5.25, 4.77, 4.75 and 4.2 million tonnes, respectively, were the five largest sorghum producing countries. India produced 4.15 million tonnes sorghum and held the sixth position in total sorghum production. Pearl millet accounts to approximately 40% production of total millets (Sharma and Niranjan, 2017). It is cultivated on approximately 30 million hectares of land in the arid and semi-arid tropical regions of Asia and Africa (Raheem et al. 2021; Satyavathi et al. 2021). India holds the first rank in the cultivation area and production of pearl millet (Satyavathi et al. 2021). Finger millet is cultivated in more than 25 nations across Africa and Asia for food as well as fodder purposes (Gebreyohannes et al. 2021; Thakur 2023). Approximately 3.7 million tons of finger millet is cultivated (majorly in tropical regions) on around 6.1 million hectares of land, accounting to nearly 12% of the global millet cultivation area (Thakur 2023). Barnyard millet is cultivated primarily in India, China and Japan (Sood et al. 2021) while major producers of foxtail millets are China, India, Afghanistan, Japan, Korea and Georgia (https://www.fao.org/millets-2023/about/en). Kodo millet cultivation is concentrated in India, Philippines, Indonesia, Vietnam, Thailand and West African regions (Ravikesavan et al. 2023). Specifically, in India, kodo millet thrives in the Deccan region, reaching as far as the foothills of the Himalayas (Ranjan et al. 2023). Proso millet is cultivated in countries such as India, Japan, China, Egypt, Arabia, and Western Europe (Prabhakar et al. 2017). Little millet production, on contrast, comes almost entirely (98%) from India (Paschapur et al. 2021).

11.3 Nutritional Properties and Health Benefits

11.3.1 Carbohydrates

Carbohydrate is the most abundant chemical constituent in millets (Table 11.1). It comprises starch and non-starch polysaccharides as major components, accounting for 60–75% and 15–30%, respectively, of grain weight. Free sugars also occur in millets, but as minor constituents accounting to only 1–2% of grain weight. Glucose, fructose and sucrose constitute the majority of free sugars in different millets (Juhaimi et al. 2019).

Similar to other cereals, starch occurs in millet grains in the form of discrete particles or granules. Millet starch granules are mostly polygonal to spherical and differ in size from 0.3 to 25 μm depending on species and variety (Annor et al. 2014; Zhu 2014; Li et al. 2019). They are also characterized by round edges and the presence of pores on the surface (Gao et al. 2022; Chao et al. 2014; Wu et al. 2014;

Table 11.1 Composition of different millets in comparison to major cereals

Cereal	Protein content (%)	Ash content (%)	Fat content (%)	Total carbohydrates (%)	Dietary fibre content (%)			Moisture content (%)
					Total	Insoluble	Soluble	
Pearl millet	10.96–11.8	1.37–2.2	4.8–5.43	61.78–67.0	11.49	9.14	2.34	8.97–12
Sorghum	9.97–10.40	1.39–1.6	1.73–3.1	67.68–70.7	10.22	8.49	1.73	9.01–12
Finger millet	7.16–7.7	2.04–2.6	1.5–1.92	66.82–72.6	11.18	9.51	1.67	10.89–12
Little millet	9.7–10.13	1.34–5.4	3.89–5.2	60.9–65.55	7.72	5.45	2.27	11.36–12
Foxtail millet	8.92–11.2	1.72–3.3	2.55–4.7	63.2–72.7	6.39–20.8	4.29–19.7	1.1–2.11	9.4–14.2
Proso millet	13.6	3.3	3.8	70	19.4	18.4	1.1	9.9
Kodo millet	9.8	3.3	3.6	66.6	–	5.2*	–	12
Barnyard millet	11	4.5	3.9	55	–	13.6*	–	12
Rice, milled	7.94	0.56	0.52	78.24	2.8	1.99	0.82	9.93
Wheat	10.59–11.6	1.42–1.6	1.47–2.0	64.72–71.0	11.23	9.63	1.6	10.58–12
Maize	8.8–9.2	1.17–1.2	3.77–4.6	64.77–73	12.24	11.29	0.94	9.26–12

Data from: Longvah et al. (2017); Saleh et al. (2013); * = crude fibre

Kim et al. 2009). Structurally, millet starches are made up of amylose and amylopectin and exhibit a semicrystalline structure. Amylose constitutes 0–34% of millet starches depending on species and variety (Mahajan et al. 2021; Sang et al. 2008). Amylose occurs as essentially linear chains of glucose units linked together by α-1,4-glycosidic bonds. Amylopectin, in comparison, is characterised by a highly branched chain structure in which glucose molecules are linked together by α-1,4- and α-1,6-glycosidic bonds. Amylose molecules exhibit an average chain length of 180–270 glucose units/chain while amylopectins of pearl millet, finger millet, proso millet and foxtail millet are characterized by an average chain length of 8–21 units/chain (Annor et al. 2014; Gaffa et al. 2004). Crystallinity in millet starches varies between 24 and 31% and it is related to amylose content and amylopectin chain length (Wu et al. 2014). Starches with a higher proportion of amylose and short amylopectin chains generally exhibit lower crystallinity.

Based on digestibility, starches are categorized into rapidly digestible starch (RDS; the portion of starch that hydrolyses within 20 min by amylolytic enzymes), slowly digestible starch (SDS; starch requiring 20–120 min for enzymatic digestion) and indigestible/resistant starch (RS; starch that resists amylolytic hydrolysis) (Englyst et al. 1992). Foods high in RDS are not considered suitable for diabetics as they cause peaks in blood glucose levels whereas SDS contributes to reduced spikes in postprandial blood glucose levels because of its slow-releasing nature, hence, improves insulin sensitivity and reduces insulin sensing dysfunction (Johnston et al. 2010). RS, in contrast, benefits the host by functioning as prebiotics (Shevkani et al. 2017), promoting probiotics (*Bifidobacterium* and *Lactobacilli*) that improve gut health in turn by (i) competing with pathogens, (ii) secreting antimicrobial peptides, (iii) promoting immunoglobulins synthesis, (iv) increasing phagocytosis and the activity of natural killer cells, (v) prevent inflammation, (vi) promoting neurotransmitters synthesis and (vii) producing short-chain fatty acids with ability to improve intestinal motility and overall functionality (Ríos-Covián et al. 2016; Qiang et al. 2009).

Slowly digestible starch generally constitutes the largest portion of millet starches, making them particularly suitable for diabetics. RDS, SDS and RS occur in different millet starches in the proportions of 13.1–50.1%, 19.9–75.4% and 1.9–51.2%, respectively (Kaimal et al. 2021). Finger millet starch has the highest proportion of SDS (75.0–75.4%) while little millet, barnyard millet and kodo millet are rich in RS (46.5–47.7%, 41.45–44.6% and 37.5–51.2%, respectively). Pearl millet, in comparison, exhibits the highest RDS content (46.3–50.1%) (Kaimal et al. 2021). Also, millet-based foods with a glycaemic index of 37–53 are categorised as low glycaemic foods (Geetha et al. 2020; Sharma and Gujral 2020). The lower rate of starch digestibility and low glycaemic index of millets is attributed to the characteristics of starches as well as the presence/absence of non-starch constituents. For example, the lower digestibility of finger millet starch was attributed to a rigid starch granule architecture and the absence of pores on granules that could serve as the entry point for hydrolysing enzymes while non-starch constituents such as lipids, proteins or phenolic compounds reduced starch digestibility by making complexes with greater resistance to amylolytic enzymes (Annor et al. 2017).

The total dietary fibre content in different millets has been shown to range between 12.1 and 33.1% (Vali Pasha et al. 2018; Pushparaj and Urooj 2011). With mean dietary fibre content of 32.5%, kodo millet is on the higher side of this range and is followed by barnyard millet (25.2%), little millet (24.1%), proso millet (22.8%), foxtail millet (19.1%), finger millet (15.6%) and pearl millet (12.6%) for total dietary fibre content (Vali Pasha et al. 2018; Pushparaj and Urooj 2011). The dietary fibre in foods is composed of indigestible substances including cellulose, lignins and other non-starch polysaccharides. Total dietary fibre is categorized into insoluble and soluble types (IDF and SDF, respectively) based on their interaction with water. SDF makes gel-like dispersions with water and can improve insulin sensitivity, lower post-prandial glucose levels and delay gastric emptying rate while IDF contributes to faecal bulk by absorbing large amounts of water (Shevkani 2023). SDFs also are fermented readily in the human gut hence promote probiotics to produce physiologically active constituents (e.g. short chain fatty acids) along with gases. Millets are particularly high in IDF content and have significant quantities of SDF. They contain 4.3–19.7% IDF and 1.1–2.3% SDF against 2.0–11.3% and 0.8–1.6%, respectively, for milled rice, wheat and maize (Table 11.1). Millets also contain β-glucans (up to 1.1%), a type of SDF that is composed of glucopyranosyl polymers linked together β-1,4 and β-1,3 glycosidic linkages (Serna-Saldivar. 2010). β-Glucan is considered to have nutraceutical effects ranging from obesity management to reduced risks of hyperglycaemia, dyslipidaemia and hypertension, improved insulin resistance and cell immunomodulation (Bozbulut and Sanlier 2019; Cloetens et al. 2012; Rahmani et al. 2019). In addition, β-glucan also exerts antioxidative effects and acts as a prebiotic (Thondre et al. 2011). In general, millets have greater amounts of β-glucans than cereals of common usage like milled rice, wheat and maize while kodo millet and sorghum are relatively higher (1.0–1.1%) in β-glucan content amongst different millets (Serna-Saldivar. 2010).

11.3.2 Proteins

Protein content of millets ranges between 7.2 and 13.6% (Table 11.1). Amongst different millets, proso millet has the highest protein content (13.6%) and is followed by foxtail millet, barnyard millet and pearl millet. Proteins constitute the most fundamental component of animal body. They are involved in almost all physiological/metabolic processes that take place in humans. Proteins ingested through foods serve as the source of amino acids pivotal for body protein synthesis and the regulation of metabolic processes.

Millet proteins contain most essential amino acids in amounts adequate for meeting human requirements (Table 11.2). Similar as other cereals, millets are deficient in lysine. However, pearl millet has relatively higher content of lysine than other millets, which is attributed to grain anatomical factors (higher ratio of

Table 11.2 Essential amino acid content (%) of various millets in comparison to major cereals

Cereal	Histidine	Isoleucine	Leucine	Lysine	Methionine	Phenylalanine	Threonine	Tryptophan	Valine
Pearl millet	2.15	3.45	8.52	3.19	2.11	4.82	3.55	1.33	4.79
Sorghum	2.07	3.45	12.03	2.31	1.52	5.1	2.96	1.03	4.51
Finger millet	2.37	3.7	8.86	2.83	2.74	5.7	3.84	0.91	5.64
Little millet	2.35	4.14	8.08	2.42	2.21	6.14	4.24	1.35	5.31
Foxtail millet	2.14	4.55	11.96	1.42	2.69	6.27	3.89	1.32	5.49
Proso millet	2.1	4.1	12.2	1.5	2.2	5.5	3.0	0.8	5.4
Kodo millet	2.23	4.44	10.84	1.91	2.73	9.56	3.85	1.32	6.78
Barnyard millet	2.42	6.04	12.46	2.13	3.08	6.86	4.72	1.31	6.59
Rice, milled	2.45	4.29	8.09	3.7	2.6	5.36	3.28	1.27	6.06
Wheat	2.65	3.83	6.81	3.13	1.75	4.75	3.01	1.4	5.11
Maize	2.7	3.67	12.24	2.64	2.1	5.14	3.23	0.57	5.41

Data from: Longvah et al. (2017); Sachdev et al. (2021)

embryo to endosperm) and protein composition differences (higher proportion of soluble protein fractions, albumins and globulins, high in lysine) (Sharma et al. 2021). As the intake of all essential amino acids in adequate quantities is essential for proper growth and maintenance, the deficiency of lysine suggests that millets alone are not sufficient for optimal growth. Hence, the consumption of millets with other protein source is recommended to fulfil the requirements of all essential amino acids. In addition to essential amino acids, millets also contain arginine, aspartic acid, glycine, glutamic acid and glutamine (Sachdev et al. 2021) which perform crucial functions in intestinal health, immunity, wound-healing, reproduction and neurological system (Brosnan and Brosnan 2013; Wu et al. 2017; Yi et al. 2018).

Digestibility is an important aspect of protein quality as it decides the bioavailability of food proteins and constituting amino acids. Finger millet, foxtail millet, barnyard millet, kodo millet, little millet and proso millet exhibit protein digestibility of 72.0–75.4% (Sharma and Gujral 2020). In this regard, millet proteins are less digestible than animal proteins (94–97%) (Damodaran 2017); however, they compare well with legume proteins (63.4–83.3%) (Vadivel and Janardhanan 2005). Moreover, studies have shown improvement of the digestibility of millet proteins with processing, e.g. germination (Hassan et al. 2020; Sharma and Gujral 2020).

In addition, millet proteins could also serve as a source of bioactive peptides with antioxidant and antimicrobial properties. For example, the hydrolysis of pearl millet proteins released peptides (including the one with a sequence of SDRDLLGPNNQYLPK) with strong antioxidant activity due to the ability to scavenge hydroxyl radicals and chelate metals (Agrawal et al. 2016). Similarly, two novel bioactive peptides (TSSSLNMAVRGGLTR and STTVGLGISMRSASVR) on interaction with the free radicals via serine and threonine residues exhibited antioxidative activity (Agrawal et al. 2019; Majid and Poornima Priyadarshini 2020). Certain acid-soluble peptides of molecular weight 2 kDa from pearl millet showed antiviral activity against the herpes simplex virus type 1 (Camargo Filho et al. 2008).

11.3.3 Lipids

Lipids occur in different millets in amounts varying between 1.5 and 5.43% (Table 11.1). Amongst different millets, pearl millet (4.8–5.4%), little millet (3.9–5.2%) and foxtail millet (2.6–4.7%) exhibit higher lipid content (Table 11.1). Millet lipids are composed of polar, nonpolar and nonsaponifiable components. Nonpolar lipids constitute the largest portion (up to 80%) of total lipids. They are comprised primarily of triglycerides (85–90%) and contain sterols, diglycerides, monoglycerides and free fatty acids in relatively lesser proportions (10–15%). The less abundant polar lipids consist of phospholipids (17–25%) and glycolipids (2.5–6.2%) while nonsaponifiable fraction (3–5%) includes carotenoids, phytosterols and tocopherols (Rao et al. 2017). Millet lipids are rich in unsaturated fatty acids (Bora et al. 2019; Longvah et al. 2017). Linoleic acid (18:2), oleic acid (18:1) and linolenic acid (18:3) are major unsaturated fatty acids in millets. Linoleic acid is generally the most abundant fatty acid (38.2–65.8%) and it is followed by oleic acid (19.4–37.8%) and linolenic acid (0.7–3.2%). However, finger millet is an exception in this regard as it contains oleic acid (47.5%) as the most abundant fatty acid. Palmitic acid (16:0) and stearic acid (18:0) are saturated fatty acids that occur in millets in amounts ranging from 8.2 to 18.4% and 1.7 to 5.5%, respectively (Bora et al. 2019). Millets also contain α-linolenic acid (ALA), an important omega-3 fatty acid in plants, which is known to have several health benefitting effects in animals. Omega-3 fats have been found useful in the prevention/treatment of inflammatory diseases (inflammatory bowel disease, Crohn's disease, rheumatoid arthritis, asthma, psoriasis, etc.), cardiovascular diseases, neurodegenerative (Alzheimer's disease) disorders and gastrointestinal cancers (Gogus and Smith 2010). In general, the ALA content of millets ranges from 16.5 to 140 mg/100 g, being the highest for pearl millet (140 mg/100 g), followed by finger millet (68.6 mg/100 g) and little (47.2 mg/100 g) millet (Longvah et al. 2017).

11.3.4 Vitamins and Minerals

Millets also serve as a source of B vitamins, particularly riboflavin, niacin, and folic acid (Kumar et al. 2018; Longvah et al. 2017; Sharma et al. 2021). Finger millet is particularly rich in thiamine (0.37–0.42 mg/100 g) while pearl millet, sorghum and foxtail millet contain the highest quantities of riboflavin (0.20–0.21 mg/100 g), niacin (0.86–4.3 mg/100 g) and pantothenic acid (0.27–0.63 mg/100 g), respectively (Table 11.3). In addition, pearl millet and sorghum also contain relatively higher quantities of vitamin B6 (0.27–0.28 mg/100 g) whereas little millet is rich in biotin (6.03 μg/100 g). In the context of folates, millets with concentration of 34.6–39.5 μg/100 g can be considered superior to rice, maize and wheat (9.32, 25.8 and 30.1 μg/100 g, respectively) (Table 11.3). Additionally, millets also contain vitamin E and β-carotene (Longvah et al. 2017). Vitamins are mainly concentrated in the bran, pericarp and aleurone layer of grains (Saldivar 2016). In general, millets are devoid of vitamin B12 and deficient in vitamin C. However, germination can increase ascorbic acid content of millets. A study by Panda et al. (2020) showed increase in mean vitamin C content of various millets from 3.69 to 5.05 mg/100 g on germination.

Millets have a rich mineral profile as they contain major as well as trace mineral elements including Ca, P, Na, K, Mg, Fe, Cu, Zn, Mn, Cr, Mo and Se (Anitha et al. 2020; Vali Pasha et al. 2018). Mineral composition of different millets is shown in Table 11.4. Finger millet is one of the richest sources of Ca (350–364 mg/100 g). This millet also exhibits higher Cu and K levels than different millets and cereals of common usage. Pearl millet has appreciable quantities of Fe (6.42–11.0 mg/100 g), Zn (2.76 mg/100 g) and Cu (0.54 mg/100 g). In general, millets can be considered a source of Se (14.12–40.41 μg/100 g). They also are superior to rice and comparable to wheat and maize for Mg content. Moreover, they exhibit lower Na content than K content (Table 11.4), a property which is desirable for reducing hypertension.

Table 11.3 Vitamin composition of selected millets in comparison to major cereals

Cereal	Thiamin (mg/100 g)	Riboflavin (mg/100 g)	Niacin (mg/100 g)	Pantothenic acid (mg/100 g)	Pyridoxine (mg/100 g)	Biotin (μg/100 g)	Folates (μg/100 g)	Tocopherols (mg/100 g)	β-carotene (μg/100 g)
Pearl millet	0.25–0.38	0.20–0.21	0.86–2.8	0.5	0.27	0.64	36.11	0.24	28.23
Sorghum	0.35–0.38	0.14–0.15	2.1–4.3	0.27	0.28	0.7	39.42	0.06	8.29
Finger millet	0.37–0.42	0.17–0.19	1.1–1.34	0.29	0.05	0.88	34.66	0.16	1.53
Little millet	0.26–0.30	0.05–0.09	1.29–3.2	0.6	0.04	6.03	36.2	0.55	1.91
Foxtail millet	0.29–0.59	0.2–0.11	1.49–3.2	0.63	0.07	1.49	39.49	0.07	1.41
Rice, milled	0.05	0.05	1.69	0.57	0.12	0.6	9.32	0.06	Nil
Wheat	0.41–0.46	0.1–0.15	2.68–5.1	1.08	0.26	1.03	30.09	0.77	3.03
Maize	0.33–0.38	0.09–0.2	2.69–3.6	0.34	0.34	0.49	25.81	0.36	186
# daily requirement (adult male/female)	1.1–1.2 mg	1.1–1.3 mg	14–16 mg	5 mg	–	–	400 μg (dietary folate equivalent)	7.5–10 mg	~3–3.6 mg (500–600 retinol equivalent)

Data from: Longvah et al. (2017); Saleh et al. (2013); # = WHO (2004)

Table 11.4 Mineral composition of selected millets in comparison to major cereals

Cereal	Ca (mg/100 g)	Mg (mg/100 g)	Fe (mg/100 g)	Zn (mg/100 g)	Cu (mg/100 g)	Se (µg/100 g)	K (mg/100 g)	Na (mg/100 g)
Pearl millet	27.35–42.0	124	6.42–11.0	2.76	0.54	30.4	365	4.11
Sorghum	25–27.6	133	3.95–5.4	1.96	0.43	26.29	328	5.42
Finger millet	350–364	146	3.9–4.62	2.53	0.67	15.3	443	4.75
Little millet	16.06–17.0	91.4	1.26–9.3	1.82	0.34	40.41	105	4.77
Foxtail millet	2.8–15.27	122	2.34–2.8	1.65	0.26	14.12	94	3.35
Rice, milled	7.49	19.3	0.65	1.21	0.23	1.01	108	2.34
Wheat	30–39.36	125	3.5–3.97	2.85	0.49	47.76	366	2.5
Maize	8.91–26	145	2.7–2.49	2.27	0.45	8.69	291	4.44

Data from: Longvah et al. (2017); Saleh et al. (2013)

11.3.5 Phytochemicals

Millets are a rich source of phytochemicals with proven health benefits (antioxidative, antiinflammatory, anticancer, antihyperlipidemic, antihypertensive, antidiabetic and antiobesity effects) (Saleem et al. 2023; Bello et al. 2022; Majid and Poornima Priyadarshini 2020; Nithiyanantham et al. 2019). Phenolics, carotenoids, tocols, saponins and phytates/phytic acid represent the most significant phytochemicals in millets.

Phenolic compounds refer to plant secondary metabolites characterized by a structure comprising of at least one phenolic group with one or more carboxylic groups. They occur in plants in bound or free forms and perform several crucial functions in signalling, wound healing and protection/defence. Millet grains are generally rich in phenolics (Goudar et al. 2023; Kumari et al. 2023; Shahidi et al. 2021). Amongst various millets, kodo millet contains the highest amount of phenolic compounds both in free and bound forms (493 and 619 mg/100 g, respectively) followed by pearl millet (415 and 580 mg/100 g, respectively), finger millet (389 and 531 mg/100 g, respectively), proso millet (375 and 520 mg/100 g, respectively), barnyard millet (270 and 389 mg/100 g, respectively), foxtail millet (245 and 395 mg/100 g, respectively) and little millet (133 and 278 mg/100 g, respectively) (Goudar et al. 2023). Phenolic acids (protocatechuic acid, vanillic, syringic acid, gentistic acid, gallic acid, *p*-hydroxybenzoic acid, ferulic acid, sinapic acid, caffeic acid, *p*-coumaric acid and chlorogenic acid) and flavonoids (catechin, epicatechin, rutin and quercetin) represent majority of phenolics in millets (Bora et al. 2019; Goudar et al. 2023; Kumari et al. 2023). Ferulic acid is the most abundant phenolic acid in millets while the most prominent flavonoid is catechin. Gentistic acid, in contrast, occurs in kodo millet and finger millet only (Bora et al. 2019). Phenolic compounds have several proven health benefits. They exert antioxidative, antidiabetic, cataract-preventive and anticancer effects. Antidiabetic, antiobesity and hypochloesterolaemic effects of finger and pearl millets were due to the complex mixture of phenolics that acted as effective inhibitors of enzymes (amylases, α-glucosidase and pancreatic amylases) reducing the postprandial hyperglycaemia (Bello et al. 2022; Shobana et al. 2009). Similarly, phenolic-induced inhibition of aldose reductase was effective in preventing cataract in diabetics (Chethan et al. 2008). Oxidised low-density lipoprotein cholesterol (LDL-C) was considered to have prominent roles in the formation of atheromatous plaques responsible for atherosclerosis while kodo millet phenolics were able to reduce LDL-C oxidation by 40% (Chandrasekara and Shahidi 2011). In addition, vanillin from proso millet and barnyard millet has shown to reduce the number of viable colon cancer cells due to the antiproloferative and non-cytotoxic cell death inducing effect (Ramadoss and Sivalingam 2020). The antioxidant activity of polyphenols has been attributed to their capacity to donate hydrogen atoms to electron deficit free radicals through the hydroxyl groups on benzene rings to form a resonance stabilized phenoxyl radical (Devi et al. 2014).

Carotenoids (fat-soluble pigments responsible for the distinctive yellow-orange colour of most fruits, vegetables, grains/seeds, and some microbes) occur in millets in quantities ranging between 78 and 366 μg/100 g (Longvah et al. 2017; Asharani et al. 2010). Amongst different millets, proso millet and pearl millet are particularly rich in carotenoids (366 and 293 mg/100 g, respectively). Sorghum, foxtail millet and finger millet also exhibit appreciable quantities (154–272 mg/100 g) of carotenoids (Longvah et al. 2017; Asharani et al. 2010). Lutein, zeaxanthin and β-carotene are majorly occurring carotenoids in millets. Foxtail millet contains the highest levels of lutein (59.40 mg/100 g) followed by pearl (29.69 mg/100 g) and finger millet (25.53 mg/100 g). Pearl millet also is high in zeaxanthin (9.30 mg/100 g) as well as β-carotene (28.23 mg/100 g) (Longvah et al. 2017). Carotenoids exert several health beneficial effects on humans. For example, they can reduce oxidative stress and inflammation by quenching hydroperoxyl and oxygen radicals (Lee et al. 2019). Lutein and zeaxanthin occur in eye tissues and have crucial roles in eye health and vision. Besides this, lutein can promote cognitive functions, prevent age-related macular disease and improve cardiovascular health (Buscemi et al. 2018). In addition, carotenoids exhibit protective actions against the development of non-alcoholic fatty liver disease and can prevent the risk of certain cancers owing to the ability to modulate gene expression and inhibit cell proliferation, transformation and micronucleus formation (Perera and Yen 2007).

Tocols consist of tocopherols and tocotrienols which are monophenols. They exist in four different forms (α, β, γ and δ), which differ from each other by the number and location of methyl groups in their chemical structures (Shahidi and De Camargo 2016). Besides exhibiting vitamin E activity, tocols exert disease-preventive effects in humans. For instance, tocols have protective effects on cell membranes owing to their strong chain-breaking antioxidant activity against peroxyl radicals. In addition, they exert gene expression modulation properties, cell proliferation inhibition and bone mass regulation (Borel and Desmarchelier 2018). Millets were reported to contain tocopherols in the α, γ and δ forms (Asharani et al. 2010; Longvah et al. 2017). In general, γ isomer was the most abundant tocopherol. The α-, γ- and δ-tocopherol content in different millets ranged from 0.03 to 1.80 mg/100 g, 0.27 to 2.60 mg/100 g and 0.02 to 0.720 mg/100 g, respectively. Amongst different millets, proso millet and pearl millet were particularly high in γ-tocopherol content (1.42–2.6 mg/100 g) (Longvah et al. 2017).

Saponins refer to a diverse group of plant compounds made up of triterpene or steroid aglycone with one or more sugar moieties. Millets have also been found to contain appreciable quantities of saponins. For example, pearl millet contains 29.1–39.53 mg/g saponins (Bhuvaneshwari et al. 2020) against 8.38–10.26 mg/g for barnyard millet and 2.13–2.63 mg/g for finger millet (Panwar et al. 2016). However, sorghum, in comparison, is lower in saponin content. Hassan et al. (2020) reported saponin content of 1.5 mg/g for sorghum grains. Saponins are conventionally regarded as an antinutrient occurring in certain food grains. However, a number of studies have linked saponins with the health-benefitting effects of plant-based diets. Various health benefits associated with saponins include reduced risk of ischemic heart diseases, renal diseases, diabetes, dental problems and certain cancers because of their ability to i) enhance faecal excretion of low-density lipoprotein cholesterol, ii) lower renal stone formation, iii) reduce blood glucose levels and iv) inhibit platelet aggregation, dental caries and metastasis (Sharma et al. 2023; Song et al. 2016).

Phytic acid, an inositol hexaphosphate, is a six-fold dihydrogenphospate ester of inositol (Brouns 2021). Phytic acid/phytates content of millets range between 7.29 and 14.6 mg/g (Hassan et al. 2020; Sharma and Gujral 2020). Phytic acid/phytates content was 7.89, 7.29, 9.14, 8.30, 8.34, 9.37 and 14.6 mg/g for finger millet, foxtail millet, barnyard millet, kodo millet, little millet, proso millet and sorghum, respectively (Hassan et al. 2020; Sharma and Gujral 2020). Similar as saponins, phytic acid has long been regarded as an antinutrient owing to its association with reduced nutrients available. However, it also exhibits nutraceutical properties such as antidiabetic, antioxidative, antiinflammatory, anticancer, neuroprotective and kidney stone prevention properties (Chen et al. 2023; Chen and Xu 2023; Shevkani 2023). The presence of hydroxyl and phosphate groups endow phytic acid molecules strong ability to inhibit free radicals (Abdulwaliyu et al. 2019) while antidiabetic properties were due to enhanced insulin sensitivity, reduced rate of starch digestion and increased viscosity of digesta due to the crosslinking of the food macromolecules with phytic acid resulting in delayed gastric emptying rates (Chen and Xu 2023).

11.4 Agro-economic Advantages

Millets have been described as a "wonder crop" due to many agro-economic advantages, including a resurgence of agro-diversity, less agricultural input requirements, high carbon sequestration and the ability to check soil erosion (Table 11.5). Compared to other cereals, millets have higher pest/disease resistance (Verma and Patel 2013) and require lesser use of agricultural inputs and labour-intensive operations for cultivation (Devi et al. 2014). Millet crops can be cultivated in marginal soils and low-input agricultural conditions, where staple cereal crops often produce low yields (Habiyaremye et al. 2017). In addition, millets are 70% more water-efficient than rice and can tolerate temperatures up to 42 °C (Kheya et al. 2023). The fact that millet can tolerate high temperatures makes it an ideal choice for tropical countries looking for a drought-resistant food solution (Saxena et al. 2018). Short plants, large leaf area index despite small leaf area, straight leaf type with high radiation use efficiency, thick cell walls, and dense root system are some of the characteristics that contribute to the mitigation of abiotic stresses (Li and Brutnell 2011). Furthermore, the C4 photosynthetic pathway of millet helps plants increase water and nutrient use efficiency as well as more flexible biomass allocation, lower hydraulic conductivity per unit leaf area, better growth, and increased tolerance to high temperatures (Sage and Zhu 2011). Therefore, millets are promising in achieving food security for smallholders living in dry areas and under harsher environmental conditions.

Table 11.5 Special agronomical characteristics associated with millets cultivation

Millet	Special agronomical characteristics
Sorghum	Better tolerance ability for moisture stress and high temperature as compared to other crops
Pearl millet	Tolerate to drought, rich in folic acid, and extremely tolerant to extreme heat; may be grown in very poor soils
Finger millet	Wider agroecological adaptability, modest tolerance for drought, temperature stress, and humidity, and a rich source of calcium
Foxtail millet	Short duration, adaptable to conditions of high altitude and little rainfall, tolerant to low soil fertility and drought
Proso millet	Short duration, tolerant to heat and drought, suited to high altitude and dry conditions
Barnyard millet	Fastest-growing, voluminous fodder suited to high elevations and dry climates
Kodo millet	It is a relatively long-duration crop, suitable for both shallow and deep soil, and a rich source of folic acid
Little millet	It is a short-duration crop, resistant to drought and waterlogging

Sources: Kheya et al. (2023); Paschapur et al. (2021); Sood et al. (2021); Seetharam (1998); Jiaju (1989); Gupta et al. (2009); Rao (2021); Refatti et al. (2019); Hegde and Gowda (1989); Doggett (1989)

In addition, the cultivation of millets has a notable rate of return on investment, even when agriculture faces climate change. The climate-resilient characteristics of millet continue to support sustainable agriculture without the need for chemical fertilizers (Thilakarathna and Raizada 2015). As injudicious use of fertilizers and pesticides in agricultural production can cause serious harm to the environment, millet cultivation is environmentally friendly as it has much less dependence on agrochemicals. Traditionally, millets have been produced as rain-fed crops for thousands of years, with little or no fertilizer inputs (Devkota et al. 2016). Another advantage associated with millet cultivation is lesser requirement of water compared to wheat and rice. Sharma (2016) reported that the increase in areas of highly water-requiring crops like rice, sugarcane, and cotton had resulted in a loss of approximately 7191 litre groundwater per hectare. Hence, the prospects for increasing the production of major cereals are diminishing as the world is already facing the challenges of an increase in dry lands and the deepening of groundwater tables (ICRISAT 2013; Kumar et al. 2018; Sharma 2016). According to the National Rainfed Area Authority (NRAA) report, even after full irrigation potential is realised, about half of the net sown area will remain rain-fed (Kumar et al. 2018; NRRA 2012). Millet cultivation could offer sustainable solutions to these problems. The rainfall requirement of certain millets, like pearl millet and proso millet, is as low as 20 cm, which is several times less than that of rice, requiring an average rainfall of 120–140 cm (Kumar et al. 2018). Most millets mature in 60–90 days after sowing, making them a water-saving crop. Barnyard millet has the shortest maturity time among millets at 45–70 days, which is almost half the maturity time of rice (Kumar et al. 2018; Hulse et al. 1980). Millets are C4 cereals; they take more carbon dioxide from the atmosphere and convert it to oxygen (Kumar et al. 2018). Thus, millets are helpful in phasing out climatic uncertainties and reducing atmospheric carbon dioxide (Sadhukhan and Debangshi 2023; Senthilkumar et al. 2023).

Moreover, scientific interventions and recent advancements in terms of agricultural research have led to the release of high-yielding varieties of millet throughout the world (Kumar et al. 2018; Joel et al. 2005). Newly developed hybrids are more resistant to diseases and have increased per-hectare production as compared to their parent varieties (Kumar et al. 2018; Joel et al. 2005). There have been new technological advancements in the post-harvesting processes of millets, which have made it easier to process and add value to millets. Machines related to millets, like threshing, decortication, destoning, and polishing, are designed. Due to low investments and expenses required to produce millets, they are proving to be a sustainable and viable income source for small and marginal farmers. There was a time when millets had disappeared from the fields and diets. However, again, it has started being consumed by people in the form of different types of value-added products (Kumar et al. 2021) in many regions of the world. The year 2023 was celebrated globally as the International Year of Millets owing to the nutritional and agro-economic benefits of millets.

11.5 Conclusion

Millets are rich source of several nutrients and nutraceuticals (proteins, resistant starch, slowly digestible starch, non-starch polysaccharides, vitamins, minerals, bioactive peptides, phenolics, saponins, phytates/phytic acids, carotenoids and tocols). The consumption of millets has been associated with reduced risk of lifestyle related chronic diseases. In addition to their nutritional and health-benefitting properties, millet cultivation presents many agro-economic advantages including but not limited to (i) short crop duration, (ii) ability to grow in marginal soils under rainfed conditions, (iii) resistance to high temperatures, water stress and diseases/pests (iv) lesser requirements of agricultural inputs and (vi) lower labour-intensive farm operations. Therefore, millets are promising not only in achieving food security in dry areas and under harsher environmental conditions but also can be helpful in achieving sustainability in agriculture through phasing out climatic uncertainties and lowering carbon footprint.

References

Abdulwaliyu I, Arekemase SO, Adudu JA, Batari ML, Egbule MN, Okoduwa SIR (2019) Investigation of the medicinal significance of phytic acid as an indispensable anti-nutrient in diseases. Clin Nutr Exp 28:42–61. https://doi.org/10.1016/j.yclnex.2019.10.002

Agrawal H, Joshi R, Gupta M (2016) Isolation, purification and characterization of antioxidative peptide of pearl millet (Pennisetum glaucum) protein hydrolysate. Food Chem 204:365–372. https://doi.org/10.1016/j.foodchem.2016.02.127

Agrawal H, Joshi R, Gupta M (2019) Purification, identification and characterization of two novel antioxidant peptides from finger millet (Eleusine coracana) protein hydrolysate. Food Res Int 120:697–707. https://doi.org/10.1016/j.foodres.2018.11.028

Ajeigbe HA, Akinseye FM, Jonah J, Kunihya A (2018) Sorghum yield and water use under phosphorus fertilization applications in The Sudan savanna of Nigeria. Glob Adv Res J Agric Sci 7:245–257

Amarakoon D, Lou Z, Lee W, Smolensky D, Lee S (2021) A mechanistic review: potential chronic disease-preventive properties of sorghum. J Sci Food Agric 101:2641–2649. https://doi.org/10.1002/jsfa.10933

Anitha S, Govindaraj M, Kane-Potaka J (2020) Balanced amino acid and higher micronutrients in millets complements legumes for improved human dietary nutrition. Cereal Chem 97:74–84. https://doi.org/10.1002/cche.10227

Annor GA, Marcone M, Bertoft E, Seetharaman K (2014) Physical and molecular characterization of millet starches. Cereal Chem 91:286–292. https://doi.org/10.1094/CCHEM-08-13-0155-R

Annor GA, Tyl C, Marcone M, Ragaee S, Marti A (2017) Why do millets have slower starch and protein digestibility than other cereals? Trends Food Sci Technol 66:73–83. https://doi.org/10.1016/j.tifs.2017.05.012

Asharani VT, Jayadeep A, Malleshi NG (2010) Natural antioxidants in edible flours of selected small millets. Int J Food Prop 13:41–50. https://doi.org/10.1080/10942910802163105

Bello OM, Ogbesejana AB, Balkisu A, Osibemhe M, Musa B, Oguntoye SO (2022) Polyphenolic fractions from three millet types (Fonio, finger millet, and pearl millet): their characterization and biological importance. Clin Complement Med Pharmacol 2:100020. https://doi.org/10.1016/j.ccmp.2022.100020

Bhuvaneshwari G, Nirmalakumari A, Kalaiselvi S (2020) Impact of soaking, sprouting on antioxidant and anti-nutritional factors in millet grains. J Phytology:62–66. https://doi.org/10.25081/jp.2020.v12.6384

Bora P, Ragaee S, Marcone M (2019) Characterisation of several types of millets as functional food ingredients. Int J Food Sci Nutr 70:714–724. https://doi.org/10.1080/09637486.2019.1570086

Borel P, Desmarchelier C (2018) Bioavailability of fat-soluble vitamins and phytochemicals in humans: effects of genetic variation. Annu Rev Nutr 38:69–96. https://doi.org/10.1146/annurev-nutr-082117-051628

Bozbulut R, Sanlier N (2019) Promising effects of β-glucans on glyceamic control in diabetes. Trends Food Sci Technol 83:159–166. https://doi.org/10.1016/j.tifs.2018.11.018

Brosnan JT, Brosnan ME (2013) Glutamate: A truly functional amino acid. Amino Acids 45:413–418. https://doi.org/10.1007/s00726-012-1280-4

Brouns F (2021) Phytic acid and whole grains for health controversy. Nutrients 14:25. https://doi.org/10.3390/nu14010025

Buscemi S, Corleo D, Di Pace F, Petroni M, Satriano A, Marchesini G (2018) The effect of lutein on eye and extra-eye health. Nutrients 10:1321. https://doi.org/10.3390/nu10091321

Camargo Filho I, Cortez DAG, Ueda-Nakamura T, Nakamura CV, Dias Filho BP (2008) Antiviral activity and mode of action of a peptide isolated from Sorghum bicolor. Phytomedicine 15:202–208. https://doi.org/10.1016/j.phymed.2007.07.059

Chandra D, Chandra S, Pallavi, Sharma AK (2016) Review of finger millet (Eleusine coracana (L.) Gaertn): A power house of health benefiting nutrients. Food Sci Human Wellness 5:149–155. https://doi.org/10.1016/j.fshw.2016.05.004

Chandrasekara A, Shahidi F (2011) Inhibitory activities of soluble and bound millet seed Phenolics on free radicals and reactive oxygen species. J Agric Food Chem 59:428–436. https://doi.org/10.1021/jf103896z

Chao G, Gao J, Liu R, Wang L, Li C, Wang Y, Qu Y, Feng B (2014) Starch physicochemical properties of waxy proso millet (Panicum Miliaceum L.). Starch-Stärke 66:1005–1012. https://doi.org/10.1002/star.201400018

Chen W, Xu D (2023) Phytic acid and its interactions in food components, health benefits, and applications: a comprehensive review. Trends Food Sci Technol:104201

Chen Y, Yuan W, Xu Q, Reddy MB (2023) Neuroprotection of phytic acid in Parkinson's and Alzheimer's disease. J Funct Foods 110:105856

Chethan S, Dharmesh SM, Malleshi NG (2008) Inhibition of aldose reductase from cataracted eye lenses by finger millet (Eleusine coracana) polyphenols. Bioorg Med Chem 16:10085–10090. https://doi.org/10.1016/j.bmc.2008.10.003

Cloetens L, Ulmius M, Johansson-Persson A, Åkesson B, Önning G (2012) Role of dietary beta-glucans in the prevention of the metabolic syndrome. Nutr Rev 70:444–458. https://doi.org/10.1111/j.1753-4887.2012.00494.x

Damodaran S (2017) Amino acids, peptides and proteins. In: Damodaran S, Parkin KL (eds) Fennema's food chemistry, 5th edn. CRC Press, Boca Raton, pp 235–356

Devi PB, Vijayabharathi R, Sathyabama S, Malleshi NG, Priyadarisini VB (2014) Health benefits of finger millet (Eleusine coracana L.) polyphenols and dietary fiber: a review. J Food Sci Technol 51:1021–1040. https://doi.org/10.1007/s13197-011-0584-9

Devkota R, Khadka K, Gartaula H (2016) Gender and labour efficiency in finger millet production in Nepal. In: Transforming gender and food security in the global south. Routledge, pp 76–95

Doggett H (1989) Small millets-a selective overview. In: Seetharam A et al (eds) Small millets in global agriculture. Oxford & IBH Publishing Co, Delhi, pp 3–17. Retrieved from https://millets.res.in/books/Small_millets.pdf#page=230

Englyst HN, Kingman SM, Cummings JH (1992) Classification and measurement of nutritionally important starch fractions. Eur J Clin Nutr 46:S33–S50

Gaffa T, Yoshimoto Y, Hanashiro I, Honda O, Kawasaki S, Takeda Y (2004) Physicochemical properties and molecular structures of starches from millet (Pennisetum typhoides) and sorghum (Sorghum bicolor L. Moench) cultivars in Nigeria. Cereal Chem 81:255–260. https://doi.org/10.1094/CCHEM.2004.81.2.255

Gao F, Li X, Li X, Liu Z, Zou X, Wang L, Zhang H (2022) Physicochemical properties and correlation analysis of retrograded starch from different varieties of sorghum. Int J Food Sci Technol 57:6678–6689. https://doi.org/10.1111/ijfs.16013

Gebreyohannes A, Shimelis H, Laing M, Mathew I, Odeny DA, Ojulong H (2021) Finger millet production in Ethiopia: opportunities, problem diagnosis, key challenges and recommendations for breeding. Sustain For 13(23):13463. https://doi.org/10.3390/su132313463

Geetha K, Yankanchi GM, Hulamani S, Hiremath N (2020) Glycemic index of millet based food mix and its effect on pre diabetic subjects. J Food Sci Technol 57:2732–2738. https://doi.org/10.1007/s13197-020-04309-5

Gogus U, Smith C (2010) N-3 omega fatty acids: a review of current knowledge. Int J Food Sci Technol 45:417–436

Goudar G, Manne M, Sathisha GJ, Sharma P, Mokalla TR, Kumar SB, Ziouzenkova O (2023) Phenolic, nutritional and molecular interaction study among different millet varieties. Food Chem Adv 2:100150. https://doi.org/10.1016/j.focha.2022.100150

Gowda NAN, Siliveru K, Prasad PVV, Bhatt Y, Netravati BP, Gurikar C (2022) Modern processing of Indian millets: A perspective on changes in nutritional properties. Food Secur 11:499. https://doi.org/10.3390/foods11040499

Gupta A, Mahajan V, Kumar M, Gupta HS (2009) Biodiversity in the barnyard millet (Echinochloa frumentacea link, Poaceae) germplasm in India. Genet Resour Crop Evol 56:883–889. https://doi.org/10.1007/s10722-009-9462-y

Habiyaremye C, Matanguihan JB, D'Alpoim Guedes J, Ganjyal GM, Whiteman MR, Kidwell KK, Murphy KM (2017) Proso millet (Panicum miliaceum L.) and its potential for cultivation in the Pacific northwest, US: a review. Front Plant Sci 7:1961. https://doi.org/10.3389/fpls.2016.01961

Hassan S, Imran M, Ahmad MH, Khan MI, Xu C, Khan MK, Muhammad N (2020) Phytochemical characterization of ultrasound-processed sorghum sprouts for the use in functional foods. Int J Food Prop 23:853–863

Hegde BR, Gowda L (1989) Cropping systems and production technology for small millets in India. In: Seetharam A et al (eds) Small millets in global agriculture. Oxford & IBH Publishing Co., Delhi, pp 209–236. Retrieved from https://millets.res.in/books/Small_millets.pdf#page=230

Hulse JH, Laing EM, Pearson OE (1980) Sorghum and the millets: their composition and nutritive value. Academic

ICRISAT. Small millets. Retrieved from http://www.icrisat.org/homepage. Accessed on Dec 2013

Jiaju C (1989) Importance and genetic resources of small millets with emphasis on foxtail millet (Setaria italica) in China. In: Seetharam A et al (eds) Small millets in global agriculture. Oxford & IBH Publishing Co., Delhi, pp 93–100. Retrieved from https://millets.res.in/books/Small_millets.pdf#page=230

Joel A, Kumaravadivel N, Nirmalakumari A, Senthil N, Mohanasundaram K, Raveendran T, Mallikavangamudi V (2005) A high yielding finger millet variety CO (Ra) 14. Madras Agric J 92:375–380

Johnston KL, Thomas EL, Bell JD, Frost GS, Robertson MD (2010) Resistant starch improves insulin sensitivity in metabolic syndrome. Diabet Med 27:391–397. https://doi.org/10.1111/j.1464-5491.2010.02923.x

Juhaimi FA, Şimşek Ş, Ghafoor K, Babiker EE, Özcan MM, Ahmed IAM, Alsawmahi O (2019) Effect of varieties on bioactive properties and mineral contents of some sorghum, millet and Lupin seeds. J Oleo Sci 68:1063–1071. https://doi.org/10.5650/jos.ess19113

Kaimal AM, Mujumdar AS, Thorat BN (2021) Resistant starch from millets: recent developments and applications in food industries. Trends Food Sci Technol 111:563–580. https://doi.org/10.1016/j.tifs.2021.02.074

Kheya SA, Talukder SK, Datta P, Yeasmin S, Rashid MH, Hasan AK, Anwar MP, Islam AA, Islam AM (2023) Millets: the future crops for the tropics – status, challenges and future prospects. Heliyon 9:e22123. https://doi.org/10.1016/j.heliyon.2023.e22123

Kim SK, Sohn EY, Lee IJ (2009) Starch properties of native foxtail millet, Setaria italica Beauv. J Crop Sci Biotechnol 12:59–62. https://doi.org/10.1007/s12892-009-0073-0

Krishnamurthy L, Upadhyaya HD, Gowda CLL, Kashiwagi J, Purushothaman R, Singh S, Vadez V (2014) Large variation for salinity tolerance in the core collection of foxtail millet (Setaria italica (L.) P. Beauv.) germplasm. Crop Pasture Sci 65:353. https://doi.org/10.1071/CP13282

Kumar A, Metwal M, Kaur S, Gupta AK, Puranik S, Singh S, Singh M, Gupta S, Babu BK, Sood S, Yadav R (2016) Nutraceutical value of finger millet [Eleusine coracana (L.) Gaertn.], and their improvement using omics approaches. Front Plant Sci 7. https://doi.org/10.3389/fpls.2016.00934

Kumar A, Tomer V, Kaur A, Kumar V, Gupta K (2018) Millets: A solution to agrarian and nutritional challenges. Agric Food Secur 7:31. https://doi.org/10.1186/s40066-018-0183-3

Kumar A, Tripathi MK, Joshi D, Kumar V (eds) (2021) Millets and millet technology (p. 438). Springer, Singapore

Kumari D, Chandrasekara A, Athukorale P, Shahidi F (2020) Finger millet porridges subjected to different processing conditions showed low glycemic index and variable efficacy on plasma antioxidant capacity of healthy adults. Food Prod Process Nutr 2:13. https://doi.org/10.1186/s43014-020-00027-9

Kumari S, Bhinder S, Singh B, Kaur A (2023) Physicochemical properties, non-nutrients and phenolic composition of germinated freeze-dried flours of foxtail millet, proso millet and common buckwheat. J Food Compos Anal 115:105043. https://doi.org/10.1016/j.jfca.2022.105043

Lee Y, Hu S, Park YK, Lee JY (2019) Health benefits of carotenoids: A role of carotenoids in the prevention of non-alcoholic fatty liver disease. Prev Nutr Food Sci 24:103–113. https://doi.org/10.3746/pnf.2019.24.2.103

Li P, Brutnell TP (2011) Setaria viridis and Setaria italica, model genetic systems for the Panicoid grasses. J Exp Bot 62:3031–3037. https://doi.org/10.1093/jxb/err096

Li Y, Hu A, Wang X, Zheng J (2019) Physicochemical and in vitro digestion of millet starch: effect of moisture content in microwave. Int J Biol Macromol 134:308–315

Longvah T, Ananthan R, Bhaskarachary K, Venkaiah K (2017) Indian food composition tables. NIN, Hyderabad

Mahajan P, Bera MB, Panesar PS, Chauhan A (2021) Millet starch: a review. Int J Biol Macromol 180:61–79. https://doi.org/10.1016/j.ijbiomac.2021.03.063

Majid A, Poornima Priyadarshini CG (2020) Millet derived bioactive peptides: A review on their functional properties and health benefits. Crit Rev Food Sci Nutr 60:3342–3351. https://doi.org/10.1080/10408398.2019.1686342

Meena RP, Joshi D, Bisht JK, Kant L (2021) Global scenario of millets cultivation. In: Kumar A, Tripathi MK, Joshi D, Kumar V (eds) Millets and millet technology. Springer, Singapore, pp 33–50. https://doi.org/10.1007/978-981-16-0676-2_2

Mrema E, Shimelis H, Laing M, Bucheyeki T (2017) Farmers' perceptions of sorghum production constraints and Striga control practices in semi-arid areas of Tanzania. Int J Pest Manag 63:146–156

National Rainfed Area Authority (NRRA) (2012). http://www.indiaenvi-ronmentportal.org.in/category/28905/publisher/national-rainfed-area-authority/. Accessed 14 Dec 2023

Nithiyanantham S, Kalaiselvi P, Mahomoodally MF, Zengin G, Abirami A, Srinivasan G (2019) Nutritional and functional roles of millets-A review. J Food Biochem 43:e12829. https://doi.org/10.1111/jfbc.12859

Panda D, Sailaja NH, Padhan B, Lenka K (2020) Sprouting-associated changes in nutritional and Physico-functional properties of indigenous millets from Koraput, India. Proc Natl Acad Sci India Sect B Biol Sci 90:79–86. https://doi.org/10.1007/s40011-019-01085-x

Panwar P, Dubey A, Verma AK (2016) Evaluation of nutraceutical and antinutritional properties in barnyard and finger millet varieties grown in Himalayan region. J Food Sci Technol 53:2779–2787. https://doi.org/10.1007/s13197-016-2250-8

Paschapur AU, Joshi D, Mishra KK, Kant L, Kumar V, Kumar A (2021) Millets for life: a brief introduction. In: Millets and Millet Technology, 1–32. PK Publishers. https://doi.org/10.1007/978-981-16-0676-2_1

Perera CO, Yen GM (2007) Functional properties of carotenoids in human health. Int J Food Prop 10:201–230. https://doi.org/10.1080/10942910601045271

Prabhakar, Ganiger PC, Boraiah B, Bhat S, Nandini C, Kiran, Tippeswamy V, Manjunath HA (2017) Improved production technology for proso millet. ICAR-AICRP on Small Millets, Bengaluru

Pushparaj FS, Urooj A (2011) Influence of processing on Dietary Fiber, Tannin and in Vitro protein digestibility of Pearl Millet. Food Nutr Sci 02:895–900. https://doi.org/10.4236/fns.2011.28122

Qiang X, Yong-Lie C, Qian-Bing W (2009) Health benefit application of functional oligosaccharides. Carbohydr Polym 77:435–441

Raheem D, Dayoub M, Birech R, Nakiyemba A (2021) The contribution of cereal grains to food security and sustainability in Africa: potential application of UAV in Ghana, Nigeria, Uganda, and Namibia. Urban Sci 5:8. https://doi.org/10.3390/urbansci5010008

Rahmani J, Miri A, Černevičiūtė R, Thompson J, De Souza NN, Sultana R, Kord Varkaneh H, Mousavi SM, Hekmatdoost A (2019) Effects of cereal beta-glucan consumption on body weight, body mass index, waist circumference and total energy intake: a meta-analysis of randomized controlled trials. Complement Ther Med 43:131–139. https://doi.org/10.1016/j.ctim.2019.01.018

Ramadoss DP, Sivalingam N (2020) Vanillin extracted from Proso and barnyard millets induce apoptotic cell death in HT-29 human colon cancer cell line. Nutr Cancer 72:1422–1437. https://doi.org/10.1080/01635581.2019.1672763

Ranjan R, Singh S, Dhua S, Mishra P, Chauhan AK, Gupta AK (2023) Kodo Millet (Paspalum scrobiculatum). In: Sharma R, Nanda V, Sharma S (eds) Nutri-Cereals, 1st edn. CRC Press, pp 193–211. https://doi.org/10.1201/9781003251279-8

Rao AN (2021) Echinochloa colona and Echinochloa crus-galli. In: Biology and management of problematic crop weed species. Academic, pp 197–239

Rao BD, Bhaskarachary K, Christina GDA, Devi GS, Vilas AT (2017) Nutritional and Health Benefits of Millets. ICAR_Indian Institute of Millets Research (IIMR), Rajendranaga

Ravikesavan R, Jeeva G, Jency JP, Muthamilarasan M, Francis N (2023) Kodo Millet (Paspalum scorbiculatum L.). In: Neglected and underutilized crops. Academic, pp 279–304

Refatti JP, De Avila LA, Camargo ER, Ziska LH, Oliveira C, Salas-Perez R, Roma-Burgos N (2019) High [CO_2] and temperature increase resistance to cyhalofop-butyl in multiple-resistant Echinochloa colona. Front Plant Sci 10:529. https://doi.org/10.3389/fpls.2019.00529

Ríos-Covián D, Ruas-Madiedo P, Margolles A, Gueimonde M, De Los Reyes-Gavilán CG, Salazar N (2016) Intestinal short chain fatty acids and their link with diet and human health. Front Microbiol 7. https://doi.org/10.3389/fmicb.2016.00185

Sachdev N, Goomer S, Singh LR (2021) Foxtail millet: A potential crop to meet future demand scenario for alternative sustainable protein. J Sci Food Agric 101:831–842. https://doi.org/10.1002/jsfa.10716

Sadhukhan A, Debangshi U (2023) Millets in meeting the nutrition security: a review. Just Agric e-mag 3:35–40

Sage RF, Zhu XG (2011) Exploiting the engine of C4 photosynthesis. J Exp Bot 62:2989–3000. https://doi.org/10.1093/jxb/err179

Saldivar SO (2016) Cereals: dietary importance. In: Encyclopedia of food and health. Elsevier, pp 703–711. https://doi.org/10.1016/B978-0-12-384947-2.00130-6

Saleem S, Mushtaq NU, Shah WH, Rasool A, Hakeem KR, Seth CS, Tahir I, Rehman RU (2023) Millets as smart future food with essential phytonutrients for promoting health. J Food Compos Anal 124:105669. https://doi.org/10.1016/j.jfca.2023.105669

Saleh ASM, Zhang Q, Chen J, Shen Q (2013) Millet grains: nutritional quality, processing, and potential health benefits. Compr Rev Food Sci Food Saf 12(3):281–295. https://doi.org/10.1111/1541-4337.12012

Sang Y, Bean S, Seib PA, Pedersen J, Shi YC (2008) Structure and functional properties of sorghum starches differing in amylose content. J Agric Food Chem 56:6680–6685. https://doi.org/10.1021/jf800577x

Satyavathi CT, Ambawat S, Khandelwal V, Srivastava RK (2021) Pearl millet: A climate-resilient Nutricereal for mitigating hidden hunger and provide nutritional security. Front Plant Sci 12. https://doi.org/10.3389/fpls.2021.659938

Saxena R, Vanga SK, Wang J, Orsat V, Raghavan V (2018) Millets for food security in the context of climate change: a review. Sustain For 10(7):2228. https://doi.org/10.3390/su10072228

Seetharam A (1998) Small millets research: achievements during 1947-97. Indian J Agric Sci 68: 431–438. https://epubs.icar.org.in/index.php/IJAgS/article/view/27642

Senthilkumar A, Saliha BB, Sheeba S, Balamurugan R, Jacob G, Raj JV (2023) Millets in meeting the nutrition security: a review. Pharm Innov J 12:3554–3557. Retrieved from https://www.thepharmajournal.com/archives/2023/vol12issue3/PartAL/12-2-401-371.pdf

Serna-Saldivar. (2010) Cereal grains: properties, processing, and nutritional attributes. In: Cereal grains: properties, processing, and nutritional attributes

Shahidi F, De Camargo A (2016) Tocopherols and Tocotrienols in common and emerging dietary sources: occurrence, applications, and health benefits. Int J Mol Sci 17:1745. https://doi.org/10.3390/ijms17101745

Shahidi F, Danielski R, Ikeda C (2021) Phenolic compounds in cereal grains and effects of processing on their composition and bioactivities: A review. J Food Bioactiv 15. https://doi.org/10.31665/JFB.2021.15281

Sharma CP (2016) Overdraft in India's water banks: studying the effect of production of water intensive crops on groundwater depletion. Master Thesis-Georgetown University, Washington DC. Retrieved from https://www.proquest.com/openview/8ada408b6ad1238ef821c42a7e3f4a4e/1?pq-origsite=gscholar&cbl=18750

Sharma N, Niranjan K (2018) Foxtail millet: properties, processing, health benefits, and uses. Food Rev Intl 34(4):329–363. https://doi.org/10.1080/87559129.2017.1290103

Sharma B, Gujral HS (2020) Modifying the dough mixing behavior, protein & starch digestibility and antinutritional profile of minor millets by sprouting. Int J Biol Macromol 153:962–970. https://doi.org/10.1016/j.ijbiomac.2019.10.225

Sharma R, Sharma S, Dar BN, Singh B (2021) Millets as potential nutri-cereals: a review of nutrient composition, phytochemical profile and techno-functionality. Int J Food Sci Technol 56:3703–3718. https://doi.org/10.1111/ijfs.15044

Sharma K, Kaur R, Kumar S, Saini RK, Sharma S, Pawde SV, Kumar V (2023) Saponins: A concise review on food related aspects, applications and health implications. Food Chem Adv 2: 100191. https://doi.org/10.1016/j.focha.2023.100191

Shevkani K (2023) Protein from land-legumes and pulses. In: Future proteins. Elsevier, pp 35–68. https://doi.org/10.1016/B978-0-323-91739-1.00003-9

Shevkani K, Singh N, Bajaj R, Kaur A (2017) Wheat starch production, structure, functionality and applications-A review. Int J Food Sci Technol 52:38–58. https://doi.org/10.1111/ijfs.13266

Shobana S, Sreerama YN, Malleshi NG (2009) Composition and enzyme inhibitory properties of finger millet (Eleusine coracana L.) seed coat phenolics: mode of inhibition of α-glucosidase and pancreatic amylase. Food Chem 115:1268–1273. https://doi.org/10.1016/j.foodchem.2009.01.042

Song S, Cong P, Xu J, Li G, Liu X, Li Z, Xue C, Xue Y, Wang Y (2016) Absorption and pharmacokinetic study of two sulphated triterpenoid saponins in rat after oral and intravenous administration of saponin extracts of Pearsonothuria graeffei by HPLC-MS. J Funct Foods 25: 62–69. https://doi.org/10.1016/j.jff.2016.05.002

Sood M, Salgotra RK, Jasrotia S (2021) Underutilized crops and their value addition. Nova Science Publishers, New Delhi

Thakur S (2023) Finger millet: the grain of goodness. Just Agric: Multi-discip e-Newsl 3:446–450

Thilakarathna MS, Raizada MN (2015) A review of nutrient management studies involving finger millet in the semi-arid tropics of Asia and Africa. Agronomy 5:262–290. https://doi.org/10.3390/agronomy5030262

Thondre PS, Ryan L, Henry CJK (2011) Barley β-glucan extracts as rich sources of polyphenols and antioxidants. Food Chem 126:72–77. https://doi.org/10.1016/j.foodchem.2010.10.074

Vadivel V, Janardhanan K (2005) Nutritional and Antinutritional characteristics of seven south Indian wild legumes. Plant Foods Hum Nutr 60:69–75. https://doi.org/10.1007/s11130-005-5102-y

Vali Pasha K, Ratnavathi CV, Ajani J, Raju D, Manoj Kumar S, Beedu SR (2018) Proximate, mineral composition and antioxidant activity of traditional small millets cultivated and consumed in Rayalaseema region of South India. J Sci Food Agric 98:652–660. https://doi.org/10.1002/jsfa.8510

Verma V, Patel S (2013) Value added products from nutri-cereals: finger millet (Eleusine coracana). Emir J Food Agric 25:169–176. https://doi.org/10.9755/ejfa.v25i3.10764

Vetriventhan M, Upadhyaya HD, Anandakumar CR, Senthilvel S, Parzies HK, Bharathi A, Varshney RK, Gowda CLL (2012) Assessing genetic diversity, allelic richness and genetic relationship among races in ICRISAT foxtail millet core collection. Plant Genetic Resour 10: 214–223

WHO (2004) Vitamin and mineral requirements in human nutrition, Second edn. FAO/WHO, Geneva, Switzerland

Wu Y, Lin Q, Cui T, Xiao H (2014) Structural and physical properties of starches isolated from six varieties of millet grown in China. Int J Food Prop 17:2344–2360. https://doi.org/10.1080/10942912.2013.803119

Wu G, Bazer FW, Johnson GA, Herring C, Seo H, Dai Z, Wang X (2017) Functional amino acids in the development of the pig placenta. Mol Reprod Dev 84:870–882

Xu J, Wang W, Zhao Y (2021) Phenolic compounds in whole grain sorghum and their health benefits. Food Secur 10:1921. https://doi.org/10.3390/foods10081921

Yi D, Li B, Hou Y, Wang L, Zhao D, Chen H, Wu G (2018) Dietary supplementation with an amino acid blend enhances intestinal function in piglets. Amino Acids 50:1089–1100

Zhang A, Ji Y, Sun M, Lin C, Zhou P, Ren J, Luo D, Wang X, Ma C, Zhang X, Feng G, Nie G, Huang L (2021) Research on the drought tolerance mechanism of Pennisetum glaucum (L.) in the root during the seedling stage. BMC Genomics 22:568. https://doi.org/10.1186/s12864-021-07888-5

Zhu F (2014) Structure, physicochemical properties, and uses of millet starch. Food Res Int 64:200–211. https://doi.org/10.1016/j.foodres.2014.06.026

Chapter 12
Rooted Solutions: The Role of Plant Biology in Climate Change Mitigation

Ayushi Sharma, DikshaVaishnav, and Parul Chowdhury

Abstract In the context of human history, climate change has traditionally transpired gradually over extended temporal scales, enabling plant species to adapt to fluctuations in climatic variables such as temperature, precipitation, and atmospheric composition. The consequential impact of climate variability on global agriculture, encompassing crop and livestock production as well as fisheries and aquaculture, has been profound. This shift in climate dynamics has manifested in notable adverse effects on crop plants, including reductions in crop yields across various regions and diminished seed protein content in cereal crops. Agricultural activities contribute significantly to Greenhouse Gas (GHG) emissions, particularly through the production and application of fertilizers, leading to the release of Nitrous Oxide—a GHG with a heat-trapping potency 298-fold greater than Carbon dioxide (CO_2) (Hassan et al. 2022). The proactive management of CO_2 levels, a pivotal factor in global warming, is essential for climate regulation. Recognizing the pivotal role of plants in the carbon cycle, particularly through photosynthesis and silicate weathering, offers avenues for intervention. Notably, the cultivation of cereal crops, constituting 75% of global food production, exhibits a unique capacity to extract nitrogen from the soil through metabolic interactions with Nitrogen-fixing bacteria. Simultaneously, legumes, exemplified by bean and pea plants, engage in a symbiotic relationship with rhizobia bacteria, forming root nodules that facilitate nitrogen acquisition in exchange for carbon, thereby minimizing greenhouse gas emissions and diminishing reliance on fossil energy. This approach holds promise for sustainable and climate-resilient agricultural practices on a global scale.

Keywords Climate change · Greenhouse gas emissions · Carbon cycle · Photosynthesis · Nitrogen-fixing bacteria · Sustainable agriculture · Climate resilience

---

A. Sharma · DikshaVaishnav · P. Chowdhury (✉)
Dr B. Lal Institute of Biotechnology, 6E, Malviya Industrial Area, Jaipur, Rajasthan, India
e-mail: parul@drblal.com

S. Kanga et al. (eds.), *Transforming Agricultural Management for a Sustainable Future*, World Sustainability Series, https://doi.org/10.1007/978-3-031-63430-7_12

12.1 Introduction

Climate change is the biggest problem in the twenty-first century. Due to major changes in air-quality, changing patterns of infections, and decrement in the food products increasing the death and illness. Climate change is affecting ecosystems through changes in climatic conditions such as increased ocean acidification and atmospheric CO_2 concentration. Changes in the atmosphere and oceans affect the biosphere, the living life on earth which is coupled to the atmosphere and hydrosphere. If the planet warms beyond 1.5 that will result in climate change, including drought, flood, heat waves and sea level rise that is very dangerous for humanity and for biodiversity (Malhi et al. 2020). Climate change is caused by the emission of greenhouse gas (like methane, CO_2) in the atmosphere leading to increased global warming. Climate change is not only caused by increasing temperature and heat wave but it also include changes in rainfall, frequent storms, increased drought and increased chances of wildfire (Jägermeyr et al. 2021).

Greenhouse gas emission needs to be reduced atmospheric CO_2 removal and technologies need to be implemented in the atmosphere (National Academies of Sciences, Engineering, and Medicine 2022). Plants play a critical role in the climate system by pulling CO_2 out of the atmosphere, plants carry out photosynthesis which convert CO_2 into plant biomass that store carbon. Plants increase the rate of silicate weathering, a chemical reaction that breaks down rocks and removes CO_2 from the atmosphere (Pires and Dolan 2012). Plants change mitigation involves removing and preventing greenhouse gas from the atmosphere. By the help of plants we can reduce GHG by some perspective:

- Use viable agricultural strategies
- Protect high carbon ecosystem that store carbon in soil, plants, forests
- Select the selective biodiverse cropping system
- Eco-Architecture can store carbon and increase productivity in urban landscape
- Genebanks provide diverse plant materials (Altieri et al. 2015)

12.1.1 The Role of Plant Biology in Addressing Climate Change

The plants play an important role in mitigating and adapting to climate change (Altieri et al. 2015). The further climatic conditions affect the growing seasons and farming system (Bowerman et al. 2023). Salt stress tolerance is one of the important breeding targets for how plants respond and adapt to salt stress. Soil salinization affects a large percentage of croplands around the world. Two articles, one explaining the protein function in guard cells, related to abscisic acid (ABA) activates Ca_{2+} signalling (Tan et al. 2023). Another one explains the disassembly of microtubules regulated by a module (Tong et al. 2023). These both articles describe the important role of stomatal regulation in water loss controlling, which is important to avoid dehydration and water use efficiency (Tan et al. 2023). Plant biology plays an important role in developing crops which have the ability to tolerate the harsh conditions and engineered the plants that are capable of capturing and storing carbon (Verslues et al. 2023). Plant biology worked on stopping greenhouse gas emission to achieve net zero CO_2 emission, improved the crop and increased the food security (Verslues et al. 2023).

12.2 Carbon Sequestration

12.2.1 Terrestrial Carbon Storage

Reduction of carbon emission and improving the capacity of storage of carbon is becoming the critical issue (Zhang et al. 2021). The terrestrial ecosystem plays a major role in carbon storage in the world (Zhang et al. 2022). The terrestrial ecosystem affects the climate system by exchanging carbon between the subsystems and atmosphere. Carbon storage increases in the terrestrial ecosystem by capturing and storing carbon in the atmosphere that leads to reduction in CO_2 concentration (Liu et al. 2019b). The changes in carbon storage affected by land use and cover changes that play an important role in balance and stability in the carbon cycle in terrestrial ecosystems (Liu et al. 2019a). Agroforestry is the useful adaptation that helps in proper land use for shrubs and trees, it protects biodiversity and stores CO_2, blocks wind and provides food (Abbas et al. 2017). Forests have woody plants that absorb 15–20% of CO_2 of annual human carbon emission and helps in soil health by maintaining organic matter in soil (Udawatta et al. 2022; Le Quéré et al. 2018; Case et al. 2021). Organic agriculture increases soil fertility and supports pest management that enhances the production of farms to climate change (Loboguerrero et al. 2019). Grassland ecosystems play an important role in ecological communities (Houghton 1994; Bahadur et al. 2015). They play a major role in carbon cycle regulation in climate and maintain the biological diversity (Wang et al. 2016; Wilcox et al. 2015).

12.2.2 Afforestation and Reforestation

Afforestation is the growing of trees and forest in an area where there is no remains of forest or tree in previous time to improve the ecosystem in that area. The need of afforestation is for ecological function, conserving biodiversity. Afforestation plays a major role in temperature regulation (Vivek and Montenegro 2011; Peng et al. 2014) many other importance of afforestation like introduction of new species, carbon- sequestration, food, firewood, beekeeping, grazing area, medicinal herbs (Nilsson and Schopfhauser 1995). Tree growth in afforestation helps in reducing air pollution, decreasing the greenhouse gas, conserving water and maintaining soil health, saving energy and environment, creating wildlife and plant diversity (HWF H.W. Foundation 2017). When human civilization began the trees were cut down more globally, this increased greenhouse gas (Pedersen et al. 2022). Restoring forest is not easy but it is important to decrease global warming (Pedersen et al. 2022). Restoring the forest is costly and consumes more time in selecting the species and in seedling (Martínez-Ramos et al. 2016).

12.2.3 Crop-Based Carbon Capture

Carbon farming is to use land or farm to capture carbon naturally by vegetation and soil and remove greenhouse gas from agricultural production. Carbon farming's purpose is to reduce atmospheric CO_2. Carbon farming contributes toward climate change mitigation. Agroforestry helps in a wide range of soil functions that include regulation of nutrients and water in soil and remove atmospheric CO_2 (Li et al. 2022). Forests have woody plants that absorb 15–20% CO_2 of annual human carbon emission and helps in soil health by maintaining organic matter in soil (Udawatta et al. 2022; Le Quere et al. 2018; Case et al. 2021).

12.3 Plant Adaptation

12.3.1 Crop Breeding for Climate Resilience

Drought is the most concern in crop production and plant biology and it causes reduction in crop yield and restricts the geographical distribution and viability of the natural population of plants. Present researchers are aimed at improving carbon capacity and climate resilience in plants to study how plant science can help to mitigate climate change and increase food security (Verslues et al. 2023). The main purpose of drought tolerance breeding is to reduce the gap between production in normal conditions and in stress conditions without affecting production rate. And also improve the yield so that it can be stable in stress condition (Lunduka et al. 2019). Recombinant DNA technology is one of the most important technologies by which we can overcome the loss which is caused by genetic erosion and lack of tolerance genes. In Recombinant DNA technology transfer of genes from one relative species to another. Transgenesis is the transfer of foreign genes in plants, fertilizing the plant and regenerating the plants with desirable expression (Chauhan and Khurana 2017; Shrawat and Armstrong 2018).

12.3.2 Wild Plant Species as Genetic Resources

Rapidly increasing in population and shifting pattern the food demand is increasing (Tilman et al. 2017). Protecting the genetic resources can develop the crop which is resistant to abiotic and biotic stress and can fulfill the global food requirement (Salgotra and Chauhan 2023). The protection of these important genetic resources is very important. The one effective way to protect crop wild varieties is to understand their phenotypic and genomic indices and then select them on the phenotypic and genomic indices.

12.3.3 Plant-Microbe Interactions

The interaction and relationship between microbes and plants are essential to the health and yields of plants. Many microorganisms, such as bacteria, fungi, and viruses interact with plants (Bonfante 2010). Plants can benefit from beneficial interaction with microbes.

- Take up nutrients (such as those from mycorrhizal fungus) from the soil)
- Fix atmospheric nitrogen (rhizobia)
- Resistance against illness and pest (such as rhizobacteria, which promote plant development)
- Accept abiotic stressors such as salt and dryness (example endophytes)

The intricate molecular pathways underlying plant-microbe interaction have been uncovered by recent study. For instance, scientists have discovered microbial effector chemicals that control plant immune responses and plant receptors that identify certain microbial signals (Win et al. 2012). The symbiotic relationship between plants and microorganisms improves plant tolerance to drought, salinity, pests, and diseases. For example, mycorrhizal fungi help plants access water and nutrients in dry or saline soils (Bonfante 2010). Rhizobia fix nitrogen from the air and support plant growth in nutrients-poor soil (Glick 2012). Rhizobia, which promotes plant growth, produces hormones and other compounds that increase plant growth and resistance to pests and diseases. Endophytes can help plants withstand abiotic stresses such as drought, salinity, and heavy metal toxicity. Here are some concrete examples of symbiotic relationships that improve plant resilience. Mycorrhizal fungi form symbiotic relationships with the roots of plants. Fungi expand a plant's root system, making water and nutrients available from a larger area of the soil. Mycorrhizal fungi produce compounds that help plants withstand abiotic stresses such as drought and salinity (Bonfante 2010). Rhizobia is a bacteria that binds with the nitrogen present in air. For plants nitrogen is an essential nutrient but its presence in soil is less. Rhizobia shows symbiotic relationship with the roots of legumes plants (such as kidney beans, peas, and lentils). This symbiotic relationship gives benefit to both the host bacteria providing nitrogen to plants, and plants providing carbohydrate to bacteria (Glick 2012). Plant growth promoting rhizobia (PGPR) are a diverse group of bacteria that can promote plant growth and resistance to pests and diseases. PGPR produces hormones, antibiotics, and other compounds that benefit plants. For example, some PGPRs produce hormones that stimulate plant growth and root development. Others can produce antibiotics that can suppress plant pathogens. Scientists are working to develop ways that increase the power of symbiotic relationships to develop plant growth. A biofertilizer is developed that has microbes like mycorrhizal fungi, rhizobia, and PGPR that are beneficial to plants (Bonfante 2010; Glick 2012).

12.4 Sustainable Agriculture

Sustainable agriculture is a holistic approach to farming that seeks to produce food in a way that is environmentally sound, economically viable and socially just. It is based on the principles of soil conservation, reducing chemical inputs and using renewable resources (Khangura et al. 2023).

12.4.1 Regenerative Agriculture

Regenerative agriculture is a subset of sustainable agriculture that focuses on improving soil health and ecosystem function. It uses practices such as cover cropping, crop rotation and grazing management to build soil organic matter, increase biodiversity and improve water retention (Khangura et al. 2023). Soil conservation and restoration are essential for sustainable agriculture healthy soils are productive resilient to drought and other stresses and filter pollutants out of water (Loisel and Gallego-Sala 2022) soil conservation practices such as minimizing tillage and planting cover crops help to protect soil from erosion and nutrients loss soil restoration practices such as compost application and agroforestry can help to rebuild degraded soils (Loisel and Gallego-Sala 2022). Reducing chemical inputs is another important aspect of sustainable agriculture. Excessive use of chemical fertilizers and pesticides can pollute water and soil, harm beneficial organisms and contribute to climate change (Sharpley 2016). Sustainable farmers use integrated pest management and other practices to reduce their reliance on chemical inputs (Stenberg 2017).

12.4.2 Perennial Crops and Agroforestry

Perennial crops and agroforestry system are key components of sustainable agriculture (Kreitzman et al. 2020) perennial crops such as trees shrubs and perennial grasses have deep root systems that help to protect soil from erosion and improve water infiltration agroforestry system combine perennial crops with livestock or annual crops to create diverse and productive ecosystem (Kreitzman et al. 2020). Perennial food crops offer a number of advantages over annual crops (Kreitzman et al. 2020) they are more resilient to drought and other stresses and they require fewer inputs such as fertilizer and pesticides perennial food crops can also help to improve soil health and biodiversity. Multifunctional agroforestry systems provide a variety of benefits including food production, soil conservation, water quality improvement and carbon sequestration (Ivezic et al. 2021) they can also help to support biodiversity and provide habitat for wildlife.

12.4.3 Precision Agriculture

Precision agriculture is a farming approach that uses technology to manage crops and livestock more efficiently (Presti et al. 2022). Precision agriculture techniques can help to reduce chemical inputs, improve yields and reduce environmental impact. For example farmers can use gps and soil sensors to target fertilizer and pesticide applications to specific areas of a field. Smart farming techniques use data technology to improve farm management decisions (Abdalla and El-Ramady 2022). For example farmers can use sensors to collect data on soil moisture crop health and livestock performance this data can then be used to make better decisions about irrigation fertilization and pest management. Data-driven decision making for resource optimization is essential for sustainable agriculture by using data to track their resource use and environmental impact farmers can identify areas where they can improve efficiency and reduce waste (KrisnaWijaya et al. 2022).

12.5 Biodiversity Conservation

Biodiversity is the variety of life on earth including all plants, animals and micro-organisms it is essential for maintaining healthy ecosystems and providing us with the resources we need to survive however biodiversity is under threat from a number of factors including habitat loss climate change and invasive species (Tsioumani 2020). Carbon emission due to deforestation and due to degradation of forest contribute 12–20% of anthropogenic global greenhouse gas emission annually (Paoli et al. 2010). Protecting and restoring biodiversity is not a simply moral obligation, but a critical imperative for ensuring a sustainable future for humanity (Mace et al. 2012).

12.5.1 Preservation of Plant Diversity

Plants play a vital role in ecosystems providing food and habitat for other organisms; they also help to clean the air and water and regulate the climate, however plant diversity is declining at an alarming rate according to the global biodiversity outlook 5, one in five plant species is at risk of extinction (Tsioumani 2020). Scaling up conservation efforts investing in sustainable development, and promoting transformative changes in our relationship with nature are crucial for ensuring a future where biodiversity thrives (Steffen et al. 2015).

Botanic gardens and seed banks play an important role in preserving plant diversity (Offord and Zimmer 2023) botanic gardens collect, cultivate and display plants from all over the world; they also conduct research on plant conservation and education seed banks store seeds of rare and endangered plants which can be used to restore populations in the wild (Liu et al. 2023). Plants are much more important for ecosystem. Plants play important role in natural ecosystem and the service they provide to human and animals. Rare species are more useful than the unusual plants in future (Corlett 2016).

12.5.2 Protecting Endangered Plant Species

There are a number of ways to protect endangered plant species. One important way is to identify and protect their critical habitats. This may involve establishing protected areas such as national parks and wildlife refuges; it is also important to manage invasive species that threaten endangered plant species (Brooks et al. 2019).

Another important way to protect endangered plant species is to reintroduce them to their former habitats this is a complex and challenging process but it has been successful in some cases for example the black footed ferret was once thought to be extinct but it has been successfully reintroduced to a number of sites in the United States (Kliver et al. 2023).

12.5.3 Habitat Restoration

Habitat restoration is the process of returning a degraded ecosystem to its natural state (Banin et al. 2023) habitat restoration is important for biodiversity conservation because it provides food and shelter for wildlife. It can also help to improve water quality and reduce soil erosion.

There are a number of different approaches to habitat restoration. One common approach is to plant native vegetation native plants are adapted to the local climate and soil conditions and they provide food and habitat for a variety of wildlife (Banin et al. 2023). Another important approach to habitat restoration is to restore natural processes such as fire and flooding. These processes are essential for maintaining healthy ecosystems (Banin et al. 2023).

12.5.4 Creating Wildlife Corridors

Wildlife corridors are areas of land that connect different habitats (Uh-Navarrete et al. 2023). Wildlife corridors allow animals to move between habitats in search of food, water and mates. Wildlife corridors are important for biodiversity conservation because they help to maintain healthy populations of animals (Uh-Navarrete et al. 2023). Wildlife corridors provide safe passage between the fragmented habitats, allowing animals to access resources, find mates, and disperse to new area (Gilbert-Norton et al. 2010).

There are a number of ways to create wildlife corridors one way is to restore natural habitats along streams and rivers another way is to create wildlife bridges and tunnels that allow animals to cross roads and other barriers (Uh-Navarrete et al. 2023).

12.6 Carbon Offset and Trading

12.6.1 Plant-Based Carbon Offsets

To store CO_2 from the atmosphere some investments are done on projects to offset GHG emission (Barnett 2023). To store CO_2 there are some methods like planting trees, restoring the forests, agricultural technologies that can seize the carbon in soil (Barnett 2023).

Planting trees have a very important role to carbon quarantine, but to mature & reach their full growth plant require time due to which carbon sequestration potential takes time (Michaelowa et al. 2023). Forest is an effective process but it is expensive and time consuming (Michaelowa et al. 2023). Agricultural practices are used to store carbon in soil, this is cheaper & easy to perform, but the long-term effects for Carbon storage are still being studied (Gratz et al. 2023).

Opportunities for carbon credit generation carbon credits are generated for the projects that are used to remove or reduce the GHG (Greenhouse Gas) from the atmosphere (Gratz et al. 2023). There are many methods for plant based carbon offset credits. Landowners can generate carbon credit by planting plants on their land and restoring the forest. Farmers can use agricultural technologies to store carbon in the soil. Companies can invest in plant based carbon offset projects (Magyari 2023).

12.6.2 Carbon Markets and Trading

Carbon market is a system where carbon credits are traded (Magyari 2023). Carbon market may be voluntary or mandatory. Voluntary carbon markets allow a business or individual to buy carbon credits. In the mandatory carbon market they require companies to buy carbon credit to cover a percentage of this complete emission (Magyari 2023). There are many platforms for carbon trading Buyers and sellers can contact each other on these platforms. These platforms are regulated by the government to maintain the fairness and transparency of the market (Magyari 2023).

12.7 Challenges and Future Directions

Biodiversity is the main reason for maintaining the functioning of the ecosystem and natures importance to people (Cardinale et al. 2012). Intraspecific diversity (genetic variation within a species) has a key effect on the ecosystem. The genetic diversity increases the productivity of the plant community in herbaceous plants (Crutsinger et al. 2006; Kotowska et al. 2010). Genetic diversity promotes productivity because different species have different genotypes with diverse traits which use resources in many ways that enhance utilization of resources in the community. Genetic diversity is a variability of a population, which indicates the equilibrium between mutation & loss in variation (Hughes et al. 2008; Leffler et al. 2012).

12.8 Balancing Conservation and Agriculture

Agriculture is a main key to global biodiversity loss (Tilman et al. 2017). Terrestrial area is widely used in agriculture, which is used for livestock production (FAO 2018; Poore and Nemecek 2018). Natural land used in agricultural land which results in habitat loss, fragmentation and affects biodiversity (Joppa et al. 2016; Kleijn et al. 2009; Morton et al. 2006). Less agriculture's effect on biodiversity is one of the Aichi Targets agreed upon by the Convention of Biological Diversity. Conservation can be done by focussing on conservation of threatened species versus conservation of landscapes (EC 2011). High levels of biodiversity on agricultural land indicates lower yields (Gabriel et al. 2013). A conservation practice that delivers such a range of benefits is cover cropping, which is widely promoted to decrease erosion & improve soil quality, nutrient cycling, weed control among other ecosystem services (Schipanski et al. 2014). Better management of the agroecosystem will control the natural enemies (Hobbs et al. 2008) and provide additional benefits to wildlife conservation (pollinators, birds) soil health and quality of water.

Due to the increase in climate change innovation is a major role for national & subnational decarbonisation processes. Recents development with different technological approaches for both private and public sector from non-carbon technologies to reduce sources of GHG to capture and decrease global warming (Sovacool 2021; UNGC-Accenture 2015). Research is turning challenging problems into manageable solutions (George et al. 2016). In 2015 at the United nation climate change conference in Paris, 20 countries including UK, the US, China & India, doubled their investment in low- carbon technology for the 'Mission Innovation' (Sanchez and Sivaram 2017). Innovative technologies like renewable energy (solar photovoltaics and microinverters), smart grid & energy storage have improved in recent years due to increased deployment has led to substantial cost reduction (IRENA 2021; Viardot et al. 2013). The seriousness of global warming has driven some companies to start important innovation in manufacturing the less carbon & energy- efficient process and technologies (Lee 2013).

Environmental problems need urgent solutions. Environmental problems like ozone depletion, climate change and marine pollution have increased in recent past years and need focus at the international level. International environmental agreements have been reached in some areas, like Montreal Protocol on ozone layer depletion and number of agreements on Marine pollution. Current negotiations following the UN Conferences on environment & Development in Rio de Janeiro & Kyoto.

12.9 Conclusion

Climate change is the biggest problem in the twenty-first century. Climate change is caused by GHG emissions like CO_2, methane which leads to global warming. Plants play an important role in climate change by removing CO_2 from the atmosphere. Plants show symbiotic relationships with microbes that help in nitrogen fixation. Interaction between plants & their herbivores symbionts and competitors. Plants carry out photosynthesis which converts CO_2 in plant biomass & stores in the plant. Plants increase the weathering of rocks from which chemical reactions take place that remove CO_2 from the atmosphere. By the help of plants GHG can be removed from the atmosphere. Climatic conditions affect the farming and growing season to adapt from the stressful conditions. Plants have the ability to tolerate stress conditions like salt stress, drought stress. Plant biology helps in stopping the emission of GHG and achieve net zero CO_2 emission & improved crop variety to increase food production. Forest management for carbon maintenance is a low cost, low technology, relatively easy way to maintain carbon conditions and global climate change. The USA is developing a climate change policy that recognizes forestry as a good source for carbon emission and offers other opportunities for afforestation, reforestation and grassland. Forests give current opportunities in mitigating climate change (Charnley et al. 2010). Climate change is projected to exaggerate biotic & abiotic stresses that potentially can threaten the sustainable crop production. Adoption of organic agricultural practices like integrated pest management, use of bioinoculants & botanicals can help in enduring climate change & will reduce dependence of chemicals (Kumar et al. 2022).

References

Abbas F, Hammad HM, Fahad S et al (2017) Agroforestry: a sustainable environmental practice for carbon sequestration under the climate change scenarios—a review. Environ Sci Pollut Res 24: 11177–11191. https://doi.org/10.1007/s11356-017-8687-0

Abdalla ZF, El-Ramady H (2022) Applications and challenges of smart farming for developing sustainable agriculture. Environ Biodiver Soil Secur 6(2022):81–90

Altieri MA, Nicholls CI, Henao A, Lana MA (2015) Agroecology and the design of climate change-resilient farming systems. Agron Sustain Dev 35(3):869–890. https://doi.org/10.1007/s13593-015-0285-2

Bahadur B, Rajam MV, Sahijram L, Krishnamurthy K (2015) Plant biology and biotechnology: volume I: plant diversity, organization, function and improvement. Springer, New Delhi

Banin LF et al (2023) The road to recovery: a synthesis of outcomes from ecosystem restoration in tropical and sub-tropical Asian forests. Philos Trans R Soc B 378(1867):20210090

Barnett M (2023) Climate change and uncertainty: an asset pricing perspective. Manag Sci 69: 7562–7584

Bonfante P (2010) Plant–fungal interactions in mycorrhizas. eLS

Bowerman AF, Byrt CS, Roy SJ, Whitney SM, Mortimer JC, Ankeny RA, Gilliham M, Zhang D, Millar AA, Rebetzke GJ, Pogson BJ (2023) Potential abiotic stress targets for modern genetic manipulation. Plant Cell 35(1):139–161. https://doi.org/10.1093/plcell/koac327

Brooks TM, Pimm SL, Akçakaya HR, Buchanan GM, Butchart SHM, Foden W, Hilton-Taylor C, Hoffmann M, Jenkins CN, Joppa L, Li BV, Menon V, Ocampo-Peñuela N, Rondinini C (2019) Measuring terrestrial area of habitat (AOH) and its utility for the IUCN Red List. Trends Ecol Evol 34(11):977–986. https://doi.org/10.1016/j.tree.2019.06.009. Epub 2019 Jul 16. PMID: 31324345

Cardinale B, Duffy J, Gonzalez A et al (2012) Biodiversity loss and its impact on humanity. Nature 486:59–67

Case MJ, Johnson BG, Bartowitz KJ, Hudiburg TW (2021) Forests of the future: climate change impacts and implications for carbon storage in the Pacific Northwest, USA. For Ecol Manag 482:118886. https://doi.org/10.1016/j.foreco.2020.118886

Charnley S, Diaz D, Gosnell H (2010) Mitigating climate change through small-scale forestry in the USA: opportunities and challenges. Small Scale For 9:445–462. https://doi.org/10.1007/s11842-010-9135-x

Chauhan H, Khurana P (2017) Wheat genetic transformation using mature embryos as explants. Methods Mol Biol 1679:153–167

Corlett RT (2016) Plant diversity in a changing world: status, trends, and conservation needs. Plant Diver 38(1):10–16

Crutsinger GM, Collins MD, Fordyce JA, Gompert Z, Nice CC, Sanders NJ (2006) Plant genotypic diversity predicts community structure and governs an ecosystem process. Science 313: 966–968

EC (2011) The EU biodiversity strategy to 2020. European Commission, Brussels

FAO (2018) FAOSTAT database: land use. Food and Agricultural Organization of the United Nations, Rome

Gabriel D, Sait SM, Kunin WE, T.G. (2013) Benton food production vs. biodiversity: comparing organic and conventional agriculture. J Appl Ecol 50:355–364. https://doi.org/10.1111/1365-2664.12035

George G, Howard-Grenville J, Joshi A, Tihanyi L (2016) Understanding and tackling societal grand challenges through management research. Acad Manag J 59(6):1880–1895

Gilbert-Norton L, Wilson R, Stevens JR, Beard KH (2010) A meta-analytic review of corridor effectiveness in facilitating plant dispersal for restoration. Restor Ecol 18(2):110–121

Glick BR (2012) Plant growth-promoting bacteria: mechanisms and applications. Scientifica 2012(1):963401

Gratz D, Cornish A, Goeringer P (2023) Carbon markets 101: understanding how carbon contracts operate. Drake J Agric L 28:3

Hassan MU, Aamer M, Mahmood A, Awan MI, Barbanti L, Seleiman MF et al (2022) Management strategies to mitigate N2O emissions in agriculture. Life 12(3):439

Hobbs PR, Sayre K, Gupta R (2008) The role of conservation agriculture in sustainable agriculture. Philos Trans R Soc Lond B 363:543–555

Houghton RA (1994) The worldwide extent of land-use change. Bioscience 44:305–313. https://doi.org/10.2307/1312380

Hughes A, Inouye BD, Johnson MT, Underwood N, Vellend M (2008) Ecological consequences of genetic diversity. Ecol Lett 11:609–623

HWF H. W. Foundation (2017) Benefits of a Forest…like Hitchcock Woods. Hitchcock Woods Foundation (HWF). https://www.hitchcockwoods.org/our-work/education/benefits-of-forests/#Trees-Add-Beauty-and-Improve-P

IRENA (2021) Renewable power generation costs in 2020. International Renewable Energy Agency, Abu Dhabi

Ivezić V, Yang Y, van der Werf W (2021) Crop yields in European agroforestry systems: a meta-analysis. Front Sustain Food Syst 5:606631

Jägermeyr J, Müller C, Ruane AC, Elliott J, Balkovic J, Castillo O et al (2021) Climate impacts on global agriculture emerge earlier in new generation of climate and crop models. Nature Food 2(11):873–885

Joppa LN, O'Connor B, Visconti P, Smith C, Geldmann J, Mann MH, Watson JEM, Butchart SHM, Sawmy MV, Halpern BS, Ahmed SE, Balmford A, Sutherland WJ, Harfoot M, Hilton-Taylor C, Foden W, DiMinin E, Pagad S, Genovesi P, Hutton J, Burgess ND (2016) Filling in biodiversity threat gaps. Science 352(6284):416–418

Khangura R, Ferris D, Wagg C, Bowyer J (2023) Regenerative agriculture—a literature review on the practices and mechanisms used to improve soil health. Sustain For 15:2338. https://doi.org/10.3390/su15032338

Kleijn D, Kohler F, Báldi A, Batáry P, Concepción ED, Clough Y, Díaz M, Gabriel D, Holzschuh A, Knop E, Kovács A, Marshall EJP, Tscharntke T, Verhulst J (2009) On the relationship between farmland biodiversity and land-use intensity in Europe. Proc R Soc B Biol Sci 276:903–909. https://doi.org/10.1098/rspb.2008.1509

Kliver S et al (2023) Chromosome-length genome assembly and karyotype of the endangered black-footed ferret (Mustelanigripes). J Hered 114:esad035

Kotowska AM, Cahill JF, Keddie BA (2010) Plant genetic diversity yields increased plant productivity and herbivore performance. J Ecol 98:237–245

Kreitzman M et al (2020) Perennial staple crops: yields, distribution, and nutrition in the global food system. Front Sustain Food Syst 4:216

Krisnawijaya NNK et al (2022) Data analytics platforms for agricultural systems: a systematic literature review. Comput Electron Agric 195:106813

Kumar P, Singh S, Pranaw K, Kumar S, Singh B, Poria V (2022) Bioinoculants as mitigators of multiple stresses: a ray of hope for agriculture in the darkness of climate change. Heliyon 8(11):e11269. https://doi.org/10.1016/j.heliyon.2022.e11269. PMID: 36339753; PMCID: PMC9634370

Le Quéré C, Andrew RM, Friedlingstein P, Sitch S, Hauck J, Pongratz J, Pickers PA, Korsbakken JI et al (2018) Global carbon budget 2018. Earth Syst Sci Data 10:2141–2194. https://doi.org/10.5194/essd-10-2141-2018

Lee S (2013) Existing and anticipating technology strategies for reducing greenhouse gas emissions in Korea's petrochemical and steel industries. J Clean Prod 40:83–92

Leffler EM, Bullaughey K, Matute DR, Meyer WK, Ségurel L, Venkat A, Andolfatto P, Przeworski M (2012) Revisiting an old riddle: what determines genetic diversity levels within species? PLoS Biol 10:e1001388

Li S, Gong S, Hou Y, Li X, Wang C (2022) The impacts of agroforestry on soil multi-functionality depending on practices and duration. Sci Total Environ 847:157438. https://doi.org/10.1016/j.scitotenv.2022.157438. Epub 2022 Jul 23. PMID: 35882333

Liu X, Wang S, Wu P, Feng K, Hubacek K, Li X, Sun L (2019a) Impacts of urban expansion on terrestrial carbon storage in China. Environ Sci Technol 53:6834–6844. https://doi.org/10.1021/acs.est.9b00103

Liu X, Li X, Liang X, Shi H, Ou J (2019b) Simulating the change of terrestrial carbon storage in China based on the FLUS-InVEST model. Trop Geogr 39:397–4091

Liu U et al (2023) Conserving useful plants for a sustainable future: species coverage, spatial distribution, and conservation status within the millennium seed Bank collection. Biodivers Conserv 32:1–49

Loboguerrero AM, Campbell BM, Cooper PJM, Hansen JW, Rosenstock T, Wollenberg E (2019) Food and earth systems: priorities for climate change adaptation and mitigation for agriculture and food systems. Sustain For 11:1372. https://doi.org/10.3390/su11051372

Loisel J, Gallego-Sala A (2022) Ecological resilience of restored peatlands to climate change. Commun Earth Environ 3:208. https://doi.org/10.1038/s43247-022-00547-x

Lunduka RW, Mateva KI, Magorokosho C, Manjeru P (2019) Impact of adoption of drought-tolerant maize varieties on total maize production in south eastern Zimbabwe. Clim Dev 11(1): 35–46

Mace GM, Norris K, Fitter AH (2012) Biodiversity and ecosystem services: a multilayered relationship. Trends Ecol Evol 27(1):19–26

Magyari J (2023) Decarbonization challenges and opportunities in the central European energy sector: implications for management. Soc Econ 45:451–471

Malhi Y, Franklin J, Seddon N, Solan M, Turner MG, Field CB, Knowlton N (2020) Climate change and ecosystems: threats, opportunities and solutions. Philos Trans R Soc Lond Ser B Biol Sci 375(1794):20190104. https://doi.org/10.1098/rstb.2019.0104. Epub 2020 Jan 27. PMID: 31983329; PMCID: PMC7017779

Martínez-Ramos M, Pingarroni A, Rodríguez-Velázquez J, Toledo-Chelala L, Zermeño-Hernández I, Bongers F (2016) Natural forest regeneration and ecological restoration in human-modified tropical landscapes. Biotropica 48(6):745–757

Michaelowa A et al (2023) International carbon markets for carbon dioxide removal. PLoS Clim 2(5):e0000118

Morton DC, DeFries RS, Shimabukuro YE, Anderson LO, Arai E, Del Bon Espirito-Santo F, Freitas R, Morisette J (2006) Cropland expansion changes deforestation dynamics in the southern Brazilian Amazon. Proc Natl Acad Sci USA 103:14637–14641. https://doi.org/10.1073/pnas.0606377103

National Academies of Sciences, Engineering, and Medicine (2022) New directions for chemical engineering. National Academy of Sciences, Washington, DC (United States)

Nilsson S, Schopfhauser W (1995) The carbon-sequestration potential of a global afforestation program. Clim Chang 30:267–293. https://doi.org/10.1007/BF01091928

Offord CA, Zimmer HC (2023) Home gardens contribute to conservation of the critically endangered Wollemi Pine: evaluation of a botanic garden-led horticultural release programme. Plants People Planet 6:116

Paoli GD, Wells PL, Meijaard E et al (2010) Biodiversity conservation in the REDD. Carbon Balance Manag 5:7. https://doi.org/10.1186/1750-0680-5-7

Pedersen JTS, van Vuuren D, Gupta J, Santos FD, Edmonds J, Swart R (2022) IPCC emission scenarios: how did critiques affect their quality and relevance 1990–2022? Glob Environ Chang 75:102538

Peng SS et al (2014) Afforestation in China cools local land surface temperature. In: Proceedings of the National Academy of Sciences, pp 2915–2919

Poore J, Nemecek T (2018) Reducing food's environmental impacts through producers and consumers. Science 360(6392):987–992. https://doi.org/10.1126/science.aaq0216

Presti DL et al (2022) Current understanding, challenges and perspective on portable systems applied to plant monitoring and precision agriculture. Biosens Bioelectron:115005

Pires ND, Dolan L (2012) Morphological evolution in land plants: new designs with old genes. Philos Trans R Soc B Biol Sci 367(1588):508–518

Sanchez DL, Sivaram V (2017) Saving innovative climate and energy research: four recommendations for Mission innovation. Energy Res Soc Sci 29:123–126
Salgotra RK, Chauhan BS (2023) Ecophysiological responses of rice (Oryza sativa L.) to drought and high temperature. Agronomy 13(7):1877
Schipanski ME, Barbercheck M, Douglas MR, Finney DM, Haider K et al (2014) A framework for evaluating ecosystem services provided by cover crops in agroecosystems. Agric Syst 125: 12–22
Sharpley A (2016) Managing agricultural phosphorus to minimize water quality impacts. Sci Agric 73:1–8
Shrawat AK, Armstrong CL (2018) Development and application of genetic engineering for wheat improvement. Crit Rev Plant Sci 37:335–421. https://doi.org/10.1080/07352689.2018.15147181
Sovacool BK (2021) Who are the victims of low-carbon transitions? Towards a political ecology of climate change mitigation. Energy Res Soc Sci 73:Article 101916
Steffen W et al (2015) Planetary boundaries: guiding human development on a changing planet. Science 347:1259855. https://doi.org/10.1126/science.1259855
Stenberg JA (2017) A conceptual framework for integrated pest management. Trends Plant Sci 22(9):759–769
Tan YQ, Yang Y, Shen X, Zhu M, Shen J, Zhang W et al (2023) Multiple cyclic nucleotidegated channels function as ABA-activated Ca2+ channels required for ABA-induced stomatal closure in Arabidopsis. Plant Cell 35(1):239–259
Tilman D, Clark M, Williams DR, Kimmel K, Polasky S, Packer C (2017) Future threats to biodiversity and pathways to their prevention. Nature 546(7656):73–81. https://doi.org/10.1038/nature22900
Tong CS, Su M, Sun H, Chua XL, Guo S, Ramaraj RSO et al (2023) Collective dynamics of formin and microtubule and its crosstalk mediated by FHDC1. bioRxiv, 2023-07
Tsioumani E (2020) Convention on biological diversity: a review of the post-2020 global biodiversity framework working group negotiations. Environ Policy Law 50(1/2):55–59
Udawatta RP, Walter D, Jose S (2022) Carbon sequestration by forests and agroforests: a reality check for the United States. Carbon Footpr 1:8. https://doi.org/10.20517/cf.2022.06
Uh-Navarrete AE et al (2023) Discovering the fish fauna of a lagoon from the southeast of the Yucatan peninsula, Mexico, using DNA barcodes. PeerJ 11:e16285
UNGC-Accenture (2015) A call to climate action. Available at: https://www.unglobalcompact.org/library/3551
Verslues PE, Bailey-Serres J, Brodersen C, Buckley TN, Conti L, Christmann A et al (2023) Burning questions for a warming and changing world: 15 unknowns in plant abiotic stress. Plant Cell 35(1):67–108
Viardot E, Wierenga ET, Friedrich B (2013) The role of cooperatives in overcoming the barriers to adoption of renewable. Energy Pol 63:756–774
Vivek AK, Montenegro A (2011) Small temperature benefits provided by realistic afforestation efforts. Nat Geosci 4:514–518. https://doi.org/10.1038/ngeo1182
Wang JY, Li AN, Bian JH (2016) Simulation of the grazing effects on grassland aboveground net primary production using a DNDC model combined with time-series remote sensing data-a case study in Zoige Plateau, China. Rem Sens 8:168. https://doi.org/10.3390/rs8030168
Wilcox KR, Blair JM, Smith MD, Knapp AK (2015) Does ecosystem sensitivity to precipitation at the site-level conform to regional-scale predictions? Ecology 97:561–568
Win J, Chaparro-Garcia A, Belhaj K, Saunders DGO, Yoshida K, Dong S et al (2012, January) Effector biology of plant-associated organisms: concepts and perspectives. In: Cold Spring Harbor symposia on quantitative biology, vol 77. Cold spring harbor laboratory press, pp 235–247
Zhang Y, Shi X, Tang Q (2021) Carbon storage assessment in the upper reaches of the Fenhe River under different land use scenarios. Acta Ecol Sin 41:360–373
Zhang P, Li Y, Yin H, Chen Q, Dong Q, Zhu L (2022) Spatial-temporal variation and dynamic simulation of ecosystem carbon storage in the north-south transitional zone of China. J Nat Resour 37:1183–1197

Chapter 13
Crop Yield Prediction in Diverse Agricultural Landscapes

Devesh Kumar Bandil and Sunil Kumar

Abstract This chapter explores the dynamic challenges and opportunities associated with crop yield prediction in diverse agricultural landscapes. As agriculture varies across regions, incorporating a multidisciplinary approach becomes crucial for developing accurate and context-specific yield prediction models. The chapter begins by highlighting the significance of tailoring predictive methodologies to the unique characteristics of different agricultural environments. The initial sections provide an overview of the diverse factors influencing crop yield in various landscapes, including soil types, climate patterns, and cropping systems. Understanding the complexity of these factors is essential for designing effective prediction models that can accommodate the heterogeneity present in different agricultural settings. The chapter then delves into the integration of remote sensing technologies, machine learning algorithms, and precision agriculture techniques tailored to diverse landscapes. Emphasis is placed on the adaptability and scalability of these technologies to address the specific needs of smallholder farmers, large-scale commercial agriculture, and everything in between. Case studies from a range of agricultural landscapes worldwide are presented to illustrate successful applications of predictive models. These case studies highlight the practical implications of tailoring crop yield prediction approaches to the unique challenges and opportunities present in specific regions. Lessons learned from these real-world implementations contribute to the development of best practices for diverse agricultural contexts. Furthermore, the chapter addresses the importance of incorporating local knowledge and engaging with stakeholders in the development and implementation of predictive models. Recognizing the role of farmers, agronomists, and local communities in the process enhances the relevance and acceptance of these technologies in diverse landscapes. Challenges associated with data availability, technology adoption, and resource constraints are discussed, along with potential strategies to overcome these hurdles. The chapter concludes by outlining future research directions and the need for ongoing collaboration between researchers, policymakers, and practitioners to refine and adapt crop yield prediction models for the ever-evolving tapestry of global

D. K. Bandil (✉) · S. Kumar
Arya College of Engineering & IT, Kukas, Jaipur, Rajasthan, India

S. Kanga et al. (eds.), *Transforming Agricultural Management for a Sustainable Future*, World Sustainability Series, https://doi.org/10.1007/978-3-031-63430-7_13

agriculture. This chapter serves as a comprehensive guide for professionals and researchers seeking to navigate the complexities of predicting crop yields in diverse agricultural landscapes.

Keywords Crop yield prediction · Remote sensing · Machine learning · Precision agriculture · Soil variability · Climate patterns · Cropping systems · Data integration · Geographic information systems (GIS)

13.1 Introduction

The agricultural sector plays a pivotal role in sustaining global food needs, and accurate crop yield predictions are essential for efficient resource management. In this chapter, we delve into the fascinating world of crop yield prediction using a diverse dataset representing various crops and countries. Each entry in the dataset contains information about the crop type, country of cultivation, and the corresponding yield in metric tonnes. The exploration will involve applying machine learning techniques to forecast crop yields and draw insights from the data.

13.2 Understanding the Dataset

13.2.1 Dataset Overview

The dataset comprises information on diverse crops, including Cassava, Maize, Plantains and others, Potatoes, Rice, Sorghum, Soybeans, Sweet potatoes, Wheat, and Yams. Each entry provides details on the crop, the country where it is cultivated, and the reported yield in metric tonnes.

Dataset Description
Variable Description
Area -----> Country names that plant crops
Item-----> Types of crops planted
Year----> Time of planted crops in years 1990–2013
average_rain_fall_mm_per_year------> Average rain fall per year
Pesticides_tonnes-----> Pesticides used in tonnes
avg_temp----> Average temperature
hg/ha_yield----> Crops yield production value in hectogram per hectare (Hg/Ha) (https://www.kaggle.com/code/mahmoudmagdyelnahal/crop-yield-prediction-99/input)

13.2.2 Exploratory Data Analysis

Exploring the dataset involves understanding the distribution of crops, identifying potential outliers, and gaining insights into the global distribution of crop cultivation and yields. This section provides a comprehensive analysis of the dataset to inform subsequent modeling efforts.

Inferences

Number of Entries: ---> The dataset consists of 28,242 entries, ranging from index 0 to 28,241.

Columns: ---> There are 8 columns in the dataset.

Data Types

Most of the columns (6 out of 8) are of the int64 & float64 data type.

Only the Item and Area columns are of the object data type.

Missing Values: ---> There doesn't appear to be any missing values in the dataset as each column has 28,242 non-null entries.

	Unnamed: 0	Year	hg/ha_yield	average_rain_fall_mm_per_year	pesticides_tonnes	avg_temp
count	28242.000000	28242.000000	28242.000000	28242.000000	28242.000000	28242.000000
mean	14120.500000	2001.544296	77053.332094	1149.055980	37076.909344	20.542627
std	8152.907488	7.051905	84956.612897	709.812150	59958.784665	6.312051
min	0.000000	1990.000000	50.000000	51.000000	0.040000	1.300000
25%	7060.250000	1995.000000	19919.250000	593.000000	1702.000000	16.702500
50%	14120.500000	2001.000000	38295.000000	1083.000000	17529.440000	21.510000
75%	21180.750000	2008.000000	104676.750000	1668.000000	48687.880000	26.000000
max	28241.000000	2013.000000	501412.000000	3240.000000	367778.000000	30.650000

Numerical Features: --->

Average_rain_fall_mm_per_year: The average rainfall per year is approximately 1149, with the least rainfall being 51 and the most 3240.

Pesticides_tonnes: The mean pesticides used in tonnes is a whopping 37,077 tonnes, with minimum as little as 0.04 and maximum as huge as 367,778 tonnes.

hg/ha_yield: The average crop production yield is 77053.3. Ranging from 50 all the way to 501,412 hectograms per hectare.

Categorical Features (Object Data Type)

Area: There are 101 unique countries/areas, with India being the most frequent category, occurring 4048 times out of 28,242 entries.

Item: 10 unique types of crops are present. The most common type is "Potatoes", occurring 4276 times.

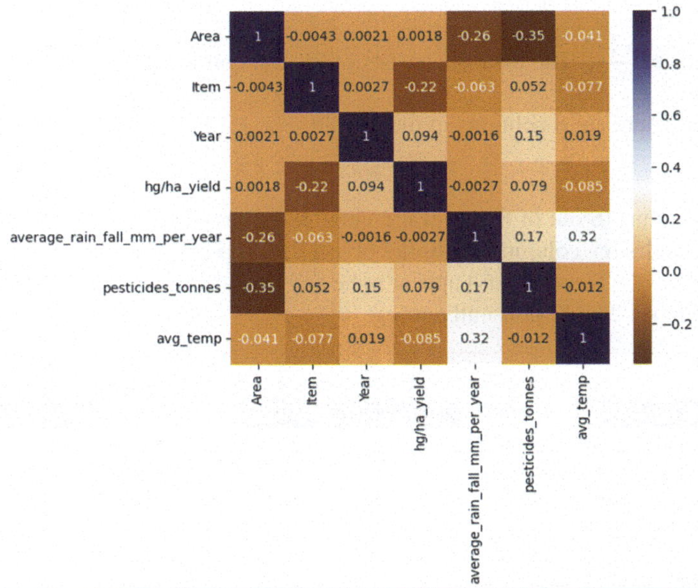

Inferences: --->

Area: There's a strong correlation between Area and pesticides_tonnes, along with Area and average rainfall

Items: There's a strong correlation between Item and hg/ha_yield

Inferences

rainfall: Most rainfall is between 0–1000 and very few is around 3000

pesticides_tonnes: The majority of used pesticides is little to zero

avg_temp: Most average temperatures are around 25. *hg/ha yield: There is a vast majority of yield production around 0.

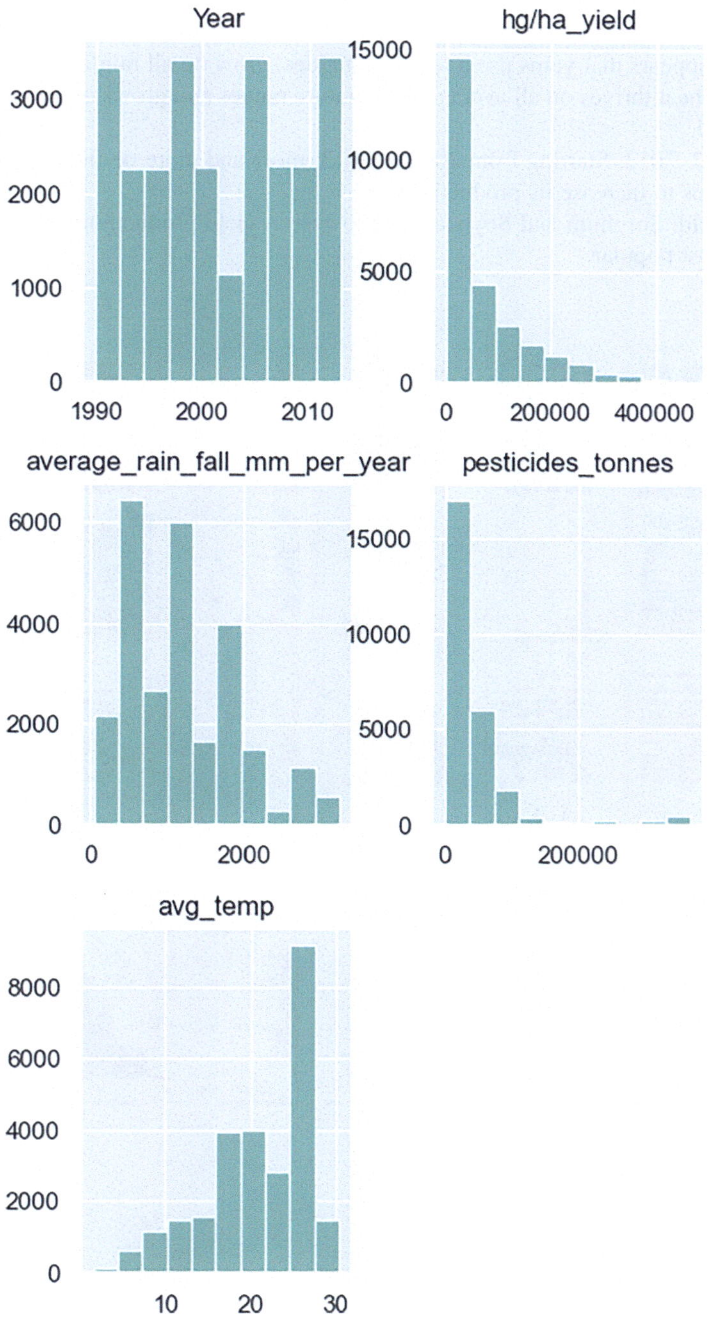
Year
3000
2000
1000
0
1990
2000
2010
hg/ha_yield
15000
10000
5000
0
0
200000
400000
average_rain_fall_mm_per_year
6000
4000
2000
0
0
2000
pesticides_tonnes
15000
10000
5000
0
0
200000
avg_temp
8000
6000
4000
2000
0
10
20
30

Inferences: Items

yams: It appears that yams use a lot of pesticides, grow on all rainfall ranges.

wheat: Wheat thrives on all average temperature ranges and grows on average to no rainfall.

year: 2003–2013: Starting from 2003 to 2013, more and more pesticides were used on yams to increase its productivity.

hg/ha_yield: Sorghum and Soybeans are lowest in yield production while potatoes are most popular.

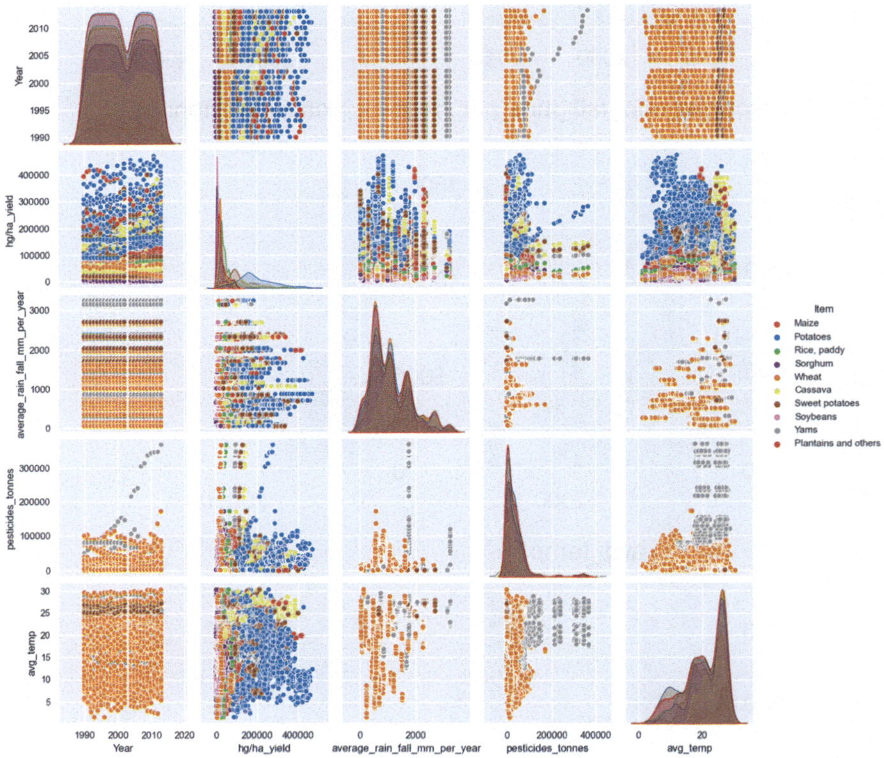

Inferences

Group 0: Australia produced the largest amount of yield harvesting potatoes while Angola has the least yield harvesting sorghum, soybeans and maize.

Group 1: Egypt was producing most yield in this group growing sweet potatoes and potatoes while Ecuador was struggling in harvesting wheat.

Group 2: France and Germany are at the top of yield production both harvesting potatoes while Honduras wasn't in luck harvesting wheat

Group 3: India soared in growing Cassava and Japan was at the top growing potatoes while Madagascar struggled to grow soybeans and sorghum.

Group 4: Morocco and Mexico shined in yield production of potatoes while Niger failed to have high productions of wheat

Group 5: Pakistan was struggling in growing sorghum while South Africa Spain, and Saudi Arabia shined in growing Potatoes

Group 6: Tajikistan failed to harvest large amounts of soybeans while United Kingdom and Turkey produced a myriad of potatoes Collectively:

Top producing countries: United Kingdom, France, Germany, Australia, and Japan

Top produced Item: potatoes

Least producing countries: Zimbabwe, Azerbaijan, Angola, Niger, Tajikistan

Least produced item: sorghum and soybeans

Inferences

Top rainfall countries: Bangladesh, Colombia, Guyana, Indonesia, Nicaragua, Papua New Guinea, Ecuador, and Suriname have rainfall with over 2000 mm

Least rainfall countries: Saudi Arabia, Pakistan, South Africa, Mali, Mauritania, Morocco, Niger, Libya, Iraq, Egypt, Azerbaijan, and Algeria have rainfall with less than 500 mm.

Inferences

Top pesticides using countries: Brazil, Italy, Argentina

France: For a high yield producing country, using over 80,000 tonnes of pesticides could be harmful.

Japan: Another high producing country using over 60,000 tonnes of pesticides.

Least pesticides using countries: Algeria, Angola, Azerbaijan, Bulgaria, Burkina Faso, Burundi, Cameroon, Central African Republic, Croatia, Egypt, El Salvador, Greece, Guinea, Guyana, Haiti, Honduras, Hungary, Indonesia, Iraq, Jamaica, Kenya, Kazakhstan, Libya, Madagascar, Malawi, Mali, Mauritania, Mauritius, Mozambique, Nepal, Niger, Papua New Guinea, Rwanda, Senegal, Saudi Arabia, Sri Lanka, Suriname, Tajikistan, Uganda, Zambia, Zimbabwe, Uruguay have been using pesticides with less than 10,000 tonnes

All poorly producing yield countries are included in least pesticides usage countries

Germany, Australia, and United Kingdom are top producing countries and they use an average amount of pesticides around 30,000 tonnes

Now this begs the question; do pesticides affect yield production?

Inferences

Group 0: Brazil had low yield production and as it used more and more pesticides their production went higher. However, Argentina, Australia, and Algeria kept their pesticides usage low and they had more yield production that Brazil ever could.

Group 1: Colombia kept having low yield production despite using more pesticides. It was almost constant despite the increased usage of pesticides. Canada had one high yield production then went lower while using more pesticides until it had the same fate as Colombia Egypt, El Salvador, Dominican Republic, and Chile on the other hand, got very high yield productions while using a significantly lower number of pesticides.

Group 2: France has high yield production despite its high pesticide usage as it is one of the leading yield production countries. Although, it lessened a little with more pesticide's usage. Germany, Guatemala, and Guyana have high yield production while using less pesticides. Even Guatemala had higher production than France at some point.

Group 3: India and Japan have high yield productions while using less amounts of pesticides. India even beat Japan in some yield productions while using less pesticides as well. Italy however, is one of the countries that has the highest pesticides usage and its yield production wasn't as high as India and Japan.

Group 4: Mexico uses a significant number of pesticides and its yield production kept increasing as more pesticides were used. Mali, Mauritius, and Niger had some high yield productions while using almost no pesticides which is impressive. Morocco kept using slightly more pesticides until it got a soaring 350k+ production then decreased slightly again. Nicaragua used more pesticides got to a maximum yield production then decreased again to an amount even lesser than it used to have before additional pesticides.

Group 5: Senegal got to a soaring 350k+ production while using almost no pesticides. Saudi Arabia got high yield productions while using a small number of pesticides. Portugal got more yield productions while utilizing a little more pesticide until it reached a certain threshold and decreased again. South Africa has a steady increase of yield production while adding more and more pesticides Spain on the other hand, has some yield fluctuations but overall, it is a high yield production. Although Senegal and South Africa beat it using less pesticides.

Group 6: Tajikistan and Suriname use almost no pesticides and they reached high amounts of yield productions. Even though Tajikistan is one of the least producing countries but it has very high potential of being close to the top producing countries. United Kingdom has high yield production and uses a fair number of pesticides it's no surprise as it's one of the leading producing countries. Turkey has lesser yield productions than United Kingdom and Suriname while utilizing more pesticides. *Thailand and Ukraine used way more pesticides and still their productions were lower than the above countries.

Conclusion While using pesticides increases yield production, there was an obvious pattern happening. Yield production is at a considerable amount -> use pesticides -> yield production increases -> comes to maximum production -> yield gradually decreases -> yield becomes even less than the starting yield amount no matter how many more pesticides the country uses.

However, there were some exceptions to this pattern such as France and Japan as they are top leading producing countries and other countries. You can increase production by using more pesticides but it is temporary nonetheless and in the long run it is better to use less pesticides. There were a lot of countries that showed impressive yield productions using almost no pesticides and this amount was never reached by other countries who do use pesticides. Those countries were Senegal, Suriname, Tajikistan, Mali, Mauritius, Guyana, Algeria, and El Salvador. They should focus more on growing more crops and businessmen should invest in growing crops in these countries as they have a very high potential in yield production of crops. Other countries who use a tiny number of pesticides and still have promising yield productions are Egypt, Chile, Morocco, South Africa, and Guatemala. It is also recommended to invest in these countries as they have reassuring yield productions. The perfect number of pesticides to use without having to go through the tedious pattern is 30k and below.

Inferences

Top pesticides used on items: Yams
Least pesticides used on items: Plantains and others

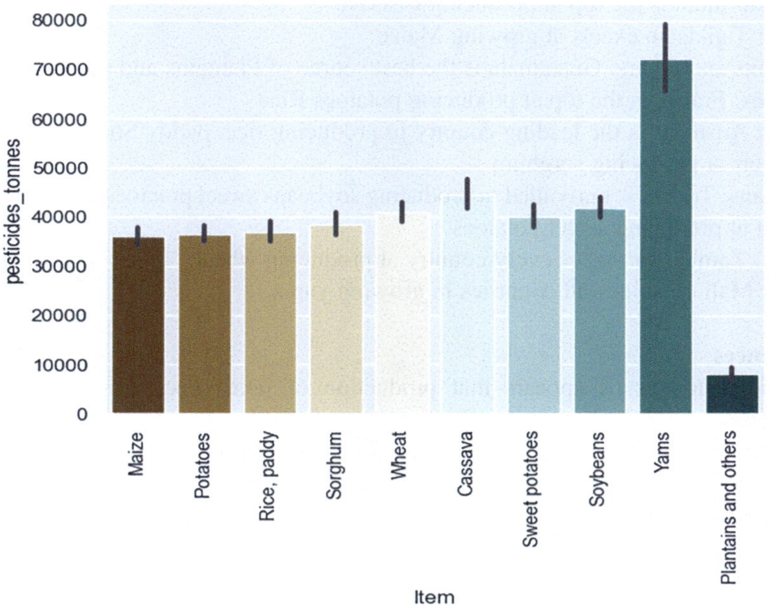

Inferences Top producing items: potatoes, cassava
Least producing items: soybeans, sorghum, wheat, maize, and rice, paddy

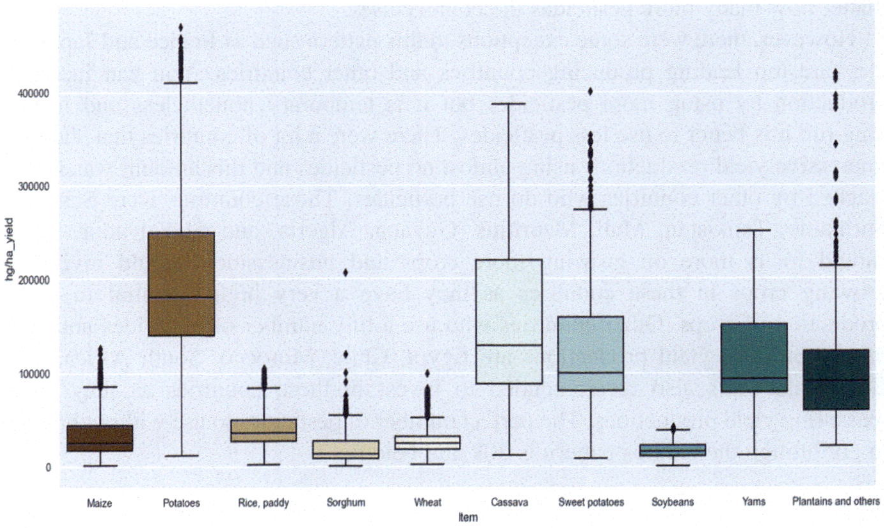

Inferences
Cassava: India is the top at producing Cassava
Maize: Tajikistan excels at growing Maize
Plantains and others: Guatemala is the best source of Plantains and others
Potatoes: France is the top at producing potatoes Rice
Paddy: Australia is the leading country in producing rice, paddy Sorghum: Algeria excels at producing sorghum
Soybeans: Turkey is unrivalled in producing soybeans sweet potatoes: Senegal is the best at producing sweet potatoes
Wheat: Zambia surpasses every country at producing wheat
Yams: Mali outshines all countries in growing yams

Inferences
Pesticidies_tonnes: it appears that production of items decreases as pesticides increases
Rainfall: it appears that production of items decreases as rainfall increases for more than 2500 avg_temp: it shows that average yield decreases as average temperature decreases below 5

13.3 Crop Yield Prediction Models

13.3.1 Data Preparation (https://www.kaggle.com/code/mahmoudmagdyelnahal/crop-yield-prediction-99/input)

Preparing the data involves cleaning, encoding categorical variables, and splitting it into training and testing sets. Special attention is given to handling any missing or anomalous data points.

13.3.2 Gradient Boosting Regressor Model for Crop Yield Prediction

Linear Regression models are applied to predict crop yields based on the given features. The chapter discusses the assumptions, model training, and evaluation using the provided data.

```
# 4. Gradient boosting Regressor
# Checking R2 score for GBR
Gb= GradientBoostingRegressor(n_estimators=400,  random_state=maxRS, learning_rate=0.1, max_depth=3)
Gb.fit(x_train,y_train)

#prediction
predGb=Gb.predict(x_test)
print('R2_Score:',r2_score(y_test,predGb))
print('Mean abs error:',mean_absolute_error(y_test, predGb))
print('Mean squared error:',mean_squared_error(y_test, predGb))
print("Root Mean Squared Error: ", np.sqrt(mean_squared_error(y_test,predGb)))

R2_Score: 0.9117295432106156
Mean abs error: 16207.094811583229
Mean squared error: 611762141.4701954
Root Mean Squared Error:  24733.825855904204
```

13.3.3 Random Forest Regression Model for Crop Yield Prediction

Random Forest, as an ensemble learning method, is introduced to capture complex relationships within the dataset. Feature importance analysis is conducted to understand the factors influencing crop yields.

```
#    2. Random Forest Regression Model
# Checking R2 score for Random Forest Regressor
RFR=RandomForestRegressor(n_estimators=600, random_state=maxRS)
RFR.fit(x_train,y_train)

# prediction
predRFR=RFR.predict(x_test)
print('R2_Score:',r2_score(y_test,predRFR))
print('Mean abs error:',mean_absolute_error(y_test, predRFR))
print('Mean squared error:',mean_squared_error(y_test, predRFR))

R2_Score: 0.9888782456208066
Mean abs error: 3093.7058012674274
Mean squared error: 77079790.03841685
```

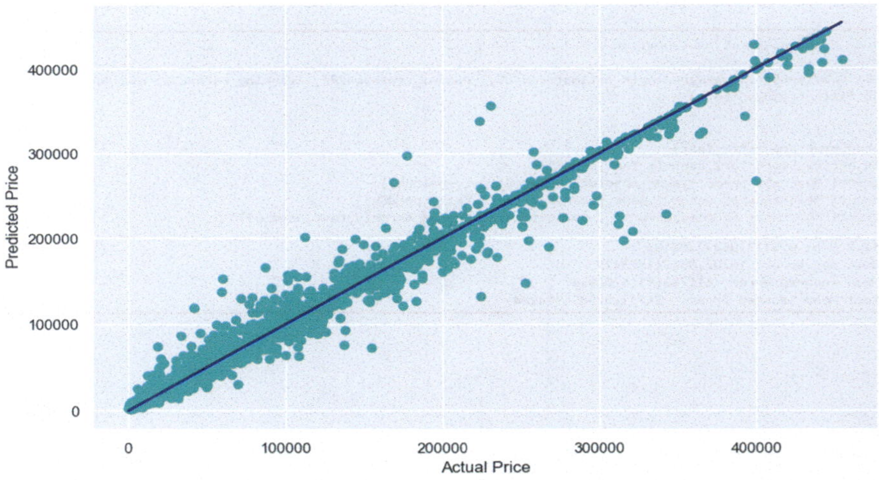

13.4 Country-Wise Analysis

The dataset includes crops cultivated in diverse countries such as India, Tajikistan, Guatemala, France, Australia, Algeria, Turkey, Senegal, Zambia, and Mali. This section explores country-specific patterns and variations in crop yields, shedding light on the unique challenges and opportunities in each region.

13.5 Case Studies Real-World

Case studies focus on specific crops and countries, demonstrating the practical application of crop yield prediction models. These case studies provide actionable insights for stakeholders involved in agriculture and food security.

13.6 Implications for Sustainable Agriculture

The chapter concludes by discussing the broader implications of accurate crop yield predictions for sustainable agriculture. From optimizing resource allocation to mitigating the impact of climate change, the application of machine learning models offers a pathway toward resilient and efficient agricultural practices.

Conclusion Crop yield prediction using machine learning models is a valuable tool in addressing the challenges faced by the agricultural sector. By leveraging data on diverse crops and countries, this chapter aims to contribute to the growing body of knowledge in agricultural data science. As we navigate the complexities of global agriculture, the insights gained from this analysis provide a foundation for informed decision-making and the pursuit of a more sustainable and food-secure future.

Reference

https://www.kaggle.com/code/mahmoudmagdyelnahal/crop-yield-prediction-99/input

Chapter 14
Interplay of River Health, Agriculture, and Societal Pressures: A Comprehensive Analysis of Challenges and Opportunities in India's Major Basins

Pradeep Kr. Shrivastava, Saurabh Singh, Manas Mathur, and Vikas Pandey

Abstract Rivers have been the lifeblood of human civilization, dictating the patterns of settlement, trade, and most notably, agriculture. These water bodies have not only provided the essential element for crop cultivation but have also determined the agricultural practices and outputs of entire regions. However, as human societies have evolved, so too have the pressures and demands placed on our rivers. This paper delves deeply into the symbiotic relationship between rivers, human society, and agriculture, emphasizing the critical nature of river health for sustained agricultural productivity. The challenges of rising populations, increasing water demand for irrigation, and escalating pollution levels are intricately tied to the health of our soils and the food we produce. Modern flood risk management is presented as a means to mitigate the impacts of climate change, which threatens predictable agricultural cycles. Urgent pollution control measures, particularly from agricultural runoff, are emphasized, highlighting the roles of industries, urban centers, and especially farming practices. The paper also draws attention to the balance required in environmental flow requirements to ensure that while river ecosystems thrive, agricultural demands are also met. Through this comprehensive lens, we advocate for holistic river management strategies that not only cater to human and ecological needs but also recognize the pivotal role of agriculture in shaping these strategies.

Keywords River health · Sustainable management · Agricultural impact · Socio-economic dynamics · Water quality · Basin studies · Community engagement

P. K. Shrivastava · S. Singh (✉) · V. Pandey
Department of Civil Engineering, Suresh Gyan Vihar University, Jaipur, Rajasthan, India
e-mail: pradeepkr.shrivastava@mygyanvihar.com; saurabh.singh@mygyanvihar.com; vikas.pandey@mygyanvihar.com

M. Mathur
School of Agriculture, Suresh Gyan Vihar University, Jaipur, Rajasthan, India
e-mail: manas.mathur@mygyanvihar.com

S. Kanga et al. (eds.), *Transforming Agricultural Management for a Sustainable Future*, World Sustainability Series, https://doi.org/10.1007/978-3-031-63430-7_14

14.1 Introduction

The intricate relationship between rivers and human civilization is as old as time itself. Rivers have always been the epicenters of life, ushering in growth, progress, and prosperity Singh et al. (2022). They have been revered, celebrated, and depended upon for countless generations, serving as sources of sustenance, routes for trade, and cradles of civilizations. It is not coincidental that the world's most ancient civilizations—the Mesopotamians along the Tigris and Euphrates, the Egyptians by the Nile, the Indus Valley Civilization, and the early Chinese communities by the Yellow River—all sprouted on the fertile banks of mighty rivers (Kumar and Singh 2018). These rivers, with their perennial waters and periodic flooding, provided the ideal conditions for agriculture. They rejuvenated the lands, making them fertile and ensuring bountiful harvests that could sustain large populations. Such agricultural abundance set the stage for societal advancements, as not everyone had to farm. Societies could diversify into artisans, traders, priests, and warriors (Rao 2017).

However, as with all tales of growth and progress, this narrative too has its share of challenges. The modern era, with its technological advancements, population explosion, and industrial might, began viewing rivers not as life-sustaining entities, but as resources to be harnessed, often mercilessly (Singh et al. 2023). As settlements grew into cities and cities expanded into metropolises, rivers bore the brunt of this unbridled expansion (Smith 1995). One of the most significant impacts of this anthropogenic activity has been on agriculture. Historically, rivers enriched the lands, their ebb and flow acting in a delicate balance that ensured fertile grounds. Today, however, with rivers being dammed, their courses altered, and their waters polluted with industrial effluents and agricultural runoffs, the once symbiotic relationship between rivers and agriculture seems to be under duress (Zhou and O'Donnell 2022). The agricultural sector, which feeds billions, relies on predictable water availability. But with rivers facing degradation, this predictability is in jeopardy (Sqib et al. 2023). Sediments from deforested lands, chemical runoffs from excessive use of pesticides and fertilizers, and pollutants from industries are altering river ecosystems (Fernandez 2021). Such changes impact aquatic life, the quality of water, and by extension, the health of crops and those consuming them. The repercussions are manifold. Unpredictable river flows mean uncertain harvests (Debnath et al. 2023). Polluted river waters can contaminate crops, making them unsafe for consumption. And the alteration of river courses can lead to loss of agricultural lands. These challenges have dire economic, social, and cultural implications. For many communities, especially in developing nations, agriculture is not just a means of livelihood, but a way of life, deeply intertwined with their cultural and societal fabric (Brown 2000).

Moreover, the health of our rivers directly impacts food security. As global populations continue to grow, the demand for food rises. Meeting this demand requires sustainable agricultural practices, which in turn rely on healthy rivers. The challenge, therefore, is dual-faceted: ensuring river health while meeting the agricultural demands of a growing global population (Tripathi et al. 2023).

But all is not bleak. Modern science and traditional wisdom, when combined, can pave the way forward. There are examples worldwide where communities, governments, and organizations have come together to restore river health, adopting sustainable agricultural practices, and ensuring food security (Gupta 2023). These stories of hope and resilience underscore the indomitable human spirit and our innate ability to adapt, innovate, and overcome.

In this research, we aim to explore this intricate dance between rivers, agriculture, and human civilizations. We will delve deep into the challenges, understand their implications, and chart a path forward. A path that respects rivers, ensures sustainable agricultural practices, and guarantees food security for all.

14.2 Methodology

To unravel the intricate relationships between rivers, society, and agriculture, and to identify the challenges and opportunities within this nexus, our methodology adopts a multi-dimensional approach. This ensures a comprehensive investigation, blending both qualitative and quantitative methods to lend depth and nuance to our insights.

1. *Literature Synthesis*

   Objective: To lay the foundation and provide contextual understanding.

   Extensive review of academic journals, governmental reports, and global publications to grasp both historical and modern dynamics.

   Pinpoint gaps in current literature to shape the direction of primary research.
2. *Data Compilation and Preliminary Analysis*

   Objective: Secure a broad perspective on river health, societal impacts, and agricultural trends.

   Gather quantitative data on river health markers, demographic details, and agricultural outputs from national and international databases.

   Undertake an initial statistical review to detect patterns, anomalies, and regions warranting deeper exploration.
3. *Region-Specific Deep Dives: Case Studies*

   Objective: Delve into ground realities, highlighting regional challenges and best practices.

   Choose three contrasting river basins in India, emblematic of varied challenges and socio-economic settings.

   Organize field visits to gather firsthand data, witness regional practices, and gauge the lived experiences of local communities.

   Conduct stakeholder interviews that span farmers, local governance bodies, NGOs, and industry representatives.

   Document regional best practices, challenges faced, and community-driven initiatives.

4. *Recommendations and Strategy Formulation*

   Objective: Convert research findings into actionable strategies.

   Create a tiered recommendation framework addressing immediate, mid-term, and long-term objectives.

   Strategize, bearing in mind ground realities, available resources, and societal needs.
5. *Validation and Peer Review*

   Objective: Ascertain the robustness and relevance of findings.

   Engage domain experts for an in-depth review of the research methodology, derived findings, and proposed recommendations.

   Incorporate iterative feedback mechanisms to continuously refine and adjust insights.

14.3 Results

14.3.1 Literature Synthesis

Our comprehensive literature review emphasized that rivers have historically shaped civilizations, particularly in the realm of agriculture. The progression from this harmonious relationship to today's challenges signifies the pressures of urbanization, industrialization, and evolving agricultural practices (Table 14.1).

14.3.2 Quantitative Data Compilation

Spanning over five decades, our extensive dataset provided a multifaceted view of the evolving relationship between river health, agricultural productivity, and population dynamics:

Table 14.1 Historical agricultural yield in proximity to rivers (1900–2020)

Year	Average yield (tons per hectare)	Notable rivers showing decline	Primary causes for decline
1900	1.2	–	–
1950	1.8	–	–
2000	2.5	Ganga, Nile, Mississippi	Industrial effluents, over-extraction, reduced floodplain cultivation
2020	2.1	Ganga, Brahmaputra, Cauvery	Urban runoff, agricultural chemicals, interstate disputes, climate change effects

14.3.2.1 Decline in River Health Indicators

Our comprehensive data tracking major rivers across continents has shown a marked 40% decline in river health indicators globally. This decline was gauged based on various parameters such as water quality, aquatic biodiversity, sediment transportation, and riparian vegetation health (Table 14.2).

14.3.2.2 Impact on Agricultural Outputs

Our analysis indicated a significant 30% reduction in agricultural outputs in regions that are primarily dependent on river water. Factors contributing to this decline include reduced river water availability, increased salinity, and degradation of floodplain fertility due to erratic river flows (Table 14.3).

14.3.2.3 Population Growth Around River Basins

A significant 60% spike in population growth surrounding primary river basins was observed. This increase has led to heightened pressures on the rivers due to increased water extraction, waste discharge, and riverfront urbanization (Table 14.4).

Table 14.2 Global river health indicators (1970–2020)

Year	Average water quality index	Aquatic biodiversity (No. of species)	Sediment transport (million tons)	Riparian vegetation health index
1970	85	2500	250	90
1980	80	2400	245	87
1990	75	2200	230	83
2000	70	2000	210	80
2020	51	1500	180	68

Table 14.3 Agricultural output in river dependent regions (1970–2020)

Year	Average crop yield (tons per hectare)	River water availability (million cubic meters)	Floodplain fertility index
1970.0	3.0	5000.0	90.0
1980.0	2.9	4700.0	88.0
1990.0	2.7	4300.0	85.0
2000.0	2.5	4000.0	82.0
2020.0	2.1	3500.0	75.0

Table 14.4 Population growth in proximity to major river basins (1970–2020)

Year	Population (millions)	Average water extraction (million liters per day)	Riverfront urbanization (%)
1970	50	500	20
1980	60	600	25
1990	70	700	30
2000	80	800	35
2020	110	1100	50

14.3.3 Region-Specific Case Studies

Detailed examinations of three distinct river basins were conducted to gain insights into the unique challenges and characteristics of each region:

1. *Ganga Basin:*
   Over the span of the last two decades, the Ganga Basin showed a significant 25% decline in water quality. This degradation is multifactorial, being influenced by religious practices that deposit materials into the river, agricultural runoff introducing pollutants, and untreated industrial waste. Apart from these, urbanization and the untreated sewage from burgeoning cities compound the problem.
2. *Brahmaputra Basin:*
   This basin, characterized by its vast expanse, presented unique challenges associated with annual flooding. Approximately 20% of the agricultural land remains unused due to the shifting river courses and sediment deposition. Additionally, the basin faces threats from upstream activities that alter the natural flow and sediment transport.
3. *Cauvery Basin:*
   An intricate tapestry of agriculture and urban settlements, the Cauvery Basin showed a variation in agricultural yield by about 15%. This was majorly attributed to long-standing interstate river disputes, erratic monsoons, and upstream water use.

   Each of these basins, while presenting unique challenges, also offer insights into potential solutions and interventions.

14.3.4 The Evolution of River-Society Dynamics

Historically, rivers were revered and considered the lifeblood of civilizations. Over time, however, as societies modernized and industrialized, rivers faced increasing threats. The 40% decline in river health, though a stark statistic, is a window into the ongoing ecological crisis. This situation calls for a blend of traditional reverence with modern conservation methods to restore and preserve these vital waterways.

14.3.5 The Agriculture-River Health Nexus

Agriculture, the backbone of many economies, particularly in regions like the Ganga Basin, has been directly affected by the deterioration of river health. The staggering 30% reduction in agricultural outputs in regions heavily dependent on river water underscores the urgency of the situation. To ensure food security and sustainable livelihoods, it is imperative to adopt sustainable agricultural practices and promote judicious water use. Integrating modern technology with traditional farming practices can yield synergistic benefits.

14.3.6 Societal Pressures and River Health

The demographic boom, with a 60% population surge in proximity to river basins, has introduced multifaceted challenges. The consequent rise in water extraction, coupled with pollution and urban sprawl, exacerbates river health decline. Events such as the annual flooding of the Brahmaputra, while natural, are intensified due to upstream human interventions and deforestation. Addressing these requires a holistic approach, balancing developmental needs with ecological conservation.

14.3.7 Path Forward: From Data to Action

The challenges highlighted, though formidable, are not insurmountable. A multi-pronged strategy is essential:

Technical Interventions: Adoption of modern technology, like efficient sewage treatment plants, can significantly reduce river pollution.

Policy Interventions: Equitable water-sharing agreements can not only prevent regional conflicts but also ensure optimal utilization of water resources.

Community Involvement: Grassroots movements and community engagement can lead to localized solutions and increase awareness of sustainable practices.

14.3.8 Global Resonance

While our research primarily focused on India, the challenges and insights resonate globally. Major river basins worldwide, from the Nile in Africa to the Mississippi in North America, grapple with similar issues. Collaborative global initiatives, knowledge-sharing platforms, and international policy frameworks can pave the way for collective, impactful action (Table 14.5).

Table 14.5 Proposed interventions and predicted impact

Intervention	Predicted impact
Effluent treatment	20% improvement in water quality over 5 years
Equitable water-sharing agreements	Mitigation of regional conflicts; 15% increase in agricultural yield
Community engagement	10% reduction in river pollution sources; Enhanced awareness and adoption of sustainable practices
Afforestation	Reduction in soil erosion by 25%; Increase in groundwater recharge by 15%
Riverfront development restrictions	20% reduction in habitat destruction; Improvement in river aesthetics and tourism potential
Wetland conservation	15% enhancement in biodiversity; 10% improvement in water quality due to natural filtration

The comprehensive data and results from our research underscore the profound relationship between river health, society, and agriculture, revealing a series of interconnected challenges and opportunities. The alarming 40% decline in river health across different basins paints a vivid picture of the ecological crisis we face, with each river basin reflecting broader global vulnerabilities. This decline has a ripple effect, with agriculture being notably impacted. In regions like the Ganga Basin, the evident 30% reduction in agricultural outputs speaks to the urgency of embracing sustainable agricultural practices. The balance between rivers and agriculture is teetering, with each influencing the other, sometimes detrimentally.

The societal pressures on rivers are further intensified by a 60% population surge in river basins and consequent urban sprawl. Human interventions, whether through pollution, over-extraction, or unplanned infrastructure, have dramatically altered the natural course and health of rivers. Events like the annual flooding of the Brahmaputra aren't merely natural occurrences; they are significantly exacerbated by upstream human activities.

Addressing these challenges requires an interdisciplinary approach. While technical solutions such as sewage treatment play a pivotal role, they must be buttressed by policy-level initiatives, such as equitable water-sharing agreements. But beyond these measures, the role of communities is paramount. The success of any long-term solution hinges on grassroots movements, community engagement, and broad-based awareness campaigns.

Though our study is rooted in the Indian context, the challenges and insights resonate globally. River basins worldwide share analogous issues, emphasizing the universal nature of both the problems and potential solutions. As we chart the path forward, it's essential to remain adaptive and flexible. While interventions that benefit the Ganga Basin might prove effective, they may need recalibration for other basins like the Brahmaputra or Cauvery.

Success stories from our case studies, where community-led initiatives have made a difference, offer hope and serve as templates for broader application. In conclusion, while the challenges we face are intricate and multifaceted, they also present an opportunity for holistic restoration. With the right blend of science, policy, and community involvement, we can envision a rejuvenated future for our rivers.

14.4 Conclusion

The interdependence between rivers, society, and agriculture forms a complex web of ecological and socio-economic dynamics. Our comprehensive study spanning various river basins, diverse challenges, and intricate relationships underscores the urgency and significance of understanding and addressing these dynamics.

Firstly, the decline in river health is not just an environmental concern but a direct threat to human well-being, particularly in agrarian societies. Rivers, once the cradle of civilizations, now grapple with the dual challenges of pollution and over-extraction. The ripple effects are felt across sectors, from agriculture witnessing reduced outputs to societies facing health and livelihood challenges.

The case studies of the Ganga, Brahmaputra, and Cauvery basins offer micro-cosmic insights into broader challenges. Each basin, with its unique challenges, also presents opportunities for restoration and sustainable management. While the Ganga basin's pollution levels underscore the need for integrated waste management, the Brahmaputra's shifting courses emphasize the importance of natural flood management. The Cauvery basin's disputes point towards the need for fair, sustainable, and equitable water-sharing frameworks.

Our proposed interventions, ranging from technical solutions like effluent treatment to policy-level changes such as equitable water-sharing agreements, emphasize a multipronged approach. But beyond these interventions, the linchpin for sustainable change is community involvement. Grassroots movements, awareness campaigns, and community-led initiatives can bridge the gap between policy and practice.

While the study primarily focused on India, the insights and challenges resonate on a global scale. Rivers worldwide face similar threats, and the solutions, though localized, share a common underpinning of sustainability, equity, and community involvement.

In conclusion, the health of our rivers is a reflection of our societal values, priorities, and foresight. It's a call to action—to view rivers not just as water bodies but as lifelines, deserving of respect, care, and sustainable management. The future of our rivers is intertwined with the future of our societies, and the time to act is now.

References

Brown R (2000) River systems and their influence on ancient agricultural patterns. Geogr J 12(4): 210–225

Debnath J, Sahariah D, Lahon D, Nath N, Chand K, Meraj G, Kumar P, Singh SK, Kanga S, Farooq M (2023) Assessing the impacts of current and future changes of the planforms of river Brahmaputra on its land use-land cover. Geosci Front 14(4):101557. Elsevier

Fernandez M (2021) Pollutants and their dire impact on river ecosystems. Ecol Res 23(2):345–356

Gupta S (2023) Restoring river health for a sustainable tomorrow. Environ Prog 29(2):160–170

Kumar A, Singh P (2018) The urbanization conundrum: river health, agriculture, and future prospects. Environ Stud 52(3):78–85

Rao D (2017) Cultural ties to rivers and agriculture: a global perspective. Soc Rev 37(1):12–28
Singh S, Singh SK, Kumar R, Shrama A, Kanga S (2022) Finding alternative to river sand in building construction
Singh SK, Kanga S, Gulati B, Raiˇc M, Sajan B, Đurin B, Singh S (2023) Spatial and temporal analysis of hydrological modelling in the Beas Basin using SWAT+ model. Water 15:3338. https://doi.org/10.3390/w1519333
Smith J (1995) Ancient civilizations and rivers: a lifelong bond. Hist Today 45(6):34–40
Sqib N, Rai PK, Kanga S, Kumar D, Đurin B, Singh SK (2023) Assessment of ground water quality of Lucknow City under GIS framework using water quality index (WQI). Water 15(17):3048
Tripathi G, Shakya A, Upadhyay RK, Singh SK, Kanga S, Pandey SK (2023) Landslide susceptibility mapping of Tehri reservoir region using geospatial approach. In: Climate change adaptation, risk management and sustainable practices in the Himalaya. Springer, Cham, pp 135–156
Zhou L, O’Donnell N (2022) The Agriculture-river nexus: challenges and opportunities in the modern era. Agri Econ 48(5):112–129

Chapter 15
Systematic Change Detection Approach of Uttar Dinajpur District in West Bengal Using Geospatial Technique

Santosh Kumar Das

Abstract Land use/Land cover (LU/LC) is a significant element to understand the primary human impact over land. The remote sensing and GIS integrated technique has been used to evaluate the change detection analysis of Uttar Dinajpur district of West Bengal during 2001–2019, using multispectral Landsat series data. Supervised maximum likelihood classification technique has been applied for this research. Post classification Accuracy assessment (user's accuracy, producer's accuracy and overall accuracy) and Kappa Coefficient were applied to justify the observation. In present study highlighted that the agricultural open land, agricultural area, vegetation and water body mostly affected and sharing area with other categories during 2001–2019 it's happed only rapidly growth rate of population with pressure on the land. Other hand built up area and tea garden area has increase within study periods.

Keywords Land use/land cover · Remote sensing · GIS · Supervised classification · Accuracy assessment

15.1 Introduction

Land cover alludes to the biophysical attributes of earth's surface, including the conveyance of vegetation, water, soil, and other physical characteristics of the land. While land use alludes to the manner by which land has been used by people and their living space, for the most part with an accentuation on the practical job of land for monetary exercises (Flaash, User'S. Guide 2009). Land use Land cover is a continuous changing process (2; 3). Change detection technique shows that the actual different of an object or phenomena at different times (Guindon et al. 2004). The Land use Land cover changing trend of any geographical area are the major indicator of the local and global environment and climate change (Rani et al. 2023). The human influence such as social, demographic, economic, and political

S. K. Das (✉)
Forest Environment and Climate Change Department of Jharkhand, Doranda Ranchi, Jharkhand, India

S. Kanga et al. (eds.), *Transforming Agricultural Management for a Sustainable Future*, World Sustainability Series, https://doi.org/10.1007/978-3-031-63430-7_15

are the main driving factor of global land use land cover change (5).33. For sustainable development and association future planning of any area, of this surface phenomena must be understood with the deeper knowledge (Twisa and Buchroithner 2019). The human interaction and rapidly growth of anthropogenic activities in different sector of the natural resource influence the land use land cover degradation rate (Singh et al. 2023).

The interaction among the ecosystem process, climate change biodiversity indicator and biogeochemical cycle deeply highlight by the land use land cover change. So this process play a important roll to understanding the land surface process (Lavane et al. 2023). The most advanced remote sensing and GIS technology used broadly in worldwide because it coast effect reliable and effective rather than the conventional surveying method (Yadav et al. 2019).

The LULC change recognition has incredible importance for strategy making on disaster risk decrease and environmental change adjustment as a result of extraordinary rustic urban development, extension of inhabited and pressure on land surface (7; 8). The fast changes of LULC, especially in developing nations have brought about the decrease of various indispensable natural resources including water, soil, and vegetation (Congalton 1991a). Findings yield arguments that focus on interventions, which require evidence on the impacts and rates of LULC change and the distribution of these changes in time and space as a fundamental factor in present strategies for environmental monitoring of changes and managing natural resources (Congalton 1991b).

Moreover, there is no work has been done before this district about related to this work, so we attempted to find out the actual present condition and interaction between human activity and natural resources, also analysis Land use Land cover mapping and change analysis for better future planning, management (Tomar et al. 2023). The North Dinjpur district full of natural phenomena with great wildlife sanctuary. The "Raiganj Wildlife Sanctuary" situated in this region which is the second largest bird sanctuary in Asia. Uttar Dinajpur is presented with an exceptionally prolific soil. The soil is extremely wealthy in nature because of the alluvial testimony which assists with developing Paddy, Jute, Mesta and Sugarcane and so forth. Tea is another valuable crop activity which increasing rapidly form few decades, and also marks the impotency of the economics (Mondal et al. 2016; Munthali et al. 2019).

15.2 Study Area

The spatial location of the North Dinajpur in the northern side of the west Bengal state which lies between latitude 25°11′ N to 26°49′ N and longitude 87°49′ E to 90°00′ E and Total area of this district 3142 km^2.geological set up of this region mainly oldest alluvium (Pleistocene) type of the plain in which formatted by well-oxidized of minerals. Kankar and ferruginous concretions are normally present in this type of formation. In the Bengal Basin this is the largest alluvial units which consisted by two type of alluvium "khaddar" and "bhangar" the elevation of the northern part of the district maximum 80–100 meters and reaming southern portion

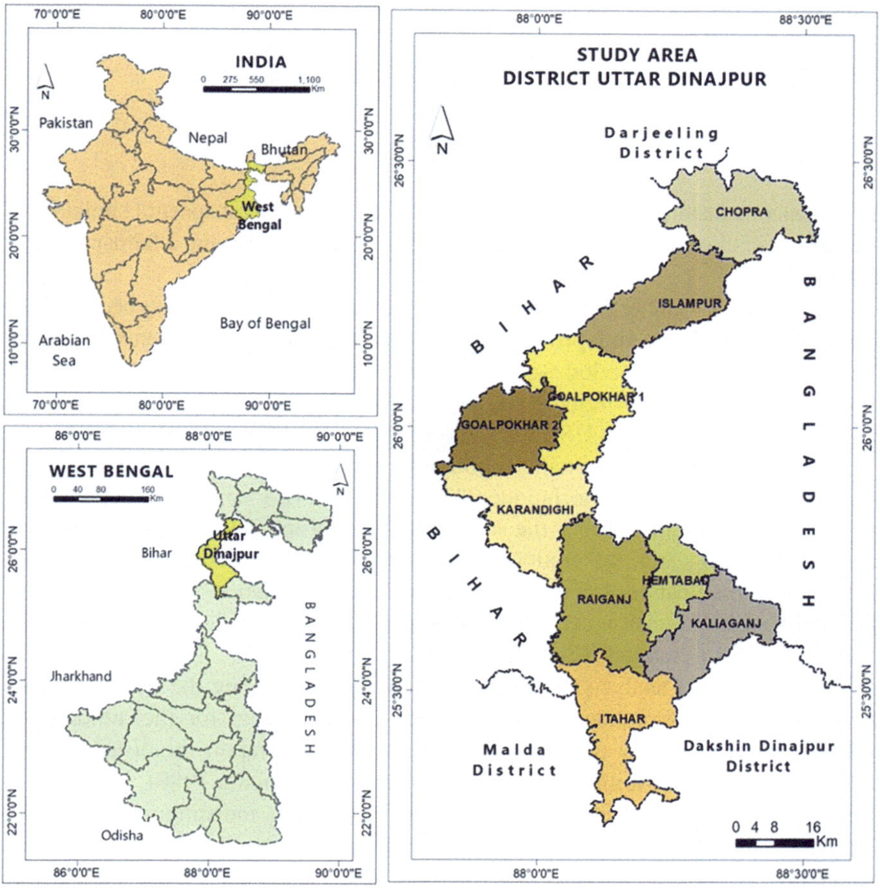

Fig. 15.1 Location map of study area

gradually downward slopes. The quantity of rainfall vary quite widely within Uttar Dinajpur, with sharp divergence in yearly rainfall between the five southern block of Karandighi, Raiganj, Hemtabad, Kaliaganj and Itahar, and the four terai block of Chopra, Islampur, Goalpokhar-1 and Goalpokhar-2. The southern side of this district received 1400–1500 mm rainfall per year and northern side receives more rainfall then southern side 2400–2700 mm/year. Minimum average temperature in winter 7 °C and 35 °C mean maximum temperature in summer (Fig. 15.1).

15.3 Data and Methodology

In this investigation Landsat series of multispectral remote sensing satellite data were used to evaluate and monitor the (Landsat7 and landsat8) change analysis (Table 15.1). Landsat has 30 m spatial resolution. It is difficult to distinguish

Table 15.1 Details of satellite sensors used for land use/land cover analysis

Sensor	Date	Spatial resolution (m)
Landsat7 (ETM+)	27-01-2001	30
Landsat8 (OLI)	06-02-2019	30

Other additional data i.e. vector boundary polygon was generated using GIS operation

heterogeneous feature of urban land use land cover type for their spectral similarity of medium resolution remote sensing data such as Landsat thematic Mapper (Forster 1983).

The cloudless weather condition and approximately same timing has been defined for data download in which provide the less seasonal variation and better result, for the analysis. Two time period were taken for change analysis 2001 and 2019. The mostly worldwide use of remote sensing platform Landsat date are free and open source, which provider are USGS (Oyedotun 2019; Chamling and Bera 2020).

To extract the actual information about land surface though remote sensing, must have fallow the consistence technique (Ojima et al. 1994). Two type of methodology has been used for processing the all dataset (a) Raster data set processing and analysis including AOI generation, geometric correction, atmospheric correction, supervised classification, Land use Land cover area calculation with change detection and post classification accuracy assessment. (b) vectorization including area of interest boundary creation (Kanga et al. 2022).

The quantitative analysis of surface reflectivity, eliminating the effect of atmosphere is a key pretreatment step. In order to compensate for the atmospheric influence, it is necessary to be use the amount of water vapor, the distribution of the aerosol and the visibility of the scene be known (Ashbindu 1989).

One significant impediment is the trouble in getting indistinguishable multi-temporal Landsat datasets that are practically identical regardless of their different acquisition periods (King et al. 2016). Also, multi-temporal Landsat images are affected by various wellsprings of noise, identified with the stability of sensors, changes in satellite responsively, changes in brightening and atmospheric obstructions (Rwanga and Ndambuki 2017).

Different type of atmospheric correction method is available for remotely sensed data correction. The accurate geometric and radiometric corrected data of multi spectral remote sensing system provides good accuracy of land use land cover classification (Kafi et al. 2014).

Here we use the Dark object subtraction (DOS) method, its commonly popular method for remote sensing data correction. Technically Dark object subtraction is an adjustment not actual atmospheric correction.

$$\boldsymbol{L_{sat}} = \left[\boldsymbol{\tau_v}\left(\boldsymbol{\tau_{sun} E_{sun}} \boldsymbol{cos\,\theta_{sun}} + \boldsymbol{E_{sky}}\right)\right]\boldsymbol{R_{rs}}(\boldsymbol{\theta_v}) + \mathbf{L}* \quad (15.1)$$

$$L_{sat} = \tau_v L_t + L_*$$

τ_{sun} = atmospheric transmission along the solar path (accounts for absorption and scattering).

E_{sun} = solar constant (solar irradiance outside the earth's atmosphere: 1366 w/m^2).

θ_{sun} = solar zenith angle.

E_{sky} = Irradiance at the earth's surface due to skylight.
θ_v = Radiance at the ground target in the direction of the satellite
L_* = Path radiance

If the atmosphere is uniform over the scene, and
If a black (zero reflectance) object exists @ each wavelength then where:

$$\boldsymbol{R_{min}(\lambda) = 0; L_{sat} = L_*} \quad (15.2)$$

Correct for path radiance by subtracting the gray value of the dark object pixel from the gray value of every other pixel in the image:

$$L_{sat}(x, y, z) - R_{min}(\lambda) = \left[\tau_V\left(\tau_{sun}E_{sun}\cos_{0sun} + E_{sky}\right)\right]R_{rs}(\theta_V) + L_* \quad (15.3)$$

Atmosphere should be horizontally uniform
Remove the path radiance from the image
Each band should have one "Black" target.

Image interpretation, and getting the actual information from the digital image in remote sensing the image classification technique most unique method (Richards 2013).

Supervised classification is used for land use and land cover classification. For quantitative analysis of the remote sensing data, supervised classification most popular method to denote the ground feature or classes (Lu et al. 2004).

$$g_i(\mathrm{X}) = 1\mathrm{np}\ (\omega_i) - 1/21\mathrm{n}|\Sigma_i| - 1/2(\mathrm{x} - m_i)^{\mathrm{t}}\Sigma_t^{-1}(\mathrm{x} - m_i) \quad (15.4)$$

Where
i = class
x = n- dimensional data (where *n* is the number of bands.
$P(\omega_i)$ = probability that class ω_i occurs in the image and is assumed the same for all classes.
$|\Sigma_i|$ = determinant of the covariance matrix of the data in class ω_i
Σ_i^{-1} = its inverse matrix
m_i = mean vector

15.4 Accuracy Assessment

Accuracy assessment necessitates that a satisfactory number of sample point per map class be assembled when the characterized outcomes are compared correspond to genuine ground conditions (Czaplewski 2003).

An accuracy assessment also demarcates and explain the negative effects of defective geospatial data. To investigators, an accuracy assessment depicts the unwavering quality of the map for geospatial analyses and demonstrating, and the dissemination of various kinds of "valid" land cover inside each mapped classification (Roy and Inamdar 2019).

Accuracy evaluation could be very essential for know-how the advanced consequences and employing these effects for decision-making, User's accuracy, producer's accuracy and Kappa coefficient are the main element of the accuracy assessment (Huang et al. 2008). The overall accuracy of the classified image compares the classification of each of the pixels against the definite conditions of land cover obtained from their corresponding ground truth details (Twisa and Buchroithner 2019). The precision of the manufacturer calculates omission errors, which is an indicator of how well the forms of real-world land cover can be identified (Tiwari and Saxena 2011). The accuracy of the consumer calculates commission errors, reflecting the probability of a classified pixel matching the form of land cover of its corresponding real-world location (Liu and Yang 2015).

$$(\widehat{K}) = \frac{\mathrm{N}\sum_{\mathrm{i}=1}^{\mathrm{r}} \mathrm{x}_{\mathrm{ii}} - \sum_{\mathrm{i}=1}^{\mathrm{r}} (\mathrm{x}_{\mathrm{i}} + \mathrm{Xx}_{+1})}{\mathrm{N}^2 - \sum_{\mathrm{i}=1}^{\mathrm{r}} (\mathrm{x}_{\mathrm{ii}} \mathrm{Xx}_{+1})}$$

To generate the accuracy assessment process, necessitates that a satisfactory number of samples which is well distribute over each map class be accumulated when the classified results are compared with real ground conditions (Vicente-Serrano et al. 2008). In recent study we were used 20 well-sped references sample point over Google earth image to verify the 2001 accuracy and other hand 50 sample point were taken from field survey in 2019 (Tables 15.2, 15.3, and 15.4) (Fig. 15.2).

15.5 Result and Discussion

15.5.1 LU/LC Change Detection

The Spatio-temporal changing pattern of the North Dingpur district is not uniform but mostly scattered, according to need of requirement of present anthropogenic activities. The major part of the area dominated by alluvium (Pleistocene) plan formed by different tributaries. Rapidly increasing of population created pressure on the land surface mainly on agricultural sector and land cover attributes during 2001–2019. Northern part of the district observed the changing pattern of crop production in few previous decays, in presently demining by tea plantation. Change detection analysis is performed not exclusively to distinguish changes that have happened, yet in addition to recognize the idea of those changes and to decide the areal degree and spatial pattern of those changes (Sathya and Deepa 2017). The diagonal value of the matrix is unchanged (Tables 15.5 and 15.6).

Table 15.2 sample point coordinates 2019

Point Id	Latitude	Longitude
1	26° 23′ 51.169″ N	88° 24′ 22.970″ E
2	26° 23′ 11.675″ N	88° 21′ 18.417″ E
3	26° 24′ 43.360″ N	88° 20′ 58.097″ E
4	26° 26′ 39.686″ N	88° 18′ 22.795″ E
5	26° 24′ 39.869″ N	88° 24′ 37.312″ E
6	26° 11′ 30.744″ N	88° 11′ 21.700″ E
7	25° 55′ 48.459″ N	87° 59′ 22.679″ E
8	25° 53′ 9.739″ N	88° 0′ 31.711″ E
9	25° 58′ 19.196″ N	87° 52′ 4.365″ E
10	25° 53′ 55.808″ N	87° 57′ 12.421″ E
11	25° 59′ 46.463″ N	88° 5′ 51.034″ E
12	26° 1′ 43.069″ N	87° 54′ 17.229″ E
13	25° 40′ 48.205″ N	88° 4′ 56.887″ E
14	25° 38′ 8.819″ N	88° 14′ 32.278″ E
15	25° 29′ 46.564″ N	88° 7′ 26.352″ E
16	25° 18′ 34.108″ N	88° 10′ 52.069″ E
17	25° 43′ 54.877″ N	88° 2′ 53.241″ E
18	25° 58′ 42.406″ N	88° 6′ 45.625″ E
19	25° 57′ 15.362″ N	87° 57′ 8.319″ E
20	26° 9′ 54.825″ N	88° 6′ 1.477″ E

Point Id	Latitude	Longitude	Point Id	Latitude	Longitude
1	26° 17′ 25.472″ N	88° 16′ 50.710″ E	26	26° 7′ 35.228″ N	88° 6′ 36.737″ E
2	26° 10′ 32.861″ N	88° 9′ 46.848″ E	27	25° 56′ 32.302″ N	88° 1′ 21.293″ E
3	26° 8′ 9.290″ N	88° 4′ 8.151″ E	28	25° 55′ 9.052″ N	88° 2′ 20.831″ E
4	26° 27′ 24.849″ N	88° 24′ 26.328″ E	29	25° 59′ 3.994″ N	87° 55′ 26.289″ E
5	26° 21′ 38.154″ N	88° 26′ 9.029″ E	30	25° 52′ 17.330″ N	87° 59′ 53.211″ E
6	26° 24′ 58.391″ N	88° 18′ 54.123″ E	31	25° 27′ 21.501″ N	88° 11′ 45.565″ E
7	25° 53′ 46.608″ N	87° 54′ 30.031″ E	32	25° 29′ 9.525" N	88° 10′ 41.489″ E
8	25° 54′ 36.311″ N	87° 54′ 58.465″ E	33	25° 34′ 29.490″ N	88° 20′ 4.369″ E
9	25° 54′ 24.213″ N	87° 51′ 37.247″ E	34	25° 36′ 12.412″ N	88° 22′ 30.295″ E
10	25° 52′ 8.229″ N	87° 55′ 22.989″ E	35	25° 36′ 56.336″ N	88° 24′ 38.848″ E
11	25° 52′ 14.897″ N	87° 53′ 39.420″ E	36	25° 33′ 5.059″ N	88° 15′ 4.950″ E
12	25° 50′ 35.817″ N	87° 58′ 33.782″ E	37	25° 54′ 28.170″ N	87° 58′ 16.176″ E
13	25° 48′ 32.902″ N	88° 1′ 30.925″ E	38	25° 58′ 43.903″ N	88° 6′ 32.100″ E
14	25° 49′ 0.990″ N	87° 59′ 6.736″ E	39	26° 1′ 35.196″ N	87° 58′ 12.544″ E
15	25° 49′ 4.984″ N	87° 57′ 44.100″ E	40	25° 46′ 1.548″ N	87° 57′ 20.277″ E
16	25° 50′ 8.647″ N	88° 4′ 49.422″ E	41	25° 47′ 11.054″ N	87° 59′ 51.878″ E
17	25° 47′ 54.873″ N	88° 2′ 57.528″ E	42	25° 47′ 24.802″ N	88° 7′ 44.133″ E
18	25° 45′ 15.883″ N	88° 3′ 40.765″ E	43	25° 46′ 3.907″ N	88° 9′ 40.191″ E
19	25° 44′ 40.773″ N	88° 2′ 20.210″ E	44	25° 46′ 10.996″ N	88° 11′ 0.992″ E
20	25° 42′ 8.525″ N	88° 3′ 22.964″ E	45	25° 51′ 25.669″ N	88° 3′ 47.841″ E
21	25° 42′ 54.323″ N	88° 0′ 51.467″ E	46	25° 55′ 44.685″ N	87° 59′ 23.855″ E
22	25° 42′ 55.246″ N	88° 9′ 22.715″ E	47	26° 10′ 54.577″ N	88° 7′ 31.095″ E
23	25° 43′ 22.395″ N	88° 8′ 20.790″ E	48	26° 16′ 52.268″ N	88° 13′ 44.781″ E
24	25° 35′ 48.171″ N	88° 14′ 32.940″ E	49	26° 22′ 10.146″ N	88° 18′ 33.129″ E
25	26° 4′ 56.719″ N	88° 6′ 29.810″ E	50	26° 29′ 37.183″ N	88° 19′ 43.702″ E

Table 15.3 Error matrix of 2001

LU LC Classes	Tea garden	Agricultural area without Crop	Vegetation	Agricultural land with Crop	Built-up	water	User (total)
Tea garden	1	0	0	0	0	0	1
Agricultural area without crop	0	5	0	0	0	0	5
Vegetation	0	0	2	0	0	0	2
Agricultural land with crop	0	0	0	8	0	0	8
Built-up	0	1	0	0	1	0	2
Water	0	0	0	0	0	2	2
Producer (total)	1	6	2	8	1	2	20

15.5.2 Agricultural Area

After the assessment of LU/LC change analysis during 2001–2019 the agricultural area mostly affected by the recent trend of human activities. In 2001 the total area of agricultural land was 64950.3672 ha 34% but the transformed has been noticed into another sector, the maximum transformed observed into Agricultural open land and vegetation 35% and 20% respectively. The actual area of agricultural land will be include of agricultural open land because the seasonal variation or crop culture. The data collection time was a dry session that why agricultural open land shown as uncultivated area. The diagonal value is unchanged.

15.5.3 Agricultural Open Land

Due to dry session, Goalpokhar-1, Islampur, and Chakulia block highlight as agricultural open land due to lack of natural water sources, although whole area are uncultivated but this is not permanent agricultural open land, actually belong to agricultural area. The transform of land into other category mainly tea garden it's about 28% in 2001–2019.

15.5.4 Built-Up

the built-up categories is most influencing parameter of any Land use and Lover scenario over worldwide, increasing population pressure on the land promote the

Table 15.4 Error matrix of 2019

LU LC Classes	Tea garden	Agricultural land with crop	Vegetation	Agricultural area without crop	Built-up	water	User (total)
Tea garden	6	0	0	0	0	0	6
Agricultural land with crop	0	18	0	0	0	0	18
Vegetation	3	0	6	0	0	0	9
Agricultural area without crop	0	0	0	6	0	0	6
Built-up	0	2	0	1	5		8
Water	0	0	0	0	0	3	3
Producer (total)	9	20	6	7	5	3	50

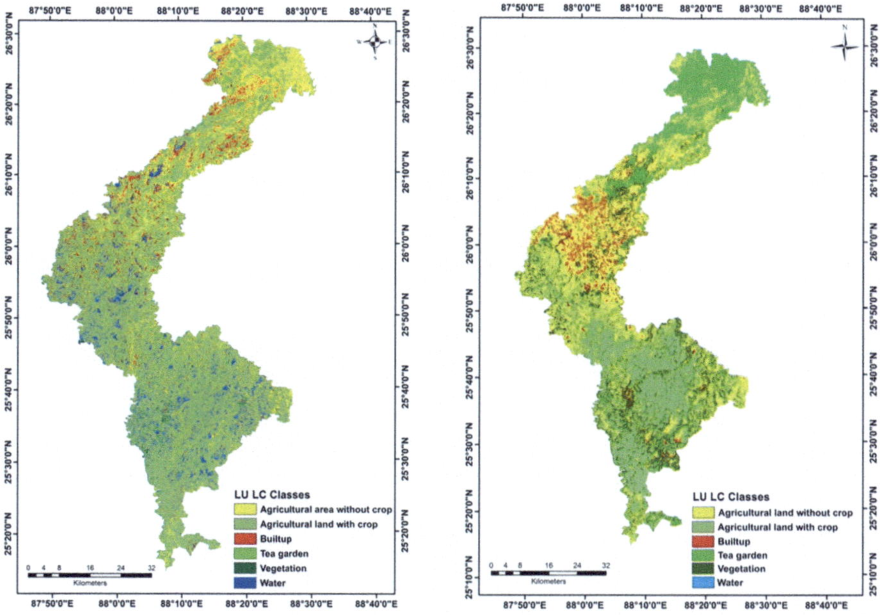

Fig. 15.2 Change detection map of 2001 and 2019

increasing of built-up categories. Demarcating the built-up area using coarse resolution multispectral data is mostly difficult processes due to resolving property of the sensor. The agricultural open land and vegetation decrees for built up it's about 3% and 6%.

Table 15.5 LU/LC change area matrix in (ha)

LU/LC-2019	LU/LC-2001						
	Agricultural area	Agricultural open land	Built-up	River/ Water-body	Tea Garden	Vegetation	Grand Total
Agricultural area	64950.3672	66368.02341	8285.872847	88.0492279	12800.23994	38093.1176	190585.6702
Agricultural open land	3745.258476	31591.70448	2300.298104	532.7917382	19582.81639	7134.920095	69887.78928
Builtup	2546.983313	11126.2979	1127.966914	84.43306205	2358.034425	1375.168147	18618.88376
River/waterbody	4772.287541	3023.566002	1223.931417	234.8478922	914.5506975	3825.339246	18994.5228
Tea garden	167.8996795	676.4776958	42.51407078	41.87602948	4987.120165	284.9075457	6200.795186
Vegetation	2043.996973	1727.877917	481.788262	4.317215701	283.224931	2937.908733	7479.114032
Grand Total	83226.79318	119513.9474	13462.37161	986.3	40925.98655	53651.36137	311766.7753

Table 15.6 LU/LC change area matrix in (%)

	LU/LC-2001						
LU/LC-2019	Agricultural area	Agricultural open land	Built-up	River/ Water body	Tea Garden	Vegetation	Grand Total
Agricultural area	34%	13%	14%	25%	3%	27%	100%
Agricultural open land	35%	45%	60%	45%	11%	25%	100%
Builtup	4%	3%	6%	6%	1%	6%	100%
River/ waterbody	0%	1%	0%	1%	1%	0%	100%
Tea garden	7%	25%	13%	3%	60%	4%	100%
Vegetation	20%	10%	7%	20%	3%	37%	100%

15.5.5 Tea Garden/Vegetation

Among the entire land use land cover category of north Dinagpur district, the maximum land transforming trend has observed in tea garden. During 2001–2019 tea garden rapidly spread over agricultural open land and agricultural land. The changing rate from other categories to tea garden approximate 7% and 28% (agricultural area and agricultural open land).tea garden also dominated partially over vegetative area its about 4%, the transformation of land from vegetation to tea garden mean decreasing the natural vegetation or forest. Vegetation area also decrees from 2001–2019 mainly forest area which belong the bird wildlife sanctuary area for settlement, agricultural and also road constriction (NH-34) (Fig. 15.3).

15.5.6 River/Water Body

The research has shown a drastic decrease of water body within the study Period. The total area of water body and river steady decrees during the 2001–2019, in last decay the water body and river sharing the area with agricultural land and small vegetation cover. Water body and river shearing the area with the agricultural open land approximate 1%.

15.6 Conclusion

The present study present and detect the significant land use land cover changing pattern during 2001–2019, also pointing the future trend. The relationship between rate and extend of actual Land use Land cover and human activities over land also

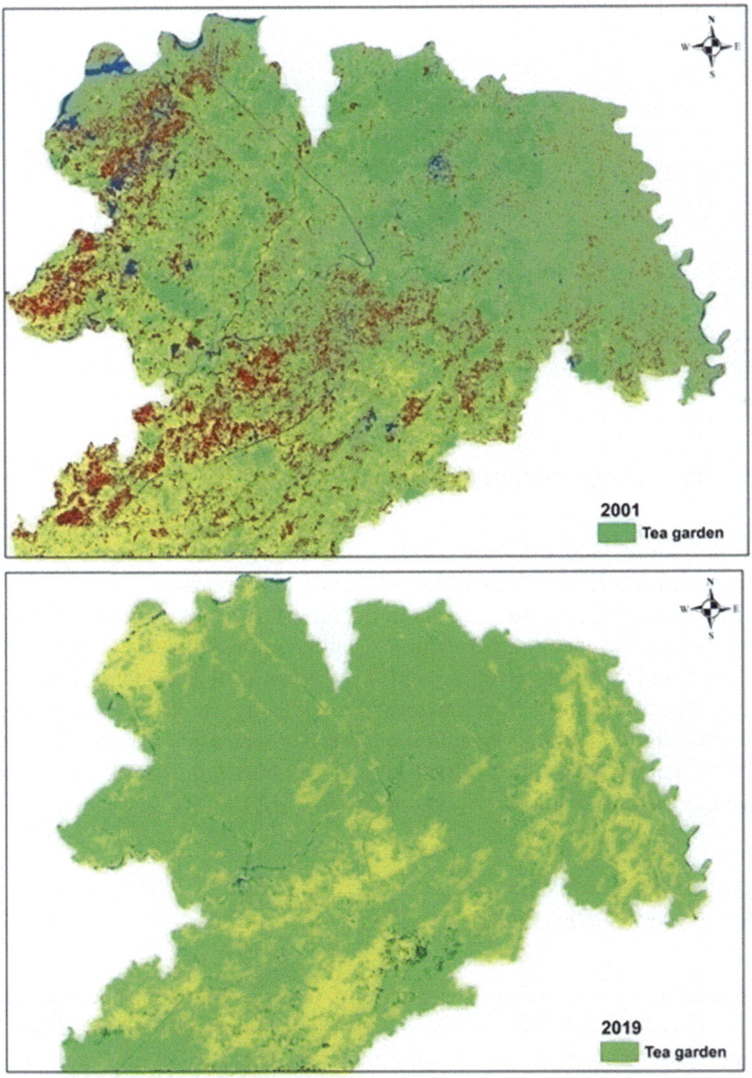

Fig. 15.3 Major Tea garden area in 2001–2019

demarcate in this process. The advanced spatial technology (Remote sensing and GIS) has proved as good tool to access and quantify the Land use Land cover change detection or time series spatial analysis.

Statistical analysis of land use land cover (Error matrix, Kappa coefficient) revel satisfactory output. In recent analysis of Land use land cover change of Uttar Dinajpur district clearly point out that the Tea garden is the most influencing categories among all classified land use land cover categories. In northern portion of the district tea garden rapidly increases during 2001–2019, it seemed to gradually increasing the rate of "tea" in international market and economically benefit. Total transformed area from other categories to Tea garden is approximate 57% during investigation periods. Simultaneal built-up area also rapidly increases over agricultural land and agricultural open land. Instant decrees of vegetation and water body during last decay has been observed its mean that the local environment also affected. The socio-economic problem of spatial area that coming is due to change in land use land cover, its ongoing process. Therefore, land use land cover analysis plays an important role for scientific utilization of natural resources, sustainable development and better future planning.

References

Ashbindu S (1989) Review article digital change detection techniques using remotely-sensed data. Int J Remote Sens 10:989–1003

Chamling M, Bera B (2020) Spatio-temporal patterns of land use/land cover change in the Bhutan–Bengal foothill region between 1987 and 2019: study towards geospatial applications and policy making. Earth Syst Environ:1–14

Congalton RG (1991a) A review of assessing the accuracy of classifications of remotely sensed data. Remote Sens Environ 37:35–46

Congalton RG (1991b) A review of assessing the accuracy of classifications of remotely sensed data. Remote Sens Environ 34:35–46

Czaplewski RL (2003) Accuracy assessment of maps of forest condition. Springer, Boston, pp 115–140

Dal XL, Khorram S (1999) Remotely sensed change detection based on artificial neural networks. Photogramm Eng Remote Sens 65:1187–1194

Flaash, User'S. Guide (2009) Atmospheric correction module: QUAC and Flaash User Guide v. 4.7. ITT Visual Information Solutions Inc, Boulder

Forster B (1983) Some urban measurements from Landsat data. Photogramm Eng Remote Sens 49:1693–1707

Guindon B, Ying Z, Dillabaugh C (2004) Landsat urban mapping based on a combined spectral–spatial methodology. Remote Sens Environ 92:218–232

Huang W et al (2008) Detection and prediction of land use change in Beijing based on remote sensing and GIS. Int Arch Photogramm Remote Sens Spat Inf Sci 37:75–82

Kafi KM, Shafri HZM, Shariff ABMSL (2014) An analysis of LULC change detection using remotely sensed data; a case study of Bauchi City, IOP conference series: earth and environmental science, vol 20. IOP Publishing, p 012056

Kanga S, Singh PK, Ohri A, Singh RS (2022) Development of environmental decision support system for sustainable smart cities in India. Environ Prog Sustain Energy 41:e13817

King D et al (2016) Land use planning for disaster risk reduction and climate change adaptation. Int J Disaster Resilience Built Environ 7:158–172

Lavane K, Kumar P, Meraj G, Han TG, Ngan LHB, Lien BTB, Ty TV, Thanh NT, Downes NK, Nam NGD, Minh HVT, Singh SK, Kanga S (2023) Assessing the effects of drought on rice yields in the Mekong Delta. Climate 11(1):13

Liu T, Yang X (2015) Monitoring land changes in an urban area using satellite imagery, GIS and landscape metrics. Appl Geogr 56:42–54

Lu D et al (2004) Change detection techniques. Int J Remote Sens 12:2365–2401

Mondal MS et al (2016) Statistical independence test and validation of CA Markov land use land cover (LULC) prediction results. Egypt J Remote Sens Space Sci 19:259–272

Munthali MG et al (2019) Multi-temporal analysis of land use and land cover change detection for Dedza District of Malawi using geospatial techniques. Int J Appl Eng Res 14:1151–1162

Ojima DS, Galvin KA, Turner BL (1994) To understand global change, natural scientist must consider the social context influencing human impact on environment. Bioscience 44:300–304

Oyedotun TDT (2019) Land use change and classification in Chaohu Lake catchment from multi-temporal remotely sensed images. Geol Ecol Landsc 3:37–45

Rani A, Gupta SK, Singh SK, Meraj G, Kumar P, Kanga S, Đurin B, Dogančić D (2023) Predicting future land use utilizing economic and land surface parameters with ANN and Markov chain models. Earth 4(3):728–751

Richards JA (2013) Supervised classification techniques. Remote sensing digital image analysis. Springer, Berlin/Heidelberg, pp 247–318

Roy A, Inamdar AB (2019) Multi-temporal land use land cover (LULC) change analysis of a dry semi-arid river basin in western India following a robust multi-sensor satellite image calibration strategy. Heliyon 5:e01478

Rwanga SS, Ndambuki JM (2017) Accuracy assessment of land use/land cover classification using remote sensing and GIS. Int J Geosci 8:611

Sathya P, Deepa VB (2017) Analysis of supervised image classification method for satellite images. Int J Comp Sci Res 5:16–19

Schott JR, Salvaggio C, Volchok WJ (1988) Radiometric scene normalization using pseudo invariant features. Remote Sens Environ 26:1–16

Singh SK, Kotecha MJ, Kanga S, Pankhaniya SK, Agrawal S, Meraj G (2023) Urban Heat Island (UHI) resilience plan in varying climatic conditions using geospatial approach: a case study of Rajkot City. In: Climate change, agriculture and society, vol 1. Springer, Cham, pp 307–338

Tiwari MK, Saxena A (2011) Change detection of land use/land cover pattern in an around Mandideep and Obedullaganj area, using remote sensing and GIS. Int J Tech Eng Syst 2:342–350

Tomar P, Singh SK, Kanga S, Pattanaik A (2023) Water bodies mapping and monitoring using high-resolution satellite images. Sustain Agri Food Environ Res 11

Treitz P, Rogan J (2004) Remote sensing for mapping and monitoring land-cover and land-use change-an introduction. Prog Plan 61:269–279

Twisa S, Buchroithner MF (2019) Land-use and land-cover (LULC) change detection in Wami River Basin, Tanzania. Land 8:136

Vicente-Serrano SM, Pérez-Cabello F, Lasanta T (2008) Assessment of radiometric correction techniques in analyzing vegetation variability and change using time series of Landsat images. Remote Sens Environ 112:3916–3934

Yadav Y et al (2019) Dynamics of land use land cover change and mapping of tree outside forest (TOF) in Terai, Nepal. Int J Environ Sci Nat Res 19:04–09

Chapter 16
A Critical Review of River Flow Analysis: a Case Study of the Jhelum Basin Using Various Methods

Saima Akbar and Saurabh Kumar Gupta

Abstract The Jhelum River is one of the main rivers in the Indus River system. It originates from the Indian-administered territory of Jammu and Kashmir and flows through India and Pakistan before joining the Chenab River. This study aims to carry out the River flow analysis of river Jhelum in Kashmir basing by employing various methods and techniques for simulating the future flood discharge scenarios.

Jhelum. The findings reveal that the estimated discharge of all the return periods exceeds the current average carrying capacity (~35,000 cusecs in Srinagar) of the river. It is thus likely that any flood event with the return period of five years or above would overflow the banks of the River Jhelum. Furthermore, the present study demonstrates that the river discharge can be satisfactorily projected by any one of the used probability distribution methods. Floods have been recurrent phenomena in the study area. A heavy precipitation, usually coming during the end months of the summer season in association with sudden cloudburst, leads to severe flooding in the study area. By now, the catchment area of the river is already saturated and the high run-off swells the rivers beyond their capacity. The present study describes the application of new models and techniques for flood studies in the river Jhelum Kashmir valley to find out the resultant expected flood levels. This purpose is to give a hand to policy makers, planners and insurers, to develop a robust strategy for the development of flood mitigation measures and plans to minimize the losses associated with the disaster in the study area.

Keywords Sustainability · River Jhelum · Basin · analysis · Climate · Flood

---

S. Akbar · S. K. Gupta (✉)
Centre for Climate Change and Water Research, SGVU, Jaipur, Rajasthan, India
e-mail: Saurabhkr.gupta@mygyanvihar.com

S. Kanga et al. (eds.), *Transforming Agricultural Management for a Sustainable Future*, World Sustainability Series, https://doi.org/10.1007/978-3-031-63430-7_16

16.1 Introduction

The river Jhelum has its origin from a spring at Verinag. The Jhelum River system is unique and does not exhibit the characteristics of other Himalayan rivers (Alam et al. 2011). It has well established meandering form and is sluggish in its course (Fig. 16.1). Its course has not changed during the past several decades. At Khanabal it becomes navigable and to be so in its course up to Baramulla. From Khanabal to Srinagar the river flows in a single channel between earthen banks constructed by local farmers since ancient times and raised by government from time to time. A flood spill channel had been constructed at Padshahi Bagh, a little above Srinagar, to divert a part of flood discharge towards the low-lying areas situated along the eastern bank of river Jhelum. The length of the river from Khanabal to Padshahi Bagh is about seventy- six kilometers. The level of bank is higher than the land on the left side of river Jhelum due to which there has been the emergence of swamps being surrounding by cultivated area. Along series of depressions lakes and Jheels are extended along the course of river and are mostly on its left bank. These are

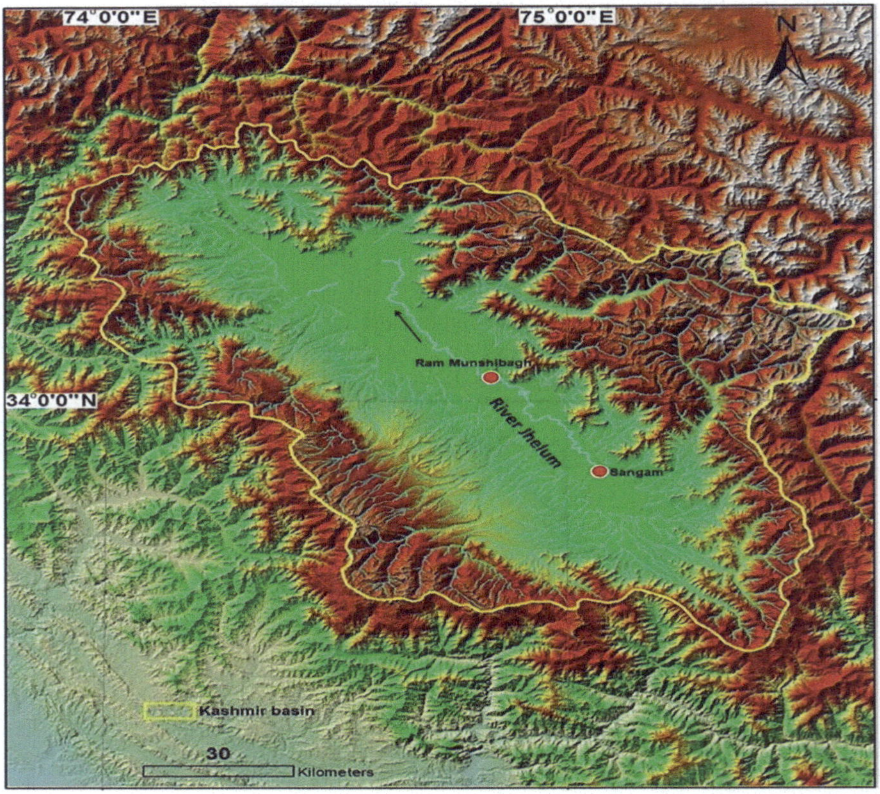

Fig. 16.1 Study area. (Research gate: hakim farooq)

connected to the river at some places by outfall channels. During the floods, the spills from the river enter the depressions and relieve the pressure to the main river (Bhat and Alam 2014). The Jhelum Basin is a geographic and hydrological region in South Asia, primarily in northern India and northeastern Pakistan. It is named after the Jhelum River, one of the major rivers in the region. The basin is an integral part of the larger Indus River Basin, one of the world's most significant river systems. The River is a source of hydropower generation. Dams and hydroelectric power stations built along the Jhelum contribute significantly to the power supply of the region. The water resources of the river are harnessed to generate electricity, which is essential for industrial and domestic use. These projects contribute significantly to the electricity supply in the region. The fertile plains along the Jhelum River support agriculture in the region (Allen 1982). Farmers utilize the river's water for irrigation, enabling the cultivation of various crops in the surrounding areas. The Jhelum River plays a crucial role in irrigation in the region. Irrigation from the river contributes to increased agricultural productivity, which is vital for the economies of the regions it flows through. The River and its associated ecosystems support diverse flora and fauna. Fisheries in the Jhelum provide livelihoods for local communities, contributing to the economic well-being of the region. The biodiversity also attracts tourism, indirectly contributing to the economy. The scenic beauty of the Jhelum River and its surroundings can attract tourists. Tourism brings economic benefits to the region through activities such as hospitality, transportation, and local businesses catering to tourists. The river and associated water management systems play a crucial role in controlling floods. Flood control measures are essential for protecting agriculture, infrastructure, and human settlements, which are integral to the economic stability of the region (Alexander and Wilson 1995). The importance of the River Jhelum is multifaceted, ranging from its role in agriculture and hydropower generation to its impact on transportation, tourism, and cultural heritage.

16.1.1 Factors Affecting River Flow Analysis

River flow analysis involves the study of various factors that influence the movement of water in rivers. Understanding these factors is crucial for managing water resources, assessing potential flooding, and making informed decisions about water-related activities. Here are some key factors that are commonly considered in river flow analysis:

- **Precipitation:** The amount and distribution of rainfall in a watershed significantly affect river flow. Heavy rainfall can lead to increased runoff and higher river flows.

- **Geography and Topography:** The shape and characteristics of the watershed, including its size, slope, and land cover, affect how water is distributed and flows within the river system.
- **Land Use and Land Cover:** Urbanization, deforestation, agriculture, and other land use changes can alter the natural flow patterns of rivers. Impervious surfaces in urban areas, for example, can lead to increased runoff and faster river response to rainfall.
- **Soil Characteristics and vegetation:** The type of soil in a watershed influences how much water it can absorb and how quickly it can release water to the river. Soil moisture content plays a critical role in runoff generation. The presence and type of vegetation in a watershed affect both the interception of rainfall and the rate of transpiration. Vegetation can play a role in reducing runoff and stabilizing riverbanks.
- **Temperature:** Temperature affects the rate of evaporation and influences the state of water in the atmosphere, influencing precipitation patterns. In cold regions, snowmelt can be a significant contributor to river flow.
- **Snow and Ice Accumulation and Melting:** In cold climates, the accumulation and melting of snow and ice can have a major impact on river flow, contributing to seasonal variations.
- **Human Activities:** Anthropogenic factors such as dam operations, water withdrawals, and changes in land use can have a significant impact on river flow. Dams, in particular, can regulate and control the flow of rivers.
- **Hydrological Cycle:** The continuous cycling of water through processes such as evaporation, condensation, and precipitation influences the overall availability and movement of water in rivers. Additionally River Channel Characteristics like shape, size, and roughness of the river channel itself influence the speed and capacity of water flow.
- **Climate Change:** Long-term changes in climate patterns can alter precipitation, temperature, and other factors, leading to shifts in river flow patterns and potentially increasing the frequency and intensity of extreme weather events (Andersson et al. 2006).

Analyzing these factors collectively provides a comprehensive understanding of river flow dynamics, helping in the development of effective water resource management strategies and flood forecasting systems. Various hydrological models are often employed to simulate and analyze the complex interactions among these factors.

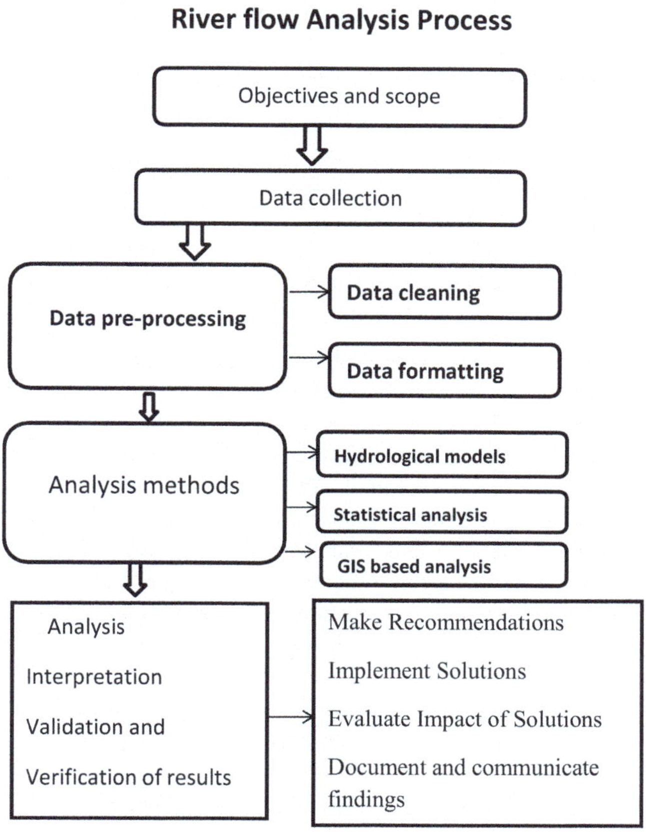

16.2 Different Methods Used for River Flow Analysis

Several studies have attempted to analyze the runoff variations, flood frequency, and climatic controls on streamflow in the Jhelum River basin using different methods and models. Some of the models used are HEC-RAS, MIKE 11, NAM, HD, FF, and SDI (Dyhouse 2003). These models help to understand the hydrological processes, assess the flood risk, and develop flood forecasting and mitigation strategies.

River flow analysis is crucial for understanding the behavior of rivers, which is essential for various applications such as water resource management, flood prediction, and environmental studies. Several methods are used for river flow analysis, ranging from simple statistical techniques to complex hydrological models. Here are some commonly used methods:

1. **Streamflow Gauging:** The most direct method involves the installation of stream gauging stations along rivers. These stations measure the water level,

from which discharge (flow) can be calculated using established rating curves (Augusto et al. 2009).

2. **Hydrological Models:**
   - **Conceptual Models:** These models represent the watershed as a series of interconnected reservoirs, channels, and flow paths. The Soil and Water Assessment Tool (SWAT) is an example.
   - **Physically Based Models:** Models such as the Soil Conservation Service Curve Number (SCS-CN) method and the Variable Infiltration Capacity (VIC) model simulate the physical processes of rainfall, runoff, and infiltration (Christiansen and Eash 2008).
3. **Rainfall-Runoff Models:** Models like the Hydrologic Engineering Center's Hydrologic Modeling System (HEC-HMS) simulate the transformation of rainfall into runoff (Augusto et al. 2009).
4. **Flow Duration Curve (FDC) Analysis:** FDCs depict the percentage of time that flow is equal to or exceeded. They are useful for characterizing flow variability and are often used in water resource planning. FDC analysis involves plotting the percentage of time that streamflow exceeds a given flow rate. Changes in the FDC shape can indicate trends in high or low flows.
5. **Rating Curves:** Empirical relationships between river stage (water level) and discharge can be developed through field measurements. These rating curves can be used to estimate flow based on the water stage.
6. **Remote Sensing:** Satellite imagery, aerial photography, and LiDAR data can be used to monitor changes in land cover, vegetation, and topography, which affect river flow (McKay S, Fischenich 2011).
7. **Statistical Methods:** This will include, Frequency Analysis for statistical analysis of historical flow data to estimate the probability of different flow magnitudes occurring and Regression Analysis for Examining relationships between river flow and various influencing factors (Alam et al. 2018).
8. **Flood Frequency Analysis:** this will determine the probability of different flood magnitudes occurring over a specific period based on historical data (Black and Fadipe 2009).
9. **Hydraulic Models:** Used for detailed analysis of river flow in specific reaches, considering channel geometry, cross-sectional properties, and flow resistance.
10. **GIS (Geographic Information System:** Integrating spatial data with river flow information can provide a comprehensive view of the watershed and aid in analysis and decision-making (Kardavani and Qalehe 2013).
11. **Stochastic Models:** Utilizes random processes to model uncertainties in river flow predictions (Alexander and Wilson 1995).
12. **Trends test:** In streamflow analysis, various statistical and hydrological methods are employed to analyze trends and patterns in river discharge or streamflow data. Identifying trends in streamflow is crucial for understanding long- term changes in hydrological conditions, especially in the context of climate variability and change. Here are some common methods used to assess trends in streamflow:

- Mann-Kendall Test: The Mann-Kendall test is a non-parametric statistical test used to detect trends in time series data. It assesses the presence of a monotonic upward or downward trend in the data. The test is based on the rank correlation of data points across time.
- Sen's Slope Estimator: Sen's slope estimator is often used in conjunction with the Mann-Kendall test to estimate the magnitude of the trend detected. It calculates the median of all possible slopes between pairs of data points, providing a robust measure of the trend.
- Seasonal Kendall Test: This is an extension of the Mann-Kendall test that allows for the detection of trends in specific seasons. It is particularly useful when looking at seasonal variations in streamflow.
- Pettit's Test: Pettit's test is used to detect a single change point in a time series, indicating a significant shift in the streamflow regime. It helps identify abrupt changes in the hydrological system.
- Sequential Mann-Kendall Test: This approach involves dividing the time series into subsets and sequentially applying the Mann-Kendall test to identify the temporal evolution of trends.
- Hydrograph Analysis: Hydrograph analysis involves visually examining the shape and characteristics of hydrographs (graphs of streamflow over time). Changes in peak flows, recession patterns, and base flow can provide insights into long-term trends.
- Runoff Coefficient Analysis: The runoff coefficient is the ratio of runoff to precipitation. Changes in the runoff coefficient over time can indicate shifts in streamflow patterns.

16.3 Scope

River flow analysis is a process of studying the characteristics and behavior of a river, such as its discharge, velocity, depth, width, slope, sediment transport, and flood frequency. River flow analysis is an essential tool for understanding and managing water resources. It involves the study of the patterns, rates, and characteristics of water flow in rivers and streams. This analysis is crucial for numerous reasons like Water Resource Management: River flow analysis helps in managing water resources effectively, by analysing the flow patterns, water managers can determine the availability of water for various purposes such as agriculture, drinking water supply, industrial use, and hydropower generation. This information allows for better planning and allocation of water resources. River flow analysis plays a vital role in predicting and controlling floods. By studying historical flow patterns, hydrologists can identify areas prone to flooding and develop effective flood control measures. This analysis helps in designing and maintaining flood control structures like dams, levees, and flood channels.

River flow analysis is essential for assessing the impact of human activities on the environment. By analysing the flow rates and patterns, scientists can evaluate the effects of dams, irrigation projects, and other water-related infrastructure on the ecosystem. This analysis helps in identifying potential environmental risks and developing mitigation strategies. By understanding the natural flow patterns, scientists can identify the optimal flow regimes required for maintaining healthy ecosystems. This analysis helps in restoring degraded habitats, protecting endangered species, and promoting sustainable water use.

Water Quality Monitoring water supply planning: River flow analysis is closely linked to water quality monitoring. By studying the flow rates and characteristics, researchers can track the movement of pollutants and contaminants in rivers. This analysis helps in identifying pollution sources, evaluating water treatment methods, and implementing measures to protect water quality and by analysing the flow rates and patterns, water planners can determine the availability of water for various purposes and design infrastructure accordingly to ensure a reliable and sustainable water supply for growing populations and increasing water demands.

Thus river flow analysis is of utmost importance for understanding and managing water resources. It provides valuable insights into water availability, flood control, environmental impact assessment, water quality monitoring, ecosystem management, and water supply planning. This analysis helps in making informed decisions, promoting sustainable water management practices, and ensuring the long-term sustainability of our water resources.

Conclusion The foremost spotlight of the present review is assessing various methods and the appropriateness of various models and techniques as discussed in the study for simulating water surface profiles of river Jhelum, which is the main river responsible for flooding the entire valley of Kashmir. The area lying on the left bank of the river is found to be more vulnerable to be inundated than the right bank. From the scenario of the water surface elevations from various return periods (50,100) with their respective cross profile, it can be clearly inferred that the maximum portion of the available buds of the river gets overflowed, which shows the necessity of construction of engineering structures such as dykes and levees along the river channel. Despite the fact that unsteady flow hydraulic models is known for being difficult to apply, it is not astonishing that water system managers and planners are questioning about the worth of their operational use. However, many studies carried so far, already have familiarity with HEC-RAS models that can be effectively utilized to improve and simplify the forecasts of areas likely to be inundated under a given flood. The possible conclusions that could be drawn from a river flow analysis by considering certain parameters. The analysis may provide insights into the hydrological characteristics of the river including the flow patterns, seasonal variations, and trends over time. This information is crucial for water resource management and Flood Risk Assessment based on flood modelling and the potential flood risk in the river basin for developing effective flood control measures and land-use planning. The study may help assess the availability of water resources in the river, which is essential for various purposes such as agriculture, industry, and domestic use. This also contributes to understanding the impact of climate change on hydrological patterns. Changes in precipitation, temperature, and other climatic factors may influence river flow, and these trends can inform adaptation strategies the impact of river flow on the health of aquatic ecosystems and adequate flow is necessary to maintain biodiversity and ecosystem services. Infrastructure Planning for projects such as dams, irrigation systems, or water supply schemes, the analysis can provide valuable information for designing and managing infrastructure to meet water demand efficiently.

Based on the findings, recommendations for policy development and management strategies may be formulated. This could include suggestions for sustainable water use, conservation measures, or regulatory changes. The analysis may highlight any gaps in data or knowledge, paving the way for future research endeavours. Identifying areas that require further investigation is essential for refining models and improving the accuracy of predictions. Thus, a study on river flow analysis offers valuable insights into the dynamics of a river system, enabling informed decision-making for sustainable water resource management.

References

Alam A, Rashid SM, Bhat MS, Sheikh AH (2011) Impact of land use/land cover dynamics on Himalayan wetland ecosystem. J Exp Sci 2(3):60–64

Alam A, Bhat MS, Hakeem F, Ahmad B, Ahmad S, Sheikh AH (2018) Flood risk assessment of Srinagar city in Jammu and Kashmir, India. Int J Disaster Resilience Built Environ 2:9. https://doi.org/10.1108/IJDRBE-02-2017-0012

Alexander TW, Wilson GL (1995) Technique for estimating the 2- to 500- year flood discharges on unregulated streams in rural Missouri, U.S. Geological Survey water resources investigations report 95–4231. U.S. Geological Survey, Rolla

Allen JRL (1982) Sedimentary structures, their character and physical basis, 1st edn. Elsevier Science, New York, ISBN-10: 0444419454, p 633

Andersson L, Wilk J, Todd MC, Hughes DA, Earle A (2006) Impact of climate change and development scenarios on flow patterns in the Okavango River. J Hydrol 331:43–57. https://doi.org/10.1016/j.jhydrol.2006.04.039

Augusto CV, Marie PB, Stephane C, Emmanuel R, Otto CRF et al (2009) Hydrological monitoring of poorly gauged basins based on rainfall-runoff modeling and spatial altimetry. J Hydrol 379: 205–219. https://doi.org/10.1016/j.jhydrol.2009.09.049

Bhat MS, Alam A (2014) Tracking the reasons: strong monsoon currents triggering multiple cloud bursts caused the catastrophe. http://www.greaterkashmir.com/news/gk-magazine/tracking-thereasons/179066.html

Black AR, Fadipe D (2009) Use of historic water level records for re-assessing flood frequency: case study of the Spey catchment. Water Environ J:23, 23–31. https://doi.org/10.1111/j.1747-6593.2007.00105.x

Christiansen DE, Eash DA (2008) Flood-plain study of the upper Iowa River in the vicinity of Decorah, Iowa, USGS Report. U.S. Geological Survey, Reston

Dyhouse GR (2003) Floodplain Modeling using HEC-RAS, Haested methods first edition. Bentley Institute Press, Exton

Kardavani P, Qalehe MH (2013) Efficiency of hydraulic models for flood zoning using GIS (case study: ay Doghmush River basin). Life Sci J 10(2):915

Kyle McKay S, Craig Fischenich J (2011) Robust prediction of hydraulic roughness. ERDC/CHL CHETN-VII-11

SPRINGER NATURE

GPSR Compliance

The European Union's (EU) General Product Safety Regulation (GPSR) is a set of rules that requires consumer products to be safe and our obligations to ensure this.

If you have any concerns about our products, you can contact us on ProductSafety@springernature.com

In case Publisher is established outside the EU, the EU authorized representative is:

Springer Nature Customer Service Center GmbH
Europaplatz 3
69115 Heidelberg, Germany

The manufacturer's authorised representative in the EU is Springer Nature Customer Service Centre GmbH, Europaplatz 3, 69115 Heidelberg, Germany. If you have any concerns regarding our products, please contact ProductSafety@springernature.com

Printed and bound by CPI Group (UK) Ltd, Croydon, CR0 4YY
01/10/2026
02225904-0006